D1626205

Textbook of Healthcare Ethics

Textbook of Healthcare Ethics

2nd Edition

Erich H. Loewy, M.D.
University of California, Davis
School of Medicine
Sacramento, California

Roberta Springer Loewy, Ph.D.
University of California, Davis
School of Medicine
Sacramento, California

KLUWER ACADEMIC PUBLISHERS
DORDRECHT / BOSTON / LONDON

A C.I.P. Catalogue record for this book is available from the Library of Congress.

ISBN 1-4020-1460-0

Published by Kluwer Academic Publishers,
P.O. Box 17, 3300 AA Dordrecht, The Netherlands.

Sold and distributed in North, Central and South America
by Kluwer Academic Publishers,
101 Philip Drive, Norwell, MA 02061, U.S.A.

In all other countries, sold and distributed
by Kluwer Academic Publishers,
P.O. Box 322, 3300 AH Dordrecht, The Netherlands.

Printed on acid-free paper

Printed in the Netherlands

This book is dedicated to our students past, present and future

And to

Dr. Thomas Frühwald of Vienna, Austria for his unflagging interest and support

Contents

Acknowledgments

There are many people we need to thank. Of course we stand, puny as we may be, on the shoulders of giants. But above all, and perhaps more importantly, we stand on the shoulders of our contemporaries: colleagues, students and laypersons who ask "naïve" questions which, hard as it is to admit, are often right on target and are the very questions which need to be asked and which we in our "conceptual ghetto" (as the late Danner Clouser used to speak of it) have failed to see.

First of all we want to thank the University of California, Davis and especially its Dean, Dr. Joe Silva, who have done all they could to support this project. It is impossible to name all of the people who by advice, criticism, "the lifted eyebrow," *etc.*, have shown their interest, pointed us to pertinent literature and deflected us from the wrong avenue of approach. Without Dr. Faith Fitzgerald again and again making major contributions by criticism, suggestions and disagreements (and above all by asking questions), this revision would not have been possible. Likewise, thanks go to our former Dean, Dr. Hibbard Williams (who is still very active in our department), for his continued and ongoing support. Our colleagues in the Division of General Medicine have uniformly given us encouragement and asked questions that helped propel the work forward. Equally, our thanks go to Dr. Tim Grennan of Kaiser Permanente for his encouragement, help and often highly incisive suggestions.

In our Bioethics Program thanks go first of all to my immediate colleagues, Dr. Ben A. Rich, and to my co-author, Dr. Roberta Loewy. Formally and informally they have helped to shape the book, its style and its basic briefness of discussion. Likewise, thanks go to our associate faculty, who are often the first to raise troubling questions. Special thanks must likewise go to Dr. Thomas Frühwald of Vienna, Austria, a geriatrician and one highly knowledgeable and interested in Bioethics. Speaking and corresponding with him has been most helpful in researching certain areas and gaining certain insights. His role has indeed been a critical one. Dr. Joachim Widder at the University of Vienna, by making suggestions and asking penetrating questions has, likewise, helped us explore regions that might otherwise have been left unexplored. My assistant, Ms. Cynthia Gomez, did much that was essential to help the work along—by filing and re-filing new versions of each chapter, keeping them in order, being protective of our time, and putting up with the sometimes not altogether pleasant personality of the senior author. In addition, her comments about the work in progress have added a lot to this edition. We owe her a great debt of gratitude. Without her it would not have been possible. Most of the revisions of this book were done in a tiny hamlet in the Haute Savoie whose proprietors (Monsieur and Madame Busson) deserve special thanks—as does, for that matter, the whole village and its people.

Last but not least we want to thank our editors Anne Ultee and Nellie Harrewijn together with Mariclaire Cloutier who shepherded this book to its conclusion. Their suggestions were invariably well taken and helpful.

Introduction to the Second Edition

The new edition of the original Textbook of Healthcare Ethics last published in 1997 has been greatly changed. In this short period of time many former questions have changed and many new approaches have been articulated. We have deleted some material which no longer forms a great deal of controversy (such as whether physicians are or are not obligated to treat AIDs patients) and have added chapters on genetics and the influences that new scientific and other knowledge have on the way we conceive and deal with ethical issues. We have tried to bring some of the previous controversies up to date. The senior author has also asked Roberta Springer Loewy to be co-author, write some of the chapters and oversee the whole way in which the book flows. Her philosophical contributions to the first edition were substantial—those to this edition are indispensable. Such a move also enables us to establish some sort of continuity for the future.

Just as when the first edition came out, our main and increasingly serious problem in health care ethics is one we are wont to shove under the rug: an increasing number of people in the United States are uninsured and most of the rest of us who are supposedly fully insured, in fact, are not. What is mainly discussed is what we have called "rich man's ethics"—the problems only those of us who can afford medical care in the first place encounter. Those of us who lack access do not worry much about the finer points of informed consent or advance directives but, rather, worry about getting our pneumonia treated or children immunized. We live with comfortable fictions: Medicaid does cover some (by no means all!) of the poor—but fewer and fewer health care organizations accept Medicaid patients. Medicare (which is supposed to "cover" those over 65) in fact allows a limited length of hospital stay after a significant out of pocket contribution. Patients who want outpatient, laboratory or x-ray coverage will have to buy part B out of their own limited funds—and part B is becoming hair-raisingly expensive and out of reach for many. Even those fortunate enough to have part B have no coverage for medications and are, therefore, in the lovely position to receive advice (in the form of the prescription) that they are often financially unable to follow. Those who are allegedly "fully insured" have co-payments that make it impossible for people on a marginal income to take advantage of that insurance—the choice is often between physician visits, lab work, x-ray and other procedures or food for the family or education for the children. The whole system (though, in fact, it is not a system but a hodge-podge of competing ones) is one of callousness, disingenuousness and hypocrisy.

Social ethics—which after all plays a critical role in health care ethics and of which health care ethics is a part—is likewise increasingly worse. The gap between the grindingly poor and those who are opulently rich has grown—indeed it is greater here than in *any* of the industrialized countries. About 25% of children (33% of black children) go hungry a good part of the time and persons may work a forty-hour week on a minimum salary and still remain beneath the unrealistically low poverty level. Our social conditions—reflected in the state of our medical care—are a national shame. This by no means makes the problems of what we have called "rich man's ethics" unimportant but, by forcing us to look at the reality of access to medical care, it puts them into a proper context and—we shall argue—imposes on health-professionals more than merely the duty to do the best for the individual patient lucky enough to be sitting in front of them. Beyond imposing this obligation on health care professionals, however, we would argue even more forcefully that those of us who pretend to do health care ethics are obligated to do all we can to rectify the conditions which today have at best limited the scope of ethical practice and have often made it impossible. It is the shame of health care ethicists and of our organizations that we have generally refused to play an active part in pushing for more justice within our health care system. Like the Nazi academics who saw, heard and spoke no evil and who, therefore, made themselves a part of that very evil, organized health care—including, and most shamefully, organized health care ethics—in this country has chosen to place itself in the same position.

Education has suffered. People are, as one of us [RSL] emphasizes, not taught how to think but what to think—or to think only within very narrow confines. Medical education has suffered as faculty are more and more forced to see more patients, earn more money for the University and receive more grant money for research. Since time is not unlimited and physicians, after all, are also people this means not only that less teaching will occur, but also that the quality of what teaching that does occur may suffer.

All of this makes teaching ethics frustrating: we teach people to do what we know they cannot do, *viz.*, get to know their patients, their values, their circumstances, etc. When a physician is only allowed fifteen minutes to see a patient this, of course, goes out the window as, indeed, does careful medical care which is predicated on a thorough history and physical in the light of an understanding of basic sciences. We will be spending considerable time articulating these problems, their geneses and their possible resolutions.

As we have been rewriting the book, the relevance of ongoing studies about the participation of the medical profession and the role of so-called bystanders in the holocaust has become more and more evident. A study of the events leading up to this tragedy and the role played by various individuals within it are highly pertinent to our problems today. We ignore them at our peril. Social injustice covers a whole spectrum of evils—the discrimination against Blacks or women is in the same continuum as is the attempt to destroy Jewry in the Holocaust. The similarities among those who would tolerate the fact that 52 million of our fellow citizens have only capricious access to health care (so long as they themselves do have access) and those who would stand by and watch Jews forbidden to enter Parks, forbidden to engage in any but menial work and ultimately beaten or shipped to Auschwitz are frightening. The most insidious forms of social evil and injustice starts in small steps and each time we accept one the likelihood of accepting the next (which is after all only a bit worse) looms large. The fact, furthermore, that so many highly trained physicians blithely participated in the holocaust (as they did in Tuskegee) is a sobering thought.

Ethicists are not, in our opinion, here to give answers. They certainly are entitled to say that some contemplated course of action would be ethically problematic and to give their reasons for such a statement. Ethicists are no more "moral" than anyone else—they are simply people who by training and daily activity are more skillful at sorting out questions and examining assumptions. Indeed, the main role of the ethicist is to examine presuppositions, inquire into precise definitions, scrutinize the logic used and, above all, to ask questions. We are all moral agents and responsible for what we do or refrain from doing.

This new edition, we hope, brings some of the issues in bioethics up to date. Given the rapidity with which new knowledge operates, they will never be totally up to date—but at least we must try. If there is one message we would like to leave with health care professionals it is that practicing "good" technical medicine with particular patients does not exhaust the duty one has: it is almost impossible to practice ethical medicine in an unethical institutional setting and it is unlikely that an unjust society will build a just institution. Thus, it is also our duty as health care professionals—whether doctors, nurses, ethicists or others—to do all we can to help improve our society and our institution.

Historical Introduction

INTRODUCTION

This chapter is mainly about the history of medicine and its ethics. As usually conceived, history is retrograde: It is what happened yesterday, and, much as we may try, it is what happened yesterday seen with a set of today's eyes. Trying to understand yesterday's culture may help us put on a pair of corrective glasses, but it fails in entirely correcting our vision. Contemporary cultural anthropology may likewise help us understand the way today's events and cultural habits shape what we call history tomorrow. Past events and the kaleidoscopic pattern of today's cultures may help guide us into a future that in at least some respects is ours to forge. Learning about ethics yesterday and thinking about ethics as it expresses itself in various cultures today can help us shape the ethics of tomorrow: This is true whether we are speaking of that part of social ethics called "medical" or of any other part of social ethics. The social aspects of medical practice—how the institution called medicine fits into and works within the greater society called culture—shape the way its ethics ultimately must play itself out. The healthcare professional–patient relationship (a relationship we generally think of as a highly personal one) and the relationships among healthcare professionals are relationships that, like any other relationships, take place within, are shaped by, and in turn help shape the social nexus of which they are merely a small part. A glance at history should teach us at least that much, and an understanding of this should help us deal with various contemporary cultural settings as well as enable us to play our part in shaping the future. No one, the conservative backlash of today notwithstanding, can possibly believe that medical practice tomorrow will not differ from medical practice today at least as radically as it differs from yesterday's. Such changes of medical practice cannot help but have a profound influence on the way that healthcare ethics (whether it is the more individual ethics of a patient–physician interaction or the more communal ethics of medical structure) is conceived. The institutional setting to a large extent determines what we recognize to be ethical problems and most certainly determines the leeway we are given to address them (managed care, which will be discussed later on, serves as one example). In turn, the institutional setting (the kind of hospital or clinic we have within a given health care system and the kind of health-care system we have within a society) depends upon the kind of society within which exists: it seems unlikely that we can craft a just institution within the context of an unjust society.

When physicians first learn about disease, they do so largely outside the context of specific patients or situations but inevitably within a very distinct cultural setting. Only after they have mastered pneumonia as a distinctive concept, only after they understand a disease's anatomic and patho-physiological underpinnings, can they turn to the specific problem of pneumonia in Mr. Tintfeather and of Mr. Tintfeather seen within his particular cultural context. Disease is more than merely a collection of clinical observations underpinned (at least frequently) by anatomical and physiological "facts": For a given state of affairs to be acknowledged as a disease, social labeling of such a state of affairs as a "disease" (instead of counting it as a "sin," as a "crime," as holy, or as an irrelevant matter) must take place. Moreover, one must make a difference between an "illness" and a "disease." One can have a disease without being ill (one may feel very well in the first stages of cancer) or can be ill without having a disease (one can feel very ill for a variety of non-pathological reasons). What counts as disease and what does not is a social construct. Epilepsy, leprosy, masturbation, and homosexuality, to name but a few, can serve as examples. To begin with Mr. Tintfeather's symptoms, findings and underlying pathology must be socially acknowledged as a "disease" instead of as a sin and sent to the priest, a crime and sent to the police, or as nothing to be concerned about and ignored. Once such symptoms, findings, and underlying pathology are socially "legitimized," one must recognize that Mr. Tintfeather (now justly labeled with this "disease") lives in a specific social setting and has personal values, attributes, and coexisting disease states peculiar to himself. All of these considerations have a profound influence on Mr. Tintfeather's diagnosis, prognosis, and, ultimately, management and outcome. Mr. Tintfeather may be an active college professor or an alcoholic living on the street, or he may be brain dead; he may be from any of a number of different and differing cultural settings, and his peculiar history may have given him a set of values peculiar to himself. He has pneumonia in that he meets certain necessary and sufficient criteria for such a diagnosis, but his particular pneumonia can be understood and dealt with only within his own peculiar circumstances. Likewise, Ms. Swidalski, who wants an abortion, cannot be dealt with outside the realization that she is, perhaps, a 15-year-old who was raped by a psychotic madman carrying a dominant gene for insanity, or a 25-year-old woman who wants an abortion so as to fit into her new spring dress. Analyzing the diagnosis and treatment of pneumonia away from Mr. Tintfeather, or the problem of abortion away from the persons involved, is perhaps helpful; but it is helpful only up to a point. It is a sterile exercise, unless the results of analysis are carefully, thoughtfully, and compassionately applied to the locus of the actual problem.

An attempt to impose one's own set of values on another when one has the power to do so is, as we shall see, a form of paternalism. Attempting to impose the values and viewpoints of one culture on another is a form of ethical imperialism. Both courses of action are at best ethically problematic. Unless one shares the belief that "might makes right"—a point of view that from an ethical perspective few would feel inclined to defend, and one that in fact would be logically indefensible—such forms of paternalism (or, on a wider scale, ethical imperialism) would have little moral standing. Attempting to pass judgment on specific actors or actions in another culture is, as we shall see in Chapter 3, something that if done at all must be done with great understanding and care. That means neither that "anything goes" nor that there are no standards: It does suggest that when such judgments are made they (1) must be made with extreme care, (2) must be made carefully to draw precise distinctions, and (3) must be argued on grounds other than one's own cultural or personal biases, traditions, or values.

——————— PRIMITIVE MEDICINE ———————

Medical ethics is an integral part of the fabric of medical decision-making and has undoubtedly always been an important consideration in medical practice. Medicine, philosophy, and religious practices had their origins in the mass of primitive beliefs and taboos that early man used to try to understand and manipulate his world. Tribes of hunter–gatherers roamed the Earth without much specialization and with little regard for individual talents or skills. No doubt it was soon evident that some were less adept at hunting and perhaps had a special knack for making flints or tools. They produced tangible objects (spears, knives, etc.), but "objects" valued by their community and were supported by their tribe in return. Specialization, if indeed it ever did not exist, was born.[1]

Humans have always feared the unknown and have always stood in terror of forces beyond their comprehension. Understanding these forces and dealing with them (even if "understanding" and "dealing" were, to our way of thinking, purely illusory) allowed humans to cope with terror. The same activity goes on today. We label a constellation of findings and call it a disease (or we attribute causality on shaky grounds), and, having so labeled it (or attributed causality), we now feel that we have, by naming, gained a certain amount of power over it; in other words, having labeled a constellation of findings with a name, we "feel better about it." It is not rare that patients with troubling symptoms who are given diagnoses that carry a very bad prognosis feel "relieved" and somehow more capable of coping. The fact that something has a name (even when it does not increase what we know about that something) gives us a feeling of being able to cope better.

Primitive man, likewise, constructed a series of stories and created a myriad of myths to explain these forces and consequently to make them appear less terrifying. To primitive humans, the world and its objects swarmed with spirits, many malevolent and all needing to be propitiated or appeased. No doubt, some members of the tribe were temperamentally better equipped to deal with such mysteries. A series of elaborate rites and customs to propitiate the unseen powers soon developed. Such persons, the "medicine men" (though recent research seems to indicate that many were, in fact, female) or shamans found in all tribes, were seen as skillful in dealing with such unseen forces.

They were the first intellectuals: persons supported by their compatriots not to produce material objects but to deal with abstractions and ideas. Philosophy, religion, and medicine have common roots in dealing with such forces. Magic was their tool.

Primitive humans, in dealing with their fellows and with the forces about them, were necessarily, even if often unconsciously, concerned with ethics. The basic ethical question, "Ought I do something that I have good reason to believe I can do?" is the basic question of ethics, and it was as valid a question then as it is now. Propitiating the unknown and mediating between it and their charges, shamans had to make judgments based on a system of values and had to evolve a set of rules eventually expressed in some sort of deliberate action or ritual. Such behavior, in turn, was in need of justification by an appeal to higher values or principles.

Shamans were not frauds. They believed (and today in many societies continue to believe) as intensely and as passionately in their capacities as we believe in penicillin or open-heart surgery. In fact, their capacities were not as minimal nor their ways as ridiculous as we often like to think: There is good evidence today that shamans knew some of the power of medicinal herbs, were quite well versed in basic surgical techniques (they set fractures and successfully—at least with the patient's long-term survival—trephined skulls), and, above all, practiced a primitive but, because of their rites and their ability to "suggest," probably a most effective type of psychotherapy. In treating their patients, therefore, they were faced with a set of moral problems (of "ought" questions) even if these questions were not labeled as such. Ethical dilemmas in treatment do not depend on the technical "rightness" or "wrongness" of the treatment; what is "right" today may well be proven "wrong" tomorrow. Such dilemmas depend on the application of a treatment sincerely believed to be the "right" thing for a patient to that patient at that particular time and upon the relationship between professional and patient. If, in the shaman's sincere belief, a certain rite can propitiate the evil spirits that cause the patient to burn and shake, the decision to use such a rite becomes a moral decision. The fact that rites rarely cure malaria is irrelevant to the issue.

———————— PRE-HIPPOCRATIC ETHICS ————————

Our knowledge of ethics in the pre-Hippocratic world is fragmentary. Hammurabi already had set some rules for medical practice and punishment for malpractice. Amundsen's claim that "ethics "is even less apt to be borrowed than is medical theory and concomitant technique"[2] is undoubtedly true, but it does not negate the influence that the philosophy of one world view has upon the philosophy (or world view) of another culture. It is now firmly established that there was widespread communication between the various cultures flourishing in the known ancient world. To believe that diverse beliefs were entirely without influence on each other would be a naïve notion. In approving or condemning another way of thinking or acting, we perforce must examine that other way, apply some standards, and pass some judgments. In so doing we are reflecting on and comparing such other ways of thinking or acting. Our own point of view is therefore inevitably put to the test, and over the long haul is apt to undergo changes inevitably influenced by such comparison and reflection.

Yet there were vast differences between the rigid rules that governed the practice of medicine in Hammurabi's Babylonia or in ancient Egypt and the freedom with which medicine was practiced in ancient Greece. In Babylonia, the *Code of Hammurabi* (1727 BCE) represents the first attempt to regulate medicine and to protect patients from incompetent (or unlucky) practitioners. In Egypt, too, medical practice was rigidly fixed by law: In the Hermetic *Book of Toth*, Egyptian physicians are enjoined not to deviate from the rigidly prescribed regimen under fear of death.[3] Aristotle, however, stated that physicians were allowed to alter treatment if, after the fourth day, the traditional approach had been found useless.[4] Persian medical ethics was "modern" in speaking not only of the cognitive but also of the characterological attributes of the good physician.[5] In Greece and in most of the later Hellenistic and Roman world, no strict laws applied.

THE ANCIENT WORLD

The fabric of healthcare ethics, like a woven cloth, has always consisted of a large number of strands combining various ground views: that of physicians, nurses, and their various "schools" and subgroups of medicine as well as other healthcare professionals, governments, and patients. Does ethics seek to benefit patients, the various professions that make up the healthcare team, or the community and its particular institutions? What is the relationship of healthcare professionals as a group and physicians, in particular, to each other as well as to the other interests to and with which they must relate? When we look at these various ways of conceptualizing ethics and what the concerns of ethics ought to be, some very obvious differences exist. The Hippocratic corpus, extending as it apparently did over several centuries and authors, is not all of one piece. Nevertheless, its main thrust is to prescribe a certain standard of decorum— "a certain etiquette, mainly to uphold a certain standard of performance, and serves to distinguish the expert from the charlatan."[6] Greek physicians, unlike their Babylonian or Egyptian colleagues, were quite unfettered by state regulations. They were itinerant craftsmen. To the Hippocratic physician the sole purpose of medicine was the application of knowledge to the treatment of disease, and his ethics consisted in doing this well. It is, as Edelstein repeatedly points out, "an ethic of outer achievement rather than one of inner intention."[7] The injunctions and enjoinders given to physicians in the Hippocratic corpus are intended to safeguard the art and guard the reputation of the profession and its practitioners. Hippocratic physicians, furthermore, were neither the only nor the most common medical practitioners in ancient Greece. The Asclepiads and many other models coexisted and, at times, freely exchanged patients with the Hippocratics. Even though patients are often thought of as having been powerless until very recent times, there is no doubt that patients could (and did) "vote with their feet." Furthermore, the way physicians dealt with various social groups (the elite among the Greeks, freedmen, or slaves) was quite different one from the other.[7] One must guard against "judging" such customs from the vantage point of one's own totally different culture and experience.

One cannot understand the medical ethics of ancient times unless one acknowledges that these ethics were informed by a wide variety of cultural and philosophical persuasions. The ancient world was a very multicultural one in which a large variety of cultures and worldviews coexisted. Even among the Greeks there was, for example, a wide variation between Homeric beliefs, which, for example, viewed chthonic personal immortality as being that of shadows in the underworld, and the beliefs of the Pythagoreans, who initiated a belief in divine personal immortality otherwise alien to Greek thought. Greek culture, Greek philosophy, and Greek medical ethics do not represent a monolithic point of view. Rather, they are composed of diverse strands forming a rich fabric that imperceptibly merges into the later Roman and early Christian practices.

The depth of the Pythagorean influence on the Hippocratic oath is debatable. Whatever its extent, a significant influence is probably beyond doubt. The Pythagoreans in many respects presaged some of the later Christian doctrines: personal immortality and an essentially life ethic, to name but two. The Hippocratic oath, when seen in its original form, prescribes the relations of student to teacher, establishes the duty to transmit knowledge as well as fixing those to whom it is and those to whom it is not to be transmitted, and sets standards of medical function and decorum. Interpreted in our light, it provides a framework of medical behavior and, perhaps, ethics to which we can relate even though we may not be able to agree with it in its entirety; viewed in the light of Greco-Roman culture, it emerges as a powerful tool seeking to safeguard the reputation of medicine and that of its practitioners rather than, in the main, seeking to promote the patient's good for its own sake. Its main emphasis is more one of etiquette than of ethics—a feature not altogether absent from most of the later oaths! It is clearly a document of outward performance, rather than one of inward intention. What matters most to the Hippocratics is how the physician's behavior is perceived and consequently what the social and material status of the profession will be.

Many physicians around the world continue to believe that upon graduating as physicians they take or took the Hippocratic oath. This is hardly, if ever, the case today. Sometimes medical students in their passage from student to physician take no oath at all. Often it is another oath altogether, and frequently it is at most one which has been very much changed. Those who oppose some debatable ways of acting by appealing to the Hippocratic oath do so on fairly flimsy grounds. Oaths or, for that matter, codes of various sorts are prescriptive instruments that seek to restrict behavior by an appeal to their own authority: While they may be (and often are) consistent with ethical medical practice, their very existence is insufficient ground for calling any practice ethically sound or not. Few if any contemporary physicians who claim to be bound by the Hippocratic oath, for example, would refrain from surgery or consider their colleagues who "cut for stone" to be acting unethically. An appeal to a given oath or code—like an appeal to the law or to religion—is an appeal that assumes the oath or code rather than the considerations that led up to it to be the arbiter of what is and what is not ethical behavior. (See also Chapter 3.)

——— THE HELLENISTIC, ROMAN, ———
AND EARLY CHRISTIAN WORLDS

The Hellenistic world spread Greek culture, modified by local custom, throughout the known world. It provided a bridge to the Roman world and to the Christian era beyond. The philosophy, the science, and the medical ethics of those times are composed of the many strands of early Western culture and, in addition, show a strong influence from the Hebrews, the Persians, and other, more Eastern nations. Hebrew medicine (and especially Hebrew hygiene) probably exerted a considerable influence. The Hebrew precept that preserving or saving life trumped all other rules (even that of one who had been an enemy) took precedence over all other religious rules probably exerted some influence on the surrounding world. Undoubtedly Pythagorean views substantially influenced Platonic ideas and ideals and to a greater or lesser extent influenced the ethics central to the writing of the Hippocratic corpus. These ideals were quite different from those of the Stoa (starting with Zeno, ca. 300 BCE) and of the Epicureans (ca. 200 BCE).

In contrast to the Pythagoreans and to Plato, who believed in personal immortality, the Stoa believed in natural personal dissolution without afterlife and without Heaven or Hell. The purpose of life was to be fulfilled on Earth. The Stoa strove for "phronesis," or practical wisdom—how to craft and lead an honorable and a "good" life. The belief in the unity of all rational beings and in the fundamental equality of all men is central to Stoic and Epicurean beliefs. Panaetius (190–109 BCE), as later represented by Cicero in *On Duties*, speaks about professional ethics. *On Duties* became "the manual of all later humanism, ancient and Christian, secular and religious alike."[7]

These different threads and views must have found expression in the late Hellenistic and early Roman world. The first expression of what we today would consider truly medical ethics is recorded in the writing of Scribonius Largus (2–52 CE).[8–9] Intriguingly, Scribonius speaks of what we would consider today to be medical humanism not as something to be argued for but rather as something "quite self-evident to himself and his readers." Influenced profoundly by the Stoa and interpreting the Hippocratic corpus in their light, Scribonius sees medicine as a "profession" and, therefore, in the view of the time, as necessarily containing a fundamental core of ethics. He introduces a textbook of pharmacology in which he argues for the use of drugs in treating patients (not by any means an established thesis then) by a chapter on what we today would clearly call "medical ethics."

At the time of Scribonius, giving pharmacological agents in and of itself was ethically problematic and was something many physicians regarded with suspicion: Giving potent extracts without any ability to standardize or even to weigh or measure accurately was not without great danger. Scribonius carefully deals with the question of what a physician is (not an easy one in the days before universities, licensing boards, or, in Greco-Roman times, state control) and what the duties of physicians are vis-à-vis patients. Humaneness, friendliness, and philanthropy, as Scribonius sees them, are not merely minor social virtues meant to enlarge medicine's reputation (as they largely were seen to be in the Hippocratic corpus and later by Galen)), but are the special obligation of the physician. In other words: Scribonius (as did Virchow very much later) recognized the practice of medicine as a social art.

Among other things, Scribonius grapples with the duty of physicians toward their patients and toward the state. When in state service (Scribonius served as physician with the Roman legions), physicians may fight against the enemy as soldiers but, as physicians, they must harm no one and treat friend and foe alike. "Medicine," Scribonius says, "is the knowledge of healing not of hurting."[9] Physicians are remiss in their duty if they do not know all that they ought to know, make use of that knowledge for the benefit of (all) the sick, and, especially, if they fail to fulfill their ethical obligations toward all, regardless of who they are, what they are, or to whom their loyalty may be. Ethics, in Scribonius' view, is intrinsic to medicine, not extrinsic to it. With Scribonius we have proceeded (some, including myself, would claim advanced) from an ethic of outward performance to one of inner intention.

Scribonius' rather progressive and, to contemporary ears, pleasing perspective (one that in its outline and humane views resembles much of Ramsey's work in this century) did not directly manage to perpetuate itself. Although Galen (131–201 CE) felt that a true physician optimally should himself be a philosopher and practice medicine out of love of humanity, he saw nothing inconsistent with other motives (love of money, love of status, etc.) underwriting, even primarily or solely underwriting, the physician's function. The matter of motive, to Galen and others, is one of personal choice and has no intrinsic connections with the practice of medicine. From the point of view of medicine, the physician's "specific morality is incidental rather than essential."[7] A physician is expected to be a technical expert in medicine and to use his skill to the best of his ability—all else is supererogatory: It might be hoped for, but it could not be expected. Once again, and less than 2000 years after Scribonius, we are back to an ethic of outward performance; inner intention is a desirable decoration, but little else. It has been said that history frequently and at least in broad outlines tends to repeat itself: Certainly many of today's views of medical ethics (especially of medical ethics as conceived in the libertarian mold) hearken back to an ethic of outward performance. The backbone of the libertarian ethic, which rests on entrepreneurialism and on "doing a good job" not because doing so has an intrinsic value but because doing so will attract more customers, certainly reminds one of such a point of view.

ARABIC INFLUENCE

Arab culture had a profound influence on European culture. One must recall that within a century of Mohammed's birth (6th century CE), both the European and African sections of the Mediterranean region were in Arab (and therefore Muslim) hands. The Arabs brought with them an extremely advanced culture: Art, music, philosophy, mathematics, science, and medicine were far ahead of what the Europe prior to the Arab conquest had to offer. Christian culture withdrew into the interior of Europe, and in many respects its own culture stagnated. In the meantime culture in the lands occupied by the Muslims flourished. Although Arab control was relatively short-lived (Christian dominance over European territories was soon reestablished, although the last Arab occupied area was not re-conquered until 1492 (the year Columbus arrived in America), its influence on subsequent intellectual history should not be underestimated. At that time, tolerance in Arab-dominated countries established a fruitful working together of various cultures and various religions. Jewish influence on philosophy, science, and the arts at that time was great.

Medical practice and knowledge, surgical as well as medical, were far advanced. This practice and knowledge, as a matter of course, it seems, included concern about ethics within that practice. The Arab–Jewish philosopher Maimonides, whose oath is frequently used even today, wrote in many respects more in the spirit of Scribonius Largus than of Galen. During the Middle Ages much of what the Arabs had accomplished was, under the influence of the Roman Catholic church, buried, lost, or entirely ignored. Only lately have we come to appreciate the importance of the contribution of Arabs and their time.[3]

━━━━━━━━━━ **MEDIEVAL TIMES** ━━━━━━━━━━

The influence of Galen permeates the medieval period. His views, adopted early on by the Church, became near dogma to be learned and memorized in medical schools, not to be challenged. Attitudes toward health and disease profoundly affected ethical positions. In one view, God sent disease as punishment (a just affliction sent in retribution for some sin) or as a test; in either case, the problem is outside man's province and jurisdiction. If we are to follow the Sermon on the Mount, are we not like the "fowl of the air" or the "lilies of the field," cared for without our efforts by our Father? Such problems had been argued in the Talmud and had been clearly adjudicated in favor of healing: God intended physicians to heal just as He expected farmers to till the soil. God no more intended the Earth to lie fallow and men to starve than He did disease to go untreated. In Christian circles, ambivalence toward medicine (the physician as opposing God's will, or the physician as instrument of God in opposing disease, pain, and death) has persisted until this day.[10]

The emphasis on Christian charity, however, worked toward the institutionalization of care for the sick. Despite the existence of hospitals in ancient Egypt and the Asclepiad temples of Greece, hospitals in the Western world first began to emerge at the beginning of the fourth century. Such "hospitals," however, were not hospitals in any modern sense. Rather, they involved a conglomerate of charitable institutions and included foundling homes, orphanages, old age homes, hostelries for the traveler, and infirmaries. Often nuns provided care and shelter to all these various groups and operated such institutions. Administration was largely in the hands of the clergy. Christian humility made service to these afflicted and troubled—even those suffering from a "vile" disease—an act of charity sure to find its reward in the hereafter.

Physicians were often priests, and the duties of physicians were, first of all, largely defined in religious terms. Healing the soul was not distinctly separated or separable from healing the body, and healing the soul had priority. Physicians were enjoined to make sure that their patients' spiritual needs had been met both because "many illnesses originate on account of sin" and because the safety of the soul was the main issue. The Lateran Council of 1225 advised that physicians admonish patients to see a priest, and Pope Pius V in 1566 asserted that after 3 days physicians could not continue to treat patients who had not confessed. Violators of this rule were to be barred from further practice.[11] Institutions granting medical licensure required graduates to take an oath promising to abide by this rule. The physician was not rarely crowded from the bedside by the priest, in part, it is said, to extract payments for relics, masses, holy candles, or appeals to the saints.[12]

Medical regulations and licensure began largely during the medieval period. In the 12th century, Roger II of Sicily decreed that all potential practitioners of medicine had to appear before judges and officers of the Crown to be examined before being licensed to practice. Roger's grandson, Frederick II of Hohenstaufen, who was the Holy Roman Emperor, confirmed and extended this decree by insisting that all those who were to be licensed must first be examined by the medical faculty of the medical college at Salerno. Before the examination could take place, candidates were required to show proof that they had undergone a rigorous course of study for a total of 8 years and had then spent an additional year (an internship of sorts) working under the direction of an experienced physician. This attempt to institute a secular license was, of course, part of the ongoing struggle between the papacy and the Holy Roman Empire: the Guelfs and the Ghibellins.[13]

Medicine, furthermore, was hardly a cohesive whole. As it emerged in the later Middle Ages, it was a pyramid. At the top stood university-trained physicians with a reputation for learning. Until after the 14th century, unfortunately, such learning largely consisted of circular memorization and scholastic quibbling. It was largely useless. Next in the hierarchical line came the surgeon (united with the physician and becoming university trained only after the 14th century). Surgeons had less training but were frequently more likely than the physicians of the day to help patients. Barber surgeons, a large step down from "surgeons," practiced phlebotomy and cautery, sometimes on their own but much of the time at the direction of physicians or surgeons. More often than not, barber surgeons were illiterate. Apothecaries mixed "Galenicals," and a host of untrained quacks pretending to things medical completed the "healthcare team" of that day.

Physicians were expected to be charitable and competent. The definition of this, of course, was in the Christian framework of the time. Euthanasia and abortion (after "animation" or "ensoulment," at any rate) were considered unethical. (This is elaborated on in Chapter 10.) As with most other aspects of life, medical ethics was determined and directly or indirectly enforced by the Church and by its secular agents in accordance with the Church's particular agenda.

——— THE PLAGUE AND ITS EFFECTS ———

The Medieval Plague (1348–1352) and the subsequent series of plagues that struck Europe had a profound influence on every aspect of material and social life. The role and duties of physicians vis-à-vis their patients were severely tested. It is amply clear that while some physicians abandoned their patients, most stayed (see also Chapter 7, on risk taking). Available documents indicate that physicians who stayed were "motivated by compassion, charity, and a sense of duty."[14–16] The Black Death of 1348 gave rise to what was then called (and is still called in many Eastern European countries) "medical deontology"—medical ethics done to examine the ethics of the profession.[15–21]

After the plague swept Europe in the 14th century, physicians began increasingly to see themselves as bound by moral duties beyond those imposed by the Church. Moreover, the formation of guilds had an influence on medicine. In institutionalizing medicine and the colleges, in keeping qualifications and licensure predominantly within medical hands, medicine shared in the medieval idea of keeping the function as well as the production of professionals within professional control. Peer review, licensure, board certification, and other aspects of modern medicine are directly derived from the basic idea of controlling the profession by and through the profession itself. The frequent preoccupation with medical etiquette, as distinct from medical ethics, can be traced to the prevailing desire to safeguard medicine's reputation, an ongoing concern of healthcare professions and something we would call "image" today. The ethic of outward performance, then as now, still played a dominant role.

As ideas of science progressed, the role of medicine changed. Francis Bacon (1561–1626) divided medicine's function into the preservation of health, the cure of disease, and the prolongation of life. The prolongation of life was seen as a new task—it had, as Amundsen has pointed out, "no classical (and also few medieval) roots."[22] The care of the "incurable," heretofore not a part of the ethical practice of medicine, now became important, initially to learn how to treat diseases previously believed incurable and, after the 17th century, for other reasons as well. Prolonging life was shortly to be seen as medicine's prime function, and keeping people alive—not necessarily the same thing—was shortly to become almost an obsession. As technology's capacities became ever greater, they tended to create their own dynamic and to become a justification for their own use. Technology, as it often does today, began to "drive itself": Doing something was and is often done not because after careful consideration it is deemed as something that, under a given set of circumstances, ought to be done, but mindlessly because it can be done.

From the early 17th century onward, works of medical ethics (as distinct from medical etiquette) began to appear. Rodericus à Castro (1546–1627), overlapping Francis Bacon, published one of the first works of medical ethics: *The Responsible Physician, or the Duties of the Physician Towards the Public*. A later work, by Johannes Bohn of Leipzig (1640–1718), deals both with the obligation of physicians toward their patients and with the physician's civic responsibilities.[22] The literature of the day, still largely rooted in Church attitudes, started to involve itself with the physician–patient relationship as well as with medicine's civic responsibilities in more modern terms. Nevertheless, the secularization of medical thinking and acting was well under way and received its final impetus with the Enlightenment.

THE ENLIGHTENMENT TO MODERN TIMES

The Enlightenment of the 17th and 18th centuries and its emphasis on human reason as well as its goals of knowledge and freedom propelled medieval thought into the modern era. Newton (1643–1727), emphasizing the application of scientific principles to the solution of problems, presages medicine's later preoccupation with science. Medicine began to view itself as a largely dispassionate scientific enterprise. Philosophers of that era, furthermore, had a profound influence on the evolution of medical ethics. Hume (1711–1776), with his emphasis on moral sentiments (the physician's character, as it were); Kant (1724–1804) examining concepts of duty, the role of autonomy in ethics, and the relationships of categories of thought; and Mill (1806–1873) and his examination of the role of utility, left a stamp on ethical thinking in medicine that is reflected in much of our thinking today. In addition, the French revolution (1789) changed the relationship of persons to one another: The concept of individual dignity and the consequent notion of personal and human "rights," despite persistent class differences, began to be emphasized. Daring to ask question of those in power was no longer unthinkable.

The social conditions in the Western world after the plague favored progressive urbanization and with that the emergence of the working class. Working one's land was the expectation prior to the violent eruption of the "Black Death." When serfdom was abolished, many of the former serfs moved into urban areas and sought jobs. Immediately after the plague working conditions favored the worker: Labor was scarce and in high demand. The first "labor laws" were, in fact, instituted to protect the work giver from the frequently rapacious demands of the worker. Soon, however, as more and more former serfs moved into the cities and as those already settled increased in number, the urban proletariat began to form out of the body of the former serfs, and conditions for workers grew ever more dismal. Workers were generally unskilled and illiterate, and their way of life and standard of living stood in stark contrast to those of artisans, merchants, businesspeople, and professionals, who tended to form a quite separate middle class. With greater emphasis on individual dignity and stirred on by the misery of the proletariat, many efforts to ameliorate their lot were made.

Marxism, developing as a result of these conditions, gave rise to the 19th- and 20th-century socialist labor movement, solidarity, and, in many respects unfortunately but inevitably, an enduring sense of class-consciousness. Social democrats eventually split from the radical communists, and the influence of social democracy on the further thrust of history was and continues to be profound. Social democrats worked not only toward a bettering of working conditions but likewise emphasized the education of the worker. Public schools developed, literacy increased, and, with this, the individual's blind adherence to what others (including physicians) prescribed tended to diminish. Laypersons began to have sufficient general education and knowledge to feel (with considerable justification) that they could and should be a party to decisions made by health professionals when it concerned their own lives. The development of social democracy, a socialist labor movement, and the increasing power of labor also had direct a direct effect on the structure of the healthcare systems in the various states. It is no accident that the European countries have well-established national healthcare systems (although their nature varies greatly from country to country), whereas the United States as of today lacks a system in which all are assured healthcare.[23] Physicians must pursue what are essentially moral ends by ever more complicated technical means. Applying these means to their patients in a more and more "Newtonian" fashion caused medical practice to lose some of its "warmer" social aspects. Traditionally paternalistic, seeking the patient's "good" on terms defined by the doctor, medicine evolved into the 19th- and early 20th-century model of "scientific" medicine. Scientific medicine, of course, had enormous benefits: It allowed many to escape disease, many to be cured, many to live with less and even without pain. It allowed physicians to understand disease process rather than conflating symptom and disease.

Unfortunately, it also allowed a new confusion: While symptom was no longer held to be disease, the patient and the social context in which disease took place were often forgotten. A military metaphor in which "batteries" of tests were utilized to help "aim" our "armamentarium" in the "conquest" of disease became universally used even when, as in chronic disease or in the care of the terminally ill, it was no longer appropriate. The patient became a battleground on which physicians waged battles with disease; the battlefield, like Verdun 70 years later, was often left devastated. In struggling for the patient's bodily health, the patient as a human being was all too often forgotten.

Physicians and physician-scientists continued to be schooled in the areas of humanism that a classical education favored and medicine, until fairly recently, could, along with law and theology, well afford to be counted among the "learned professions." In Europe, where a well-established secondary educational system ["Gymnasium," "Lycée," or "Collège" (not to be confused with the English college)] exists, physicians enter university with a firm humanistic foundation; in America, colleges (which are often forced to play "catch-up" to make up for what was not taught in high schools) only very sketchily make up for this since most students entering medical school prefer to study predominantly "hard sciences" in college. In Europe as well as in America, colleges of medicine or universities have until lately tended to teach subjects exclusively related to the technical practice and understanding of medical practice. With the increasing introduction of ethics and some of the other humanities into the curriculum of many colleges and universities, this is (hopefully) beginning to change.

A superficial examination of history would tend to support the thesis that our concern with moral issues in medicine originated in response to the possibilities raised by the proliferation of science and technology. On the other hand, some have argued that moral issues did not arise out of technology but rather that technology developed "in response to a deeper and a prior moral concern."[24] Man's fear of unknown forces, of death, and of illness—man's search, in other words, for the good life—prompted the exploration of the unknown and the development of technology to deal with it effectively. A closer examination of this relationship would indicate that it is reciprocal: Questions of morality and the development of technology are mutually reinforcing. A dialectic between moral concerns and technological options produces a synthesis facilitating the development of both. The old, paternalistic model of previous ages was incorporated in the scientific model. The patient's "good" usually continued to be defined on the physician's terms. With more and more attention given to disease, the patient tended to be neglected: The "good" was seen more often than not in terms of "conquering" a particular disease or aberration.

As we have seen, a hierarchy of medical practitioners developed during the Middle Ages. With the formal development of nursing (there had always been persons who saw their job in helping with the sick or in delivering babies) and the various other associated professions, the relationship between physician practitioner and these professions likewise developed. It forms an important consideration in healthcare ethics today. Initially, and in some respects today, physicians regarded nurses as "handmaidens" here to obey blindly. As nurses have become better trained, they have rightfully assumed more important roles. Ideally, ethically and practically one should look at today's relationship as that of a team in which each has a different, even if often overlapping, set of skills, and all serve the patient's interest.

MODERN TIMES

Not only the scientific advances of the 19th and the continuing insights of the 20th but to an equal extent by the political and social changes of that century profoundly influenced health-care ethics in the 20th century. The first world war with its carnage, its emphasis on triage (which indeed was already practiced by the Egyptians) but especially the experiences of the Nazi era led not only to an interest in medical ethics but to its incorporation into experimental process as well as into educational policy. Curiously enough physicians and nurses to a large part cooperated and often cooperated enthusiastically with the Nazi program of discrimination, extermination of the unfit, "purification" of the race ("eugenics"—a concept actually originated in the USA) and eventually the holocaust. Indeed—physicians had the highest percentage of membership in the Nazi party.[25-27] Curiously enough: whereas in Germany physicians largely cooperated with and often were instrumental in carrying out Nazi policy this was not the case in Denmark where physicians (trained frequently at the same Universities) did all they could to undermine and frustrate Nazi policy.[28]

After the "doctor's trial" in 1946 (which tried only a handful—most got away scott-free[29]) a code of ethics for human experimentation (the "Nürnberg Code") developed and has undergone several revisions ("Helsinki revisions") since. The claim that this constituted the first code of ethics for the conduct of human experimentation is false: there was a code, quite similar to what the Nürnberg Code of 1946 stipulated in effect during the Weimar republic. It was never abrogated—simply ignored.

The Nazi experience had other ramifications and is full of meaning for us today. Physicians saw themselves as obligated to help the process and to let the wishes of the government override their ethical duties—to work for the patient's good and never for his/her harm. It tried—and tried all too often successfully—to use physicians as social engineers and to make their first responsibility one to the state. In other words, it interposed a system or organization between physicians and patients just as it had interposed a system between perpetrator and victim.

> A "system"—a political, economic, or cultural system—insinuates itself between myself and the other. If the other is excluded, it is the system that is doing the excluding, a system in which I participate because I must survive and against which I do not rebel because it cannot be changed...I start to view horror and my implication in it as normalcy.[30]

There is no doubt that a system within which our activities take place is the rule rather the exception—be it a hospital, a health-care system or a system in the wider body politic. But this very fact makes it essential that those affected by the system (which in essence means all of us) play our role in making such a system acceptable. This was hardly a problem only of the Nazi state—it is a severe and perhaps the chief problem in the United States with managed care today.

This contemporary issue of human experimentation as well as the treatment of individual patients by their physician most not be reduced to the Nazi state nor to managed care today—albeit that these are two flagrant examples. From 1932 to 1972 (the very time when the doctors trials took place and the US was "beating its breasts" in righteousness) the Tuskegee experiments funded by the US Department of Health deliberately failed to treat and misled a group of Black males who had contracted syphilis so as to study the progress of their disease. The findings were widely reported and commented as to their questionable ethics in only one letter to the editor throughout this period. At other times since, experiments on prison inmates and other weak, defenseless or easily coerced persons as well as persons who did not even know that they were the subjects of such experiments have been carried out. To prevent such occurrences Institutional Review Boards (IRB's) have been established and certainly have improved matters. The problem, however, continues and we cannot lull ourselves into the convenient belief that things have been set right.

Ethicists have played a role in shaping public policy. They have, however, tended to focus, and to focus almost exclusively, on the issues involving individual practice and have largely ignored the social problems and the institutional framework in which these problems take place. Often, in our view, bioethicists have "sold out," become members of the establishment instead of its critics and allowed themselves to be used as rubber stamps by institutions, industry and government.[31] If healthcare workers, and especially physicians and ethicists, are to act responsibly, they must pay sufficient attention to the conditions their particular institutions and their health-care system provide (in today's managed care, competition, and insurance driven society patients needing medical attention are turned away at the door—a classical interposition of institutional policy between two individuals). And since, ultimately, institutions are created by society and the worldview such societies have, all of us have an interest in shaping a society that serves us all.

One cannot separate medicine and its values from the culture that nourishes it and in which it functions. Medicine's moral views can differ in emphasis and detail, but they cannot differ substantially. American society, predicated on competition and personal gain, should not be surprised if its healthcare practitioners evolve into businessmen vying for a share of the "healthcare dollar." The quality of the "product" sold to the customer (the patient) must assure customer satisfaction so that the "business," rather than the patient, may prosper. The emphasis on customer satisfaction—rather than on patient service or patient "good"—constitutes a change in moral view. An ethic of outward performance is once again given full reign in the entrepreneurial model of today's emerging medical practice.

While it cannot be entirely distinct from the culture in which it is embedded, medicine nevertheless has its own unique history and identity. Medicine and the other healthcare professions, therefore, also have their own distinctive set of values and precepts. Because of such distinct values and precepts, healthcare professionals often experience friction with the culture of which they are a part. The eventual ethos of these professions is a combination of interacting forces. Healthcare need not become entirely enmeshed in the entrepreneurial model; rather, healthcare can use its historical viewpoints and its traditions to support its own distinctive ethos. Although such an ethos perforce needs to function within specific communities and accommodate its function to them, it nevertheless does not have to have values and viewpoints that are identical when it comes to its own professional values and ways of acting. Furthermore, medicine in discharging its social function may be able to do much to ameliorate what it may see as a point of view inimical to the interest of patients in a wider sense: Social conditions have much to do with illness and health, and medicine may well see itself as obliged to try to bring about social (and, therefore, inevitably philosophical) changes in the way the community sees itself (see also Chapter 6, on the physician as citizen).

——————— CULTURAL AND SYSTEMIC ——————— CONSIDERATIONS

The way that society structures its medical system has a profound influence on medical practice. In most of the Western world today (the United States is a unique exception and one that is certainly likely to change), physicians do not practice both in the hospital and as "primary care providers" in the outpatient setting. In most societies patients choose a family physician whose practice is strictly outpatient. Such physicians will care for long-term illness (arthritis, hypertension, uncomplicated diabetes, and other chronic conditions) and take care of their panel of patients for inter-current but relatively simple illnesses. When a need for more specialized diagnostic or therapeutic interventions presents itself, the "primary care physician" will refer the patient to a hospital-based internist or other specialist who will then take care of the particular problem and eventually refer the patient back to the primary care physician. Such a difference in the medical system of course has profound implications for the patient–physician relationship and for medical ethics

One must also consider that the way illness, birth, and death occur differs with culture and society, and this too changes some of the problems in ethics. In America and in Austria, most patients are born and die in a hospital; in the Netherlands, birth and death more commonly take place in the embrace of home and family. The fact that ethical problems, while superficially similar, are quite differently experienced seems obvious.

In the United States today (and to a far lesser extent in other countries) healthcare providers and especially physicians fear lawsuits. This recent historical development has not been without influence on health professionals, hospitals, and their clients. The proliferation of malpractice actions in the United States has many roots: We are a litigious society; the profession and the media have often and for a variety of reasons exaggerated the capacity of medicine to cure or heal; physicians are (often justifiably) viewed as more than well-off; and we often have the misconception that something that goes wrong invariably has to be someone's (and most certainly not our own!) fault.

It is popular for physicians and others to claim that ethics can be reduced to law. Such a claim if acted upon would freeze the status quo: That which is, is also that which ought to be. And that is simply not true. Beyond this, physicians and increasingly other healthcare professionals have a great fear of (rather than a good deal of respect for) the law and often share a belief that the law constrains them where it clearly does not. Indeed, the law leaves (and, in our view, properly so) a good deal of latitude in the hands of physicians and other health professionals. It asks (and not unreasonably so) that physicians adhere to certain standards and guidelines within which room for decision-making occurs.

The history of the profession would not be complete without a word about healthcare systems. Since the social system of medicine forms the necessary framework of individual medical practice, the type of healthcare system a society chooses to have is of critical importance to technical as well as ethical considerations of practice. It is difficult, if not impossible, to practice ethically unless the system in which one practices is an ethical one; likewise, it is not very well possible to craft an ethically appropriate institution in the context of a society whose ethical structure will not allow this. The United States today lacks access to healthcare for many members of the society and provides only inadequate access to many more. This (to many of us) intolerable state of affairs (see Chapter 11) makes individual practice difficult and makes of the usual problems treated in this book a sort of "rich man's ethics:" problems and considerations that can only come up in those able to afford access to healthcare to begin with.

——————————— **REFERENCES** ———————————

1. Loewy EH. Clergy and physicians confront medical ethics. *Humane Med.* 1986; 3:48–51.
2. Amundsen D. History of medical ethics: ancient near east. In: Reich WT, ed. *Encyclopedia of Bioethics.* New York, NY: The Free Press, 1978:880–884.
3. Garrison FH. *An Introduction to the History of Medicine.* Philadelphia, Pa: W.B. Saunders, 1929.
4. Aristotle. Politics III, 15. In: McKeon R, ed. *The Basic Works of Aristotle.* New York, NY: Random House, 1941:1127–1324.
5. Carrick P. *Medical Ethics in Antiquity.* Boston, Mass: D. Reidel, 1985.
6. Edelstein L. The Hippocratic oath. In: Temkin O, Temkin CL, eds. *Ancient Medicine: Selected Papers of Ludwig Edelstein.* Baltimore, Md: Johns Hopkins Press, 1967:3–63.
7. Edelstein L. The professional ethics of the Greek physician. *Bull Hist Med.* 1956;30(5):391–419.
8. Deichgräber K. Professio medici: Zum Vorwort des Scribonius Largus. *Abh Akad Mainz.* 1956;9:856–862.
9. Hamilton JS. Scribonius Largus on the medical profession. *Bull Hist Med.* 1986; 60:209–216.
10. Jonsen AR. Medical ethics: Western Europe in the seventeenth century. In: Reich WT, ed. *Encyclopedia of Bioethics.* New York, NY: The Free Press, 1978:954–957.
11. Amundsen DW. History of medieval medical ethics. In: Reich WT, ed. *Encyclopedia of Bioethics.* New York, NY: The Free Press, 1978:938–951.
12. Marks G. *The Medieval Plague.* New York, NY: Doubleday, 1979.
13. Hartung EF. Medical regulations of Frederick the Second of Hohenstaufen. *Med Life.* 1934;41:587–601.
14. Gottfried RS. *The Black Death.* New York, NY: The Free Press, 1983.
15. Marks G. *The Medieval Plague.* New York, NY: The Free Press, 1979.
16. Gottfried RS. *The Black Death.* New York, NY: The Free Press, 1983.
17. Welborn MC. The long tradition: a study in 14th century medical deontology. In: Cates JL, ed. *Medieval and Historiographical Essays in Honor of James Westfall Thompson.* Chicago, Ill: University of Chicago Press, 1938:344–357.
18. Rath G. Ärztliche Ethik in Pestzeiten. *Munch Med Wochenschr.* 1957;99(5):158–162.
19. Schullian DM. A manuscript of Dominici in the Army Medical Library, *J Hist Med.* 1948;3:395–399.
20. Amundsen DW. Medical deontology and the pestilential disease in the later Middle Ages. *J Hist Med.* 1977;32:403–421.
21. Huppert G. *After the Black Death: A History of Early Modern Europe.* Bloomington, Ind: Indiana University Press, 1986.
22. Amundsen D.W. The physician's obligation to prolong life: a medical duty without classical roots. *Hastings Center Report.* 1978;8(4):23–31.
23. Navarro V. Why some countries have national-health insurance, others have a

national health-service and the U.S. has neither. *Soc Sci Med.* 1989;28(9):887–898.

24. Carse JP. The social effect of changing attitudes towards death. *Ann NY Acad Sci.* 1978;315:322–328

25. Proctor R. *Racial Hygiene: Medicine under the Nazis.* Cambridge, MA: Cambridge University Press, 1988.

26. Lifton RJ. *The Nazi Doctors: Medical Killing and the Psychology of Genocide.* New York, NY: Basic Books, 1986.

27. Kater MH. *Doctors under Hitler.* Chapel Hill, NC: University of North Carolina Press, 1989.

28. Levine E. *Darkness over Denmark: the Danish Resistance and the Rescue of the Jews.* New York, NY: Holiday House, 2000.

29. Klee E. *Was Sie Taten, Was Sie Wurden.* Frankfurt a/M, Germany: Fischer Taschenbuch Verlag;1995.

30. Quoted in Barnett V. *Bystanders: Conscience and Complicity during the Holocaust.* Westport, CT: Praeger Publishers, 1999.

31. Loewy, E. H., Loewy R.S. Bioethics at the crossroad. Health Care Analysis, 9.

2

Knowledge and Ethics

In the view we are presenting here and in the way we look at health-care ethics, ethics—like science—is not something fixed and eternal but a human activity that we in community together work to establish. This is not a sort of relativism but rather, as later sections will show, provides a commonly decided upon framework within which sufficient tolerance for a variety of views towards specific issues are acceptable and within which peaceful dialogue can occur. Furthermore, we shall insist that good ethics starts with good facts. Since these "facts" and our knowledge of the world around us changes, our viewpoint towards what constitutes an ethical problem, what constitutes the analysis of such problems and what our tentative answers will be must also change.

The way we look at ourselves—at our place within the universe, our relations with our community and with specific others—changes as our understanding of scientific, social, psychological economic and ecological "facts" changes. To claim that such insights do not have a profound effect on ethics is to claim that ethics is a revealed and immutable set of rules that we accept and "apply" to problems at hand. Although we have, as a matter of course, presented such a point of view when we briefly discussed ethical theory, our entire work is predicated on the assumption that ethics like all else is neither immutable, revealed nor simply a set of rules to be blindly applied.

For millennia humans assumed that the earth was the center of the universe and that humans formed the apex of creation separate and entirely separable from lower life forms. This led to a predominantly homocentric universe in which things and circumstances were pronounced "good" or "bad" entirely relative to how they affected human beings. Furthermore, our ability to have any long-term effects or to undertake actions whose consequences were much beyond affecting a handful of others were minimal. "*Fiat justitia, pereat mundi*" ("let justice be done even if the heaven's fall") was a debatable stance even then; but at a time when we had no way of having "the heaven's fall" it remained speculative and without profound practical consequences. Today we live in a world in which we, indeed, are capable of causing "the heavens to fall"—capable of destroying the earth as we know it. Thus, noble as it may sound, such a statement has become more than questionable today. Absolutism, whether religious, philosophical, ethical or political is no longer an ethically or for that matter practically defensible stance.

In ancient times and until recently our knowledge of ourselves and of the world about us did, of course, undergo some changes; but such changes were gradual, incremental and not profound. In the West they were, furthermore, strictly guarded against by the Catholic Church, which generally saw in change a threat to its authority and, therefore, the work of the devil. During the Muslim ascension to power much scientific, cultural and social change took place but it was still change that assumed the basic premise that man was the center of the Universe.

With the works of Copernicus and Galileo a radical change of perspective occurred: the belief that the earth was the center of the universe—that the rest of the universe rotated about it—was challenged. Rather, empirical evidence suggested it to be merely another little planet in a vast array of solar systems and planets. The Catholic Church very quickly saw the implications of such a point of view and did all it could to suppress it. But, inevitably, one cannot suppress "facts" and we began to see ourselves differently, at least those of us who had sufficient education to appreciate, absorb and integrate these new facts—the mass of mankind remained in an uneducated, religiously controlled and totally subservient state. It took a long time for such new attitudes to "trickle down" and become understood—especially against the active opposition of the parish priest.

Further changes would now come relatively quickly. Our knowledge and understanding of the world around us accelerated until today we are confronted with daily change. That, as we shall see, is not without its problems. What has been called "the scientific method" (essentially a hypothesis posing and testing rigorous inquiry) was adopted to address major problems and provide us with more than anecdotal and somewhat firmer answers. The study of human anatomy (a clandestine activity for many millennia) and a comparison of this anatomy to that of "lower" animals inevitably pointed to their enormous similarities. The work of Darwin suggested the continuity and interconnectedness of all life forms and the closeness of ourselves to what we choose to regard as "lower animals." Then, as today, fundamentalist religions (which very well understood the implications) continued to do all they could to oppose the teaching of these findings. In our view their attempts have only succeeded in making themselves look ridiculous. Darwin's findings (erroneous though some particulars may be) cannot help but suggest to us that the border between humans and other obviously sentient creatures is often more artificial and arbitrary than real. Freud, who in good part shattered the division between body and mind, showed us that our "subconscious" plays a distinctive role in our daily activities. Einstein demonstrated the arbitrariness and even questioned the non-existence of space and time. Such insights were crucial, especially in a time in which social mobility became ever more prevalent. No longer could a shoemaker, the son and grandson of shoemakers, assume that his son would follow his path. That social rigidity diminished with the French and American Revolution, with the Napoleonic wars and with ever more rapid means of communication. It continues to be affected by forces that are released, in part, by new scientific knowledge and technical capacity and in part by the work (even if often indirect and even if often not appreciated) of philosophers, social scientists, writers, artists and others in the humanities. We have come a long way from the *Nibelungenlied* to Dickens, from Corovaggio's, Vermeer's or Rembrandt's wonderful paintings to Van Gogh's "*Potato Eaters*" or Picasso's "*Guernica*," from Plato to Kant to Dewey. With and, in part, through the social sciences we have—despite horrible lapses—become more socially conscious and, with our new scientific understanding and technical developments, have opened up a vast array of opportunities as well as dangers.

The development of computers and artificial intelligence has given us new and unavoidable questions and problems: if computers can "know" and think, what is it to know and think? If computers can—as seems probable—be equipped with what we would recognize as "emotions" and if they can think and plan ahead how could and should we relate to them? Robots today can not only repair themselves but can create new and "improved" generations of robots. These are not matters that can simply be shrugged off—they will affect the way we think, live and ultimately develop as individuals, as societies and as species.

For ethics another and perhaps ultimately most critical development has occurred in the last decade. For millennia the argument as to whether emotion or reason should prevail in making judgments in ethical questions has raged. Few will doubt that emotion does in fact play a role in our recognition of and judgments about ethical problems. Some (like Plato) have argued that poor ethical judgments are the result of either poor information or poor logic: evil comes about not because men knowingly act in an evil manner but because they either know too little or commit errors in reasoning—a belief which Aristotle most certainly did not share. Much later Hume would argue that emotion not only was but in fact, should be the deciding factor—that "reason is *and of right ought to be* the slave of the passions."[1] Kant, on the other hand felt that emotion ("the passions") that he equated with inclinations (that which we want to do) should be de-emphasized as much as possible.[2] Kant, of course, recognized that on a practical level making judgments was inevitably influenced by passions, emotions and inclinations and recognized the importance of schooling the inclinations so as to make them more readily accord with our reasoned judgments. Kant's distaste for inclinations or emotions should come as no surprise: as much as he tried and, at times, succeeded in keeping his religious views separate from his ethics and philosophy, his pietistic background provided him with a distaste and suspicion of desires: things that "felt good" were more than likely to be sinful!

Scientific insights in the last decade or more have clearly demonstrated the importance of both feelings and reason in the making of judgments. The works of Damasio (Iowa), Ledoux (New York) and Roth (Bremen) have clearly and unequivocally demonstrated this interdependence. Such an interdependence however, does not answer the "ought" question—given that emotion inevitably affects the way we make decisions ought we embrace it, regard it with suspicion or as much as we can minimize its role?[3–6]

The studies of the Damasio, Ledoux and Roth as well as those of some others have shown more than mere interdependence. Higher animals are equipped with separate and separable emotional and cognitive centers—the former resides in the temporal lobe region called the "limbic system" and the other in the frontal lobes. These two areas are connected by an intricate series of neuro-chemical and electrical pathways. In patients who fully recover after their limbic system (either by surgery or by accident) has been destroyed or entirely severed from its connections with the frontal lobes or cognitive centers, cognitive function appears unimpaired: they can, for example, solve intricate problems of logic and would, at first blush, seem like fully rational beings. While emotionally "flat" they are seemingly cognitively unimpaired. Such patients can live "normal" physical lives for years. However, something has been lost. Not only are they affectively changed, their ability to make sound judgements is severely stunted: over and over again they will logically reason out a problem, implement their "rational" decision and abysmally fail to deal effectively with the problem at hand. Their judgements are not only "bad" they are often entirely self-destructive. Beyond this, patients never seem to learn from such failures and to deal with similar problems in the future in exactly the same way. Those patients studied have all led stunted, miserable lives.

The authors postulate the presence of an "emotional memory"—a memory that teaches us certain things that translate into emotive responses. "Man with big stick running after you is dangerous" is generally more an emotively learned than a cognitive and reasoned response. It appears that "emotional memory" (as distinct from cognitive memory) is predominantly laid down in infancy and early childhood, is virtually permanent (that is—it is not prone to be forgotten and is very difficult, albeit it possible, to change) and regularly interacts with our cognitive centers. As any of us who have taken examinations know, the cognitive centers are far different: memory of this sort is laid down throughout life, is easily forgotten and is malleable.

If these facts are indeed as stated (and there is no reason to doubt this) then the "ought" question is a bit nearer to an answer. Although every radical "rationalist" (or "idealist") will decry the use of empirical evidence as not providing a sound basis for answering that question and even though we may be accused of committing a "naturalistic fallacy" the fact that decisions made without input from our emotional memory turn out to be destructive cannot simply be shrugged away. Shrugging it away constitutes something even more fallacious—it suggests that scientific findings and "facts" based on them can simply be shrugged off and reason alone appealed to: something that these observations have just shown not to be the case.

We are not suggesting that emotion should be allowed to predominate. Indeed, a judgement made without careful reasoning and logic and made by emotion alone would undoubtedly prove more destructive and more dangerous. Further, it would lack all external referents, all means of retrospectively judging it as right or wrong—as is the case in the so-called "care ethic" the emotion itself becomes its own referent and right or wrong (or even better or worse) vanish.

The interrelationship between reason and emotion at the very least casts doubt on those who claim that clinical or ethical problems should be solved by purely "objective" or purely "reasoning" means and that emotions should, as far as that is possible, be kept out of either decision. What these findings do not address, however, is the extent to which either reason or emotion should be involved. In our discussion of "compassionate rationality" and "rational compassion" (see Chapters 3 and 8) we find a prototype of this problem. It may well be that the role of emotion is a hem-shoe on reason—that it may caution us that our reasonably reached conclusion may well still be faulty and in need of revision, and it may well be that reason acts in a similar manner to curb our emotion.

What is critical here is that scientific insights, technical advances and social changes alter the way we conceptualize and deal with ethical problems and that in turn the way we deal with such problems affects all others. Compartmentalizing knowledge or understanding or isolating it from other disciplines interferes with progress and ultimately stops it. There is no reason to expect that this process of change will not continue. Clinging to an unchanging ethics embodied in a few (or in many) rigid "principles" universally and unchangeably applied makes a Dodo bird of ethics—and may well presage its extinction. For—to paraphrase a section of the New Testament (something we are not apt to do usually!)—ethics is made for men, not men for ethics!

Our newer understanding of scientific "facts," our capacity to rapidly translate such insights into technical capacities and hence action and our ability to on the one hand more accurately and on the other much more poorly foretell the future have raised critical ethical questions about the role of knowledge itself. Heretofore "knowledge" (or technology) was always considered to be "value neutral"—it was the application of such knowledge or of such technology that was not. Hans Jonas in his seminal works argues that knowledge itself has assumed a moral dimension and that there may be things we ought not try to know.[7] While we do not agree either on a purely pragmatic (it is not possible to keep humans from seeking to better understand the world about them) or on an ethical (are we to be frozen into a "status quo"?) level the concerns that he raises are well worth considering.

Jonas argues that the Kantian Categorical Imperative (see Chapter 3) "act so that you could will the maxim of your action to become a universal law of nature" might well be changed to "act so that acting tomorrow remains a possibility."[7] Under some circumstances Jonas' maxim might well play havoc with some of the Kantian absolute rules! Jonas argues that the future (thanks to our far more accurate instruments) has in one sense become more readily foreseeable but that (because of the rapidity and unanticipated character of change) it has simultaneously become far more difficult to foretell. Furthermore, if new knowledge is inevitably (and for a number of good and bad reasons) turned into technical capacities, knowledge itself becomes an ethical problem. Thus he concludes that we should undertake no changes whose immediate or long-term consequences might possibly have serious unforeseen consequences—a position which unfortunately would logically mean that, since we cannot ever know what short or long term effect innovations may have with any degree of certainty, we can truly not change anything. We have called this attitude "logical negativism": it is pretty much the opposite of "logical positivism" (the belief that science is fact and can and will solve all of our problems).

While we strongly disagree with Jonas rather extreme conclusion—which would reduce our future to our present—it is deserving of a good deal of attention. In reality

Jonas' argument is a form of Hubris—it rests on the belief that humans have "gone far enough," that we are the apex of creation and that anything more we create is more likely to risk leading us down the path of destruction rather than up the difficult path to improvement. Man is capable of both—and capable of making a choice.

While the last century can be readily used to argue for Jonas' position (various genocides, a pervasive loss of a sense of community, etc.) it can with as much merit be used to argue against that position: Slavery was abolished in the 19th century and the progress (much as it leaves to be desired) made in assuring civil rights to all members of our society has not only progressed in the U.S.A. but taken on world wide proportions. The Rwandan Genocide or the various debacles in what was once Yugoslavia showed a world who had at least begun to care about such issues and who, in some instances, was prepared to take consequent action (as flawed and belated as these efforts were)—when Hitler began his persecutions the world largely yawned. Remarkable progress in public health and medicine has been made and has eased the lot of many. None of this would—without newer knowledge and its translation into appropriate technology—have been possible. Such progress has not been without its problems—anticipated and otherwise. Indeed much of this book concerns issues without which such "progress" would be mute. And yet: there is no doubt that one can prevent the abuse of new knowledge by stopping (which, in fact, we cannot hope to do) its acquisition just as one can eliminate medical malpractice by forbidding practice. But that seems a rather extreme sort of thing to do!

There are many "advances" in our knowledge which are questionable—should one really allow the search for a highly infectious and deadly micro-organism resistant to all anti-microbials or permit the development of a device which could destroy our planet? And how do we know where our search for knowledge may lead? It seems to us that the only way of providing some safeguards lies in allowing (indeed mandating) some time, thought, scrutiny and effort between the acquisition of knowledge, between its translation into technology and the application of such technology in other than a most limited fashion. Moratoriums—if they are used to think through problems, engage a wider audience and debate the issues—have distinct merit.

We live in a world that offers us astounding social and scientific opportunities as well as threatening us with extreme dangers. This makes it all the more necessary that all members of the community who may be affected by our actions (and ultimately this means all of us) are given not only a voice but the necessary conditions for participation: that is mutual respect, economic security and educational opportunity not based solely on the ability to pay. It also means that those who can*not* participate (the weak, sick, mentally disabled as well as future others) must have their interests represented. Crafting an ethical framework in which we all can live and thrive cannot—by definition—be a "top down" strategy in which at best an *ethic* **_for_** the weak and powerless is magnanimously and condescendingly promulgated; it must be an ethic **_with_** the weak and powerless. One of our first tasks, then, is to provide persons across the globe with the necessary conditions (economic security, education, personal respect for one another, freedom **_for_** thinking and expressing their thoughts) that allow them to be truly empowered. This is not a Utopian ideal but one towards which we can realize—not tomorrow, not the next day but as a clear goal towards which we ought to strive not only for ethical reasons but for reasons of preserving our very selves.[8]

In this chapter we have tried to show that ethics and of necessity that part called health-care ethics is influenced by our current knowledge of the world around us, that ethics is as dynamic as all other knowledge and that our knowledge (be it of ethics or of other things) is embedded in a social setting which modulates the way we perceive (and ultimately use) knowledge and which is, in turn, modified by that knowledge. Likewise we have pointed out the opportunities and dangers facing us in our ever-increasing knowledge, in the speed with which such knowledge is transformed into technology and with the haste with which such technology is then applied.

━━━━━━━━ **REFERENCES** ━━━━━━━━

1. Hume D. *A Treatise of Human Nature.* Oxford, England: Oxford University Press, 1968.
2. Kant I. *Grundlegung zur Metaphysik der Sitten,* ed. Wilhlem Weischedl. In: *Immanuel Kant Werkausgabe VII.* Franfurt a/M, Deutschland: Suhrkamp Verlag 1989 and Kant I: *Die Metaphysik der Sitten,* ed Wilhlem Weischedl. In: *Immanuel Kant Werkausgabe VII.* Franfurt a/M, Deutschland: Suhrkamp Verlag, 1989.
3. Roth G. *Das Gehirn und seine Wirklichkeit.* Frankfurt a/M, Deutschland: Suhrkamp Taschenbuch, 2000.
4. Damasio A. *Descartes' Error: Emotion, Reason and the Brain.* New York, NY: Avon Books, 1995.
5. Damasio AR. *The Feeling of What Happened: Body and Emotion in the Making of Consciousness.* Fort Washington, PA: Harvest Books, 2000.
6. Ledoux W. *The Emotional Brain: the Mysterious Underpinnings of Emotional Life.* New York, NY: Touchstone Books, 1998.
7. Jonas H. *Das Prinzip Verantwortung: Versuch einer Ethik für die technologische Zivilisation.* Frankfurt a/M: Suhrkamp Taschenbuch, 1984.
8. Loewy EH. *Moral Strangers, Moral Acquaintance and Moral Friends: connectedness and its condition.* Albany, NY: SUNY Press, 1977.

Theoretical Considerations

3

——— INTRODUCTION ———

What constitutes an "ethical" problem or question? In this book we shall maintain that an action (or a failure to act when action was possible) that affects other sentient beings is by its very nature ethical. We do not maintain that other factors fail to enter in and that such factors often are the more critical ones. Because of their effect on others—and ultimately we ourselves are "others"—ethical considerations are one of the most important considerations; but they are neither the only nor in all instances the deciding factors. Furthermore, the problems are usually highly complex—if only because that "other" who is affected is generally not one but many and because "affected" must not be understood merely as an immediate or direct effect. Thus ethical questions often raise questions of hierarchy (which of the various "others" affected is or are the most important and the most central to the obligation of the moral agent) as well as questions of future others and/or to the kind of situation in which our actions may place them.

We hold that the most basic thing in ethics (and especially but hardly only in Bioethics) is to be sensitive to the ethical questions lurking (and often lurking unrecognized) within a specific clinical or a wider (say allocation of beds or organs) setting. To begin reasoning about ethics and ethical questions we first need to recognize their presence. This is, it would seem, an obvious statement. Nevertheless it is astonishing and, indeed, frightening to find that many physicians when asked about ethical problems among their current patients deny that any are present.

Good ethics starts with two preconditions: (1) Good facts (or at least the best "facts" that one can get); and (2) an institutional and social setting which permits deliberating about ethics and gives enough latitude to carry out ethically acceptable decisions. Ethical speculation not grounded in good clinical and social facts is just that: speculation and a form of mysticism that can be more dangerous than it is useful. On the other hand, the institution within which the actor must act constrains not only his or her action but likewise even constrains examining a variety of options.

Managed care (which will be discussed in later chapters) may serve as a typical example. Furthermore: it is most difficult if not indeed impossible to develop a just institution in the context of an unjust society—persons or governments which can allow a state of affairs in which persons working a full week still must exist below the already unacceptably low poverty level can hardly be expected to have much interest in the plight of the uninsured whose access to health care in reality exists only in life-threatening emergencies or by the grace of capricious charity.

This chapter will be devoted to a brief sketch of ethical theory as it relates to health care ethics. It is an extremely brief sketch, and, therefore, necessarily deficient in many ways. Readers who are interested in more detailed understanding of ethics in general or health care ethics in particular are urged to refer to some of the primary sources themselves. At the outset, however, a few words need to be said about the problem of justification: How do we judge a given course of action as one ethically appropriate (or at least as ethically not inappropriate) versus judging it to be ethically inappropriate? This gets us to the question of authority: What kind of "authority" is sufficient (or put another way, what assumptions can safely be accepted) when we argue our point?

In a pluralistic world and in a pluralistic society traditional ways of appeal have proven to be to no avail when discussing such matters with persons who come from traditions, cultures, or belief systems quite different from one another. Likewise, appealing to the law of the land cannot tell us if such laws are or are not ethically appropriate—not all laws are. One hopes that most laws are ethically unproblematic and that some even help cement ethical behavior; but one also knows that throughout history some laws have ordered us to engage in some ethically very questionable or even indisputably unethical forms of activity or have intended to prevent us from acting in inarguably ethical ways. An appeal to the law is an appeal to a situation as it exists so as to justify its own existence: i.e., things simply ought to be what they are. Appealing to religious beliefs suffers from the very same narrow approach. Are religious pro- or prescriptions "good" because they are religious, or are they religiously pre- or proscribed because they are "good?" If they are considered ethically correct merely because they derive from a religious base (book, statement of higher clergy, etc.), they suffer from the same problem as an appeal to law: i.e., they become a circular way of arguing. If, however, religion only prescribes good or proscribes bad ways of acting, we are left with asking what the criteria for judging such acts good or bad are. Inevitably, we are left with using a form of justification that must lie outside religion itself.

Furthermore, we live in nations and in a world in which multiple cultural, religious and even legal systems must live side by side, and, if we are to get on with our lives, we must coexist peacefully. If appeals to religion or the law cannot guide us in making ethical judgments, we must rely on other ways of finding good reasons for justifying and arguing our propositions. Throughout this book we will assume (and will briefly argue) that it is possible to come to some general agreements about ethical issues. Such agreement relies on a framework of common human capacities and experiences and is enabled by a common sense of primitive logic sufficient to enable our dialogue. We will not and cannot (and, we think, ought not) try to build a fully contentful ethic that most of us would be willing to accept and that would allow consensus on all troubling issues. But we can forge ahead and, while seeking for broader agreement, find some content about which we would generally agree.[1]

At this point we need to mention a word about codes and oaths. Codes and oaths (which really are meant to enforce particular points of view) have been used throughout history in both religion and law. They can serve us well as guideposts if we assume that the direction in which they point is likely but not by any means necessarily the right one and if we therefore continue to question, analyze, and revise. In this book, while mindful of the importance of what religion, law, code, or oath can teach us, we shall not rely on such teachings for justifying any particular position or point of view. (See also Chapter 1.)

Medical ethics in one sense does and in another does not differ from ethics in general. In a qualitative sense, the principles of action that underwrite all moral choice underwrite the choices here; in another, more quantitative sense, medical ethics differs because of the complex and often puzzling nature of its subject material and because of the intense emotions accompanying many of its choices.[2]

Ethics in general (and applied ethics in particular), if it is to accomplish its avowed goal of helping people to live the good life, relies for its deliberations and judgments on a rich background of philosophy, psychology, anthropology, sociology, history, and the arts. Applied ethics, in addition, requires an understanding of the matter to which it is to be applied, an understanding that includes emotive and experiential considerations rather than cognitive material alone. Likewise, to be a meaningful activity and discipline, ethics requires more than merely logical reasoning: we have argued and shall continue to argue that the interest persons take in questions of ethics as well as the answers they ultimately evolve are initiated, and throughout the process leavened, by compassion. Some would deny this. They feel that practicing ethics is merely a rational activity, that ethical analysis can take place away from the context in which the problem is embedded and from the moral actors who eventually must act, that it need only be conversant with the cognitive material under immediate considerations, and that such analysis then constitutes the sufficient condition for action.

This book assumes that medical ethics cannot make a judgment about problems unless it considers both their context, their history, and their cultural setting as well as the circumstances, values, and feelings of the moral actors who are involved. A dispassionate analysis of encapsulated problems unmindful of their context may help clarify some aspects, but it cannot, in a practical sense, provide equitable, let alone compassionate, solutions. Problems taken out of their cultural and particular context, divorced from the moral actors who must act, and uninformed by history, are changed problems. Analyzing them in such an encapsulated fashion can be helpful in providing more dispassionate insights, but such insights, if they are to fulfill a meaningful role, must be carefully fitted into particular contexts and must be mindful of the moral actors involved.[3,4]

Ethics derives from the Greek *ethos*, as morals derive from the Latin *mores*. Both derive from the word for *custom, manners*, or *the disposition peculiar to a given people*. In German a third word, *sittlich*, is similarly derivable. Although these words are used interchangeably, they each carry a somewhat different flavor: morals, more a religious and often a sexual flavor; sittlich, more one of manners or convention; and ethics, one that much more clearly deals with what we shall understand by the term throughout this book. In our framework, *ethics*, although it inevitably contains the root meaning of custom, manners, or the disposition to act in certain ways, transcends it. Unless we specifically state that we are using the word *moral* in a different sense, we shall often use the word *moral* interchangeably with the word *ethical*.

It is often not easy, despite these definitions, to delineate an "ethical" from some other sort of problem. Intuitively we know that prescribing penicillin to a patient with pneumococcal pneumonia is largely (but not merely) a "technical" matter, whereas not supplying nourishment to a terminally ill patient is largely (but not solely) an "ethical" one. But we are stumped when we try to analyze the difference. As with describing what is and is not pornography, we find that we "know it when we see it."

Throughout this book, we assume that ethical problems are inseparable from other problems and that virtually all other problems likewise contain ethical questions. Medicine (or any other kind of human activity) needs to ask and first answer the "how" question: How can I (or can I) treat this disease, prolong this patient's life, restore that patient's function? Ethics asks a different albeit inevitably linked question: "Given that I can do something, ought I to do it?" Inevitably the two questions are and, above all, ought to be linked with one another. They ought to be linked especially in a day and age in which our technical capacities have, compared to former times, become not only much greater but also much more dangerous—and to far greater effect. Ethics searches for a way of life (secular ethics, Christian ethics), seeks rules of conduct (applied ethics), and inquires about ways of life and conduct (metaethics).[5] Thus, as in ancient times, ethics remains largely concerned with seeking the "good" life and, more broadly, in defining the "good." Ethics is concerned with two key concepts: the concept of "good" and the concept of "right." When we speak of the good (see Chapter 2), we refer to a goal at which we aim; when we speak of right, we have a more juridical framework in mind. Used as a noun, the concept of right denotes a justified claim, sometimes a claim justified by a particular notion of the good and meaningful only when secured by and through the community in which such a right is said to exist.

As has been stated, ethics is held to be other-directed. For an act to be considered "ethical," it must in some way, however remotely, affect another sentient being. There are those who, pointing to pure duties to oneself, would deny this.[6,7] Duties to oneself, however, are a peculiar concept. In general, when duties to self are invoked, they are justified by appealing to the harm that failure to discharge these duties would, in fact, bring to others: relatives, friends, the community, or, ultimately, God. In true duties to self and self alone, rights and duties are simultaneous rights and duties: i.e., they are owed to and by the same person. Therefore, it would seem that they are disposable by the same person. Duties owed to oneself are self-referential, i.e., they lack a referent other than oneself, and thus the arguments cannot escape a certain circularity. Under ordinary circumstances, individuals can waive their rights and can, for instance, decide that they do not want to collect a sum owed to them and thereby absolve their debtors of the duty to pay—but how do I make sense of the notion that I forgive myself a debt that I owe to myself? Forgiving myself a debt or absolving myself of an obligation (which, since I owe it to myself I theoretically should be able to do) becomes a trivial concept. For purposes of this book, problems of ethical content will be considered to be limited to problems involving another. In fact, virtually all problems—our dilemmas and, ultimately, the way we handle them—are problems for the very reason that they involve another or others.

————— DEFINING THE GOOD —————

In searching for the "good life," the "good" has largely remained elusive and indefinable. Good, in general parlance, is used as an adjective: e.g., a good knife or a good diagnosis. This, of course, invites the inquiry: what it is "good" for or in what (as Aristotle would say) does its goodness consist?

Goods may be intrinsic goods ("goods in themselves") or instrumental goods (goods that serve as the means for achieving other, usually higher, goods). It is the intrinsic good that has escaped definition. The quest for the *summum bonum*—the ultimate good, that which is good in itself—is sometimes seen as the greatest good in a hierarchy of goods (e.g., rationality, for Aristotle), sometimes as the common denominator of all other goods (e.g., pleasure, for the hedonist), and, at times, as an almost mystical, religious good (e.g., God, for the religious). The way we shall use the word is more pedestrian: the referent is the common experience of what men universally call good and what they call bad (e.g., joy as a universal good, pain as a universal evil). Other goods, for the purpose of this book, are defined largely by rational beings experiencing and enunciating that good in community.

————— THE FUNCTION OF ETHICAL ————— THEORY

Just as the principles of pharmacology are necessary if we are to prescribe properly, some understanding of ethical theory and its principles is most helpful if we are to analyze and understand moral problems. What follows is a necessarily oversimplified and therefore somewhat falsified account of ethical theory. At least this brief acquaintance with theory, however, is most helpful in understanding and working with the practical problems encountered on the ward. Too often persons unacquainted with theory and unclear about method are unable to come to grips with—or even clearly identify—such problems and end up in a discussion based on ill-founded assumptions, unexamined beliefs, and traditions held personally dear but unpersuasive to others whose assumptions, beliefs, and traditions are different.

There are, of course, a variety of ethical theories as well as many variations, interpretations and applications of each particular theory. The principles that each of these theories underwrite seem, at first glance, irreconcilably different. But such principles are fashioned by men to live by. They are instruments that should help one fashion the "good life," not narrow exercises here to constrain and oppress man. The principles we come to accept as our own vary with personal experience, innate personality, and social conditioning, but in spite of this, our decisions are often similar.[8] Ethical theories and principles are no more sacrosanct than those of physiology or pharmacology; they are merely working hypotheses with which we can hope to deal with our problems and which, in turn and by what we learn by their application, we seek to adapt and improve.

—— CONSEQUENCES AND INTENTIONS: —— UTILITARIANISM AND DEONTOLOGY

Traditionally there have been two ways of looking at the "rightness" or "wrongness" of an act. We can look either at consequences or at intentions. If we focus on the consequences of an act, we are judging the outcome we value to be good and the outcome we find bad to be bad. If we adopt this consequentialist view, we are apt to search for the "good" in any act as one that brings about the greatest amount of good. On the other hand, we can ignore the consequences of an act, judging that, in any event, the outcome is far from completely under our control, and we can then seek the good in the intention of the actor rather than in the actual consequences that are brought about by the act itself.

If we adopt the latter view, we rely on the agent's adherence to duty to judge the merits of his or her actions. (Of course, consequences are not entirely ignored: rather, consequences here are the intended and not the actualized consequences.) In both of these differing ways of looking at ethics, a standard by which to judge what is ultimately "good" is missing. Broadly speaking, the first of these two views has been called "consequentialist," or "utilitarian." It relies on the works of Jeremy Bentham (1748–1832) and on those of his student John Stuart Mill (1806–1873).[9,10] Consequentialist viewpoints, of course, preexisted Bentham and Mill just as such viewpoints have continued to find adherents and are developed further by some ethicists today (see the later section dealing with Peter Singer's "Interest Utilitarianism"). The second view is called deontological or, at times, Kantian, notwithstanding that Immanuel Kant (1724–1804), its most eloquent proponent, is not the only one. Kantian ethics has had a profound impact on ethics in general and on medical ethics in particular.

Kant puts primacy on the autonomous selection of our moral principles.[6,11,12] In essence he claims that persons are free or, what he holds to be the same thing, must act as though they were free. That is, persons are deserving of the respect that all rational beings deserve because of their capacity as self-legislating moral agents, i.e., as capable of setting their own (autonomous) moral law. Heteronomous law, in contradistinction to autonomous law, is a law extrinsic to the agent: that is, such a law is not actively thought through and embraced by the agent him or herself but has been accepted (unthinkingly) from another source such as the law of the land, custom, or religious rules. Persons who unthinkingly acquiesce to such laws, according to Kant, are not acting morally. Given their capacity to make these choices and their failure to make them they are blameworthy. Adherence to heteronomous law is not praiseworthy simply because it is the law. Adhering to heteronomous law (law that comes from outside the moral agent and is not selected by him or herself) may be morally neutral (as when we obey the law not to park in a certain place), morally blameworthy (as when we obey a law forcing us to discriminate against certain races), or praiseworthy, if our adhering to such law is motivated by more than blind obedience to the law itself.

Kant bases his rule-oriented ("deontological") moral philosophy strictly on those rules that a rational agent legislates for him- or herself (autonomously derived law), and such rules, to be moral, must be universalizable: i.e., a moral agent setting these laws for him- or herself should be able to will that such laws would apply likewise to all other rational beings. We must, Kant says, be willing for the rules we set for ourselves to become a "law of nature": i.e., we must be willing to have such rules apply universally. The categorical imperative, or "universalizability" principle is one of the fundamental struts of Kantian ethics (and, we would argue, forms one of the bases of most thinking in this subject).

Likewise, according to Kant, "imperatives" (i.e., rules) can be hypothetical or categorical. Hypothetical rules are only binding if one wishes to achieve a given end: i.e., they have the form "if you wish accomplish X, you must do Y." Categorical rules (to Kant there is only one categorical imperative from which all other rules are derived) are always binding. Such a rule simply says that logic would compel the will of all rational beings to accept such an imperative. It would be irrational to call an action moral for oneself under the same circumstances that one would consider it immoral for another—the basis of the Categorical Imperative.

Kant derives another, very critical part of his philosophy from another formulation of the categorical imperative. Since all rational beings have the capacity to set their own moral law, they are deserving of respect and cannot merely be used as means but must always also be treated as ends in themselves. This of course means that all persons—no matter their race, nationality, or station in life—are deserving of this same basic respect. In dealing with them we must respectfully consider their goals as well as our own. When we generalize such a point of view we end up with a world in which there is an ultimate "realm of ends:" in other words, all of our ends must ultimately have a common meeting point.

The rules (or "maxims," as Kant calls them) that are derived from such a categorical imperative are largely "negative" rules, and Kant believes that such rules are absolutely binding under all circumstances. Rules against lying, murders, or actively harming another, are rules that (according to Kant) one can always discharge: i.e., these are things one can always refrain from doing. Such rules are called "perfect" duties and, according to Kant, should not conflict with one another. But negative rules alone would leave many of us without some very important ethical obligations: obligations of helping one another, of beneficence, and of compassion. Kant acknowledges and, indeed, emphasizes such rules. Perfect duties (those that universalize and are binding on all rational beings at all times: the duty not to murder, for example) contrast with imperfect duties (beneficence, for example: those that are not binding at all times but cannot be done without). These imperfect duties, among other things, give content and direction to ethics.

However, since such "imperfect duties" are not as logically compelling as are the "perfect" rules, they must be argued for on somewhat different grounds. Kant does this in two ways. On the one hand, he states that while willing a world in which not acting beneficently would not contradict logic, the fact that we all at one time or another find ourselves critically in need of the help of another would make it logically impossible to "will" such a world: Doing so would "force the will to conflict with itself."[6] Beneficence and the other "imperfect" rules, moreover, can be argued for on the basis of the realm of ends. Because we must treat each other as more than merely means to our own ends, we end up in a "realm of ends" in which the interest of others plays an essential moral role. We cannot simply ignore the goals or interests of others when formulating our own or another's course of action but, if we are to act morally (that is, if we are to act in a manner we could will as a "law of nature"), we must consider the goals of others together with our own. Since we can presume that everyone is interested in his or her own welfare, such a realm of ends enjoins all of us to take at least some interest in the well being of others.

While we can, according to Kant at least, always comply (and always comply without internal conflict) with perfect duties, compliance with imperfect duties is not always possible and not always necessary. Such imperfect duties are conceived as optional: while we may not refrain from duties of beneficence altogether (and should be encouraged to do so as much as possible), we cannot always and in all situations act beneficently.

Finally, Kant holds that one cannot derive an "ought" from an "is;" that is, one cannot derive rules from things the way they are. The status quo is not right (or wrong) merely because "that's the way things are." Rather, rules must accord to reason and logic and, therefore, be universalizable. In refusing to grant the status of "ought" as equivalent to "is," Kant furthermore holds out hope for future changes. He is a true child of the time of the French revolution, although he abhorred its methods.[6,11,12]

Utilitarianism, or consequentialist ethics, can be either act or rule utilitarianism. In both, the determination of "rightness" or "wrongness" is based on the actual consequences achieved: in act utilitarianism the rightness of the act, and in rule utilitarianism the rightness of the rule, is judged by the consequences the act or the rule brings about. The trouble with either of these forms of utilitarianism, of course, is that they make any action that either produces maximal "good" in itself (act utilitarianism) or any action that conforms to a rule seen to maximize the "good" (rule utilitarianism), good only because of the consequences. Consequences, to be "good," must achieve the greatest "good" for the greatest number. One of the largest drawbacks, of course, is that the notion of the "good" remains ambiguous.

Both utilitarian and deontological ethics have problems peculiar to themselves. Neither defines the "good" except in the most general terms. To utilitarians good is a balance of pleasure over pain; for Kant the only absolute good (that which is "good in itself" and not merely an instrumental good) is the good will: i.e., the will to subject oneself to autonomously derived moral law. Utilitarianism, like all consequentialist theories, by relying purely on consequences in judging an act and its actor (or a rule and its implementation), ignores fallibility and the unpredictability of events in assessing praise or blame. An action (or rule) that turns out badly, no matter what the unforeseeable cause, is bad. In a sense, rule Utilitarianism, in that it would have you follow that rule which maximizes the good of most, is a form of deontology, insofar as it follows a set rule.

Utilitarianism, at least in its purer form, has lost much of its appeal. When applied to the realm of private relationships, it is rightly deemed dangerous: In its name, dangerous experiments on helpless people that could, however, benefit a large number would not be precluded, and a few innocent could readily be sacrificed if doing so would result in great benefit to many. In stressing outcome, utilitarianism ignores motive and the interests of the minority and fails to account for human fallibility. We shall see in later chapters, however, that when we are faced with issues of resource allocation in which we deal not with individuals but with groups, we inevitably must take consequences and the greatest good for the greatest number into account.

Deontological theories, on the other hand, are often accused of being inflexible and deficient in their ability to guide our daily decisions. An action considered to be wrong—lying, for example—is wrong under every and all circumstances. The resulting conflict is inevitable in daily life and highlighted in medical practice. Further, Kant's ethic is one of pure form and lacks specific content; the categorical imperative ("act so that you can will the maxim of the action to be a universal law") is impeccably true but provides little guidance in concrete situations. In defense of Kant, however, it must be said that he clearly did not set out to give specific answers to specific problems in concrete situations. Rather, and quite explicitly, he set out to provide a framework within which decisions about specific problems and affecting concrete situations could be understood. And, all criticism since notwithstanding, that framework retains much of its validity. It is, however, often charged that, by narrowly twisting language to accommodate a given contingency, almost any maxim could be construed to be a universal.[13] Kant, we think, might fall back on his insistence that the only thing "good in itself" (i.e., good without exception) is "the good will"—the will to formulate one's own (rationally defensible) moral law and to adhere to it—and that twisting circumstances to make what is clearly illicit permissible supports, rather than defeats, his position.

Kant stresses that adherence to moral law from duty instead of from inclination is what acting meritoriously is all about. Agents who, despite their inclination to act otherwise, act in accordance with their duty are meritorious ("praiseworthy"), whereas the agent who acts merely because of inclination is not. If one performs an act that is one's duty but does so motivated solely by inclination, one is acting neutrally and is neither praise- nor blameworthy. This intrinsic suspicion of human inclination (not surprising in one whose background is pietistic) is, as we shall see, the precise opposite of the "care-ethic" (see section to follow). While there is certainly great merit in acting morally despite one's inclination to do otherwise, it is difficult to deny praiseworthiness to the person who acts morally and takes true pleasure in doing so. It is difficult to consider Albert Schweitzer, who very obviously loved caring for disadvantaged sick people, or Mother Theresa, who evidently loves helping the poor and downtrodden, to be less praiseworthy than someone who, while helping, does so only by overcoming his inclination to ignore those in need. One can claim that overcoming one's inclination requires more energy and is more difficult than to act in accordance with it. And yet, we generally define a "good" person as one who is disposed to act ethically and kindly—not as one who, with set teeth, forces him- or herself to do so.

In dealing with patients within our vision of the patient-physician relationship, inner intention rather than merely outward performance becomes crucial. Kant's philosophy, in stressing motive and duty, does just that. Kant did not intend to produce a "cookbook" of ethics but a firm foundation upon which rational men could build the good life. Except as Kant is concerned with the "realm of ends," Kantian ethics, as with most varieties of deontological ethics, gives scant direction to solving problems of social justice that form an ever more important part of medical ethics. When it comes to questions of allocation and to questions of whether more funds for sometimes exotic crisis care or more funds for preventive or primary care should be expended, a utilitarian type of approach that seeks to produce the greatest good for the greatest number seems much more appropriate if not, indeed, inevitable.

The problem with Kantian ethics even when it comes to personal relationships is often felt to be the rigidity of the various rules. Perfect duties are always and invariably binding and, if one is to believe Kant, will never conflict. Yet this is simply not so. The classical example of hiding six Jews in the cellar but not being permitted to lie to the inquiring Gestapo man is one of a variety of difficulties with such an absolutist point of view. And yet it is evident that in this example, as in many examples taken from everyday life or from clinical practice, such absolute rules do in fact clash. When the Gestapo man is told that six Jews are hidden in the cellar—a thing he is told because of blind devotion to an absolute rule—such rules are seen to clash: the promise implicitly given when we undertake to hide another from evil would be violated by "telling the truth." Likewise, in medical ethics, lying to patients can generally be considered as unethical: persons are deserving of respect, and telling the truth is an absolute requirement. And yet there are exceptional occasions when the humanity and wisdom of such truth telling may legitimately be brought into question (legally this is called "clinical privilege:" it's the classic "if I tell him the truth he will jump out the window" sort of argument and, despite the fact that physicians have often argued otherwise, it is a very rare situation—though it does occasionally occur).

Moreover, unmodified Kantian ethics would make it difficult to choose between equally needy persons in allocating critically needed resources: The known criminal and the lifelong philanthropist could be argued to have the same standing. A view such as that of W. D. Ross may help in dealing with such problems. W. D. Ross in the earlier part of this century developed the concept of duties that are "*prima facie*,"[14] i.e.,binding unless overwhelming and compelling moral reasons to the contrary can be marshaled. Of course, such an approach fails to give direction as to what such "overwhelming moral reasons" might be, but it certainly suggests an approach. To W. D. Ross, relationships are an important source of such moral reasons. Relationships modify but do not change the basic obligations we owe to everyone. Relationships are, at least in part, defined by the obligations they entail: in explaining words like friend, student, health care professional, spouse, or even enemy to someone unfamiliar with the term, we invariably must consider what obligations such relationships entail. My general obligation to all others forms the baseline of the obligations of my special relationships. The relationship health care professionals have with their patients starts, at minimum, with the obligations all humans have to one another even though it, hopefully, will transcend that minimum. Even my obligations to my enemy are still underwritten by general human obligations: except when necessary in self-defense or to save innocent others, I cannot, and be acting in an ethical manner, simply kill or otherwise gratuitously harm an enemy. Certainly, a physician's obligation to treat a patient once a physician–patient relationship exists is arguably greater than the simultaneous obligation to treat a stranger.

One of our most prominent contemporary Utilitarians, Peter Singer—who currently holds the endowed chair for Bioethics in Princeton and formerly was at Monash University in Australia—has developed what he calls "preference utilitarianism." He and his views have been widely attacked, especially and most viciously within German speaking countries (notably Switzerland, Germany and Austria), but also in the US and elsewhere. Although we neither necessarily agree with his theory or conclusions, we believe that these attacks have largely been made by people who have either not read (or, to be less charitable, not understood) what he had to say or who have opposed him by arguing that his conclusions were *a priori* and without further discussion simply wrong and have thus attempted simply to foreclose debate. That this practice is philosophically unacceptable and violates every premise on which democracy and rational discussion is based goes without saying.

Preference Utilitarianism is based on the assumption that to act ethically is not—as it is in routine Utilitarianism—simply to maximize pleasure and minimize pain and to do so for the greatest number but, rather, that it is based on the preferences of all concerned. In other words, Singer's philosophy attempts to adjust the resolution of an issue between personal and collective interests.[15] This does not, in a sense, differ substantially from classical Utilitarianism, which holds that "good" is pleasure. Since every sentient creature has a "preference" for pursuing the greatest good and for avoiding harm the difference is arguably not very great. But Singer's philosophy extends itself far further than these underpinnings—it defends the interests of all sentient (not just human) beings, calls into doubt or at least suggests that we examine what, beyond simply "human being" we mean by person and, under some circumstances, defends abortion as well as euthanasia. Singer, for example, suggests that when medicine either has nothing more to offer (or the decision not to offer any more active treatment has been made in severely damaged newborns) actively and painlessly causing their death might arguably be a better option than "allowing" them to live longer and to die in pain. Beyond this, Singer argues that relationships are not of overwhelming importance in making ethical choices, that we are as responsible for the starving child in Bangladesh as, other things being equal, we are for our neighbours.[15,16]

The theory—and the conclusions Singer reaches through that theory (others might use the same theory to reach different conclusions)—are most certainly debatable, and his particular form of Utilitarianism suffers from the same kinds of difficulties facing other forms. But both the theory and the conclusions are not things to be simply discarded, neglected or mindlessly attacked. Singer, for example, does not say that "lower" animals have the same moral standing as do fully actualized humans; he merely argues for the claim that they do have significant moral standing and that to ignore such standing is similar to ignoring the standing of other arbitrarily disenfranchised groups (Blacks, Jews or whatever we choose to regard as "the other"). Nor does he— as he has been accused—ignore the interests of the disabled. He considers the disabled (a term which is ill defined and can range from wearing spectacles to being permanently vegetative!!) as persons with interests that must be respected and merely questions whether beings without self-awareness can have an interest in continuing to live. We would, in this regard, go further and question whether someone or something without (nor or possibly in the future again) self-awareness can be said to have interests at all. Singer illustrates what we previously said about the role of Bioethicists: to make people uncomfortable, to make them question their assumptions, to insist upon reasonable argument instead of relying on un-reflected assumptions…in other words to make them uncomfortable with the status quo and with their habitual attitudinal responses.

Throughout this book we most certainly question Singer's down-playing of relationships and hold relationships to be central to the concept of obligation—indeed one cannot explain relationships without invoking that concept. While this is especially true in the health care patient/provider relationship we would argue it is crucial in making ethical judgments. Again: this does not mean that we do not accept our relationship towards strangers as having force—indeed humans among their species and sentient beings among themselves have a relationship which forms the basis of human obligation to each other as humans and to other sentient creatures. And it certainly does not mean that we as a society or as health care providers as a profession have amply lived up to the obligation which being a human implies. But it does mean that my relationship with another plays an important role in making ethical choices.

JOHN DEWEY

John Dewey is, at least in our estimate, one of the most important figures in social ethics, and his views are, therefore, extremely important for its sub-field, health care ethics. John Dewey denies the existence of an absolute good, rejects the notion of an "intrinsic good" (or, for that matter, an intrinsic right) outside of any particular experiential context, and feels that judgments about ethical matters are judgments made about the appropriateness of specific actions to achieve specific goals. The goals (or "ends") themselves are context-dependent and, thus, cannot be immutable or valid for all times. The methodology used in moral inquiry is the same as the method used so successfully in scientific investigation: hypotheses are made and tested in the crucible of experience. The alleged strict difference between praxis and theory vanishes to be replaced by a dynamic and mutually corrective relationship. Truth is neither immutable nor the pure product of human willing, but emerges out of the complex interrelationships between persons and their physical and social environments, is tested in praxis, and is apt (indeed, is bound) to change and develop, depending on circumstances—including both positive and negative feedback. There is no absolute answer nor are there absolute solutions to problems. We start with an indeterminate situation and together to reduce the number of indeterminate aspects of the situation until it, and its implications (its possible resolutions and their possible consequences) become progressively clearer, i.e., more determinate, for all of us. Such a process improves our situation in a number of ways:

1. It helps to "get us all on the same page" by revealing tacit personal and social assumptions—whether erroneous or not—that may otherwise influence our perceptions of the situation and its implications.
2. It enables us to marshal a range of resolutions to the problematic situation and to become better equipped to recognize the strengths and weaknesses of each.
3. In creating novel ways to respond to indeterminate situations, it also creates novel possibilities—i.e., what must play out initially as new forms of indeterminacy—which, in turn, advance further development, learning, and growth.

Conventional ethical theory, as it is generally interpreted today, lacks this dynamic, homeostatic aspect. Unlike Dewey's approach, it would "apply" certain rules to a situation that occurs in a particular context without becoming personally involved—much as a cookie-cutter cuts out pre-determined patterns. In Deweyan ethical inquiry and analysis, the moral agent is not only mindful of situation and context but also becomes personally involved in the process so that he or she personally undergoes fundamental evolution and change.[17]

Dewey seeks to incorporate the concept of growth, change, and context into our moral reasoning. He is determined to preserve the importance of experience and empirical evidence in the formulation of our ethical choices rather than to base choice on predetermined ethical "principles." In doing this, he has profoundly influenced much contemporary work, especially as it relates to the importance of context and character. Not only is his work important from a practical point of view in medical ethics, in which context and specifics assume such great importance, but also it likewise has greatly influenced ethics in general, and especially social ethics. The way specific problems as well as problems of moral worth, blameworthiness, and community are examined owe a heavy debt to John Dewey.

Dewey's viewpoint of community as the basis of all human endeavor and understanding has been critical to the way social ethics is conceived. To Dewey (as, in many respects, to Rousseau and Kant), intellectual activity (like all other activity) is basically social: we need a community of others if we are to think effectively or lead successful lives Such a point of view should not be conflated with garrulousness: researchers or thinkers must, by the nature of the task, do much of their work alone.

But such aloneness is not possible unless it, in turn, exists in the matrix of a community, which inspires and can, eventually, evaluate, correct, and enrich the work at hand. Such a point of view is basic to democracy itself and to the way we as individuals see ourselves enmeshed in it.[17,18]

—— VIRTUE ETHICS AND CASUISTRY ——

Rules we choose to follow, principles we select, and theories we elect to guide us are a reflection of our character. Virtue ethics, as old as the Homeric tradition and as young as contemporary thinkers, tries to base itself on an appreciation not so much of the rightness or wrongness of given acts depending on duties and obligations as on the goodness of the persons who select such obligations and rules.[19-21] It inquires into what attributes are characteristic of persons we consider "virtuous" rather than selecting rules and then deciding that persons who follow such rules are, by virtue of rule following, virtuous.

"Virtue" is used here in the classic sense of competence in the pursuit of moral excellence. In common usage it carries an unfortunate baggage of moralism and comes across almost as mealy-mouthed. That is not the way it is used here, nor is it what people generally mean when they speak of "virtue ethics." To Plato, virtue was synonymous with excellence in living a good life, and such excellence could be attained by practice. Vice, Plato believed, was not so much caused by moral turpitude as it was the result of simple ignorance: one either lacked knowledge or lacked the ability to reason properly. To Aristotle and, later, Aquinas, virtue was a disposition to act in the right way. Aristotle strove for balance. Whereas Plato saw virtue as an intellectual trait, Aristotle saw that, in practical terms, virtue was the result of a balance among intellect, feeling, and action. "Virtue" was a state of character and the result of practice. In turn, practice resulted in habit so that the "virtuous" man could be counted on to act justly. Thinkers from then on have explicitly or implicitly considered virtue and the virtuous man (the man practiced and adept at finding moral goodness in real situations) to be an intrinsic part of ethical behavior. MacIntyre saw that more than internal qualities were involved: goodness is shaped by a social vision of the good.[19]

To judge concrete situations, virtue ethics suggests the use of certain "rules of thumb:" rules that are derived from the practice of "virtuous" practitioners in similar cases. Casuistry—very similar to "virtue ethics" and more of a method than a theory— seeks to develop "index cases," i.e., cases that have been adjudicated before and that can serve as a model when confronted by future actual cases.

The problem with virtue ethics as a single ethic to adhere to, of course, is that "goodness" and "virtue" are defined in terms of each other: the virtuous man does good things, and good things are those acts a virtuous man does. Nevertheless, other ethical theories, in just as circular a fashion, link the good with their theory of ethics. Legalistic ethical systems tend to define right action purely in terms of rules. When used by themselves and with nothing else to guide them, they are very likely to become straitjackets rather than guideposts of the moral life. Ultimately, and if carried to their logical extreme, they may interfere with decisions that, in human terms, are humane and good. Virtue ethics alone, on the other hand, suffers from an imprecision and "fuzziness" that, in defining the good in terms of the "virtuous person," runs the danger of a thoroughgoing paternalism. The two, it seems, can complement and enhance each other. If being virtuous, as Aristotle viewed it, entails striving for balance, virtue ethics may entail the balanced application of principles and rules in a thoughtful and humane fashion. Problems of ethics cannot be separated from the moral sensibilities that shape them or isolate them from the moral actors that make up their context.[22] Virtue and rules shape each other and permit moral growth and learning. In that sense and in many others, the insights of John Dewey are critical to ethics.

There is no doubt that casuistry and virtue ethics have their place in moral education. Seeking to inquire and understand the way such problems were dealt with in the past most certainly can serve as an example—but preferably as an example of inquiry (why were they dealt with in a given manner?) rather than with the way a case was handled itself. Education (be it moral or medical) seeks to instill good habits—above all good habits of inquiry, which includes skepticism about how things were dealt with in the past.

─────────── SITUATION ETHICS ───────────

A more recent attempt to enunciate a system of ethics was made by Professor Joseph Fletcher. Situation ethics, as he calls this type of approach, would judge each situation purely on its own merits, aiming for the most "loving" result that could be brought about. Initially Fletcher (who was a theologian) speaks of this as "Christian love;" in his later works he speaks of it purely as "love." Situationism (or, as it has sometimes been called, "agapism" from the Greek term for "love," which encompasses more than its mere translation denotes) is, of course, a form of act utilitarianism in which each act is judged by its outcome, which, here, instead of being pure utility, must accord with what is most "loving."[23-25]

A prior insight, of course, as to what constitutes the most loving result is needed. "Love" as a concept on which to ground morality seems at least as ill defined as notions of the "good." Situationism, Fletcher agrees, must take place in a framework of general rules, "rules of thumb," as he calls them. But to be useful, it would seem, such rules of thumb must conform to some prior insight of the "good" or of "love." Such prior insight, in turn, either is the product of rigidly conceived "truth" and, thus, will be seen to vary from society to society in its definition, or will be conceived as something persons working and thinking together craft in community and test in praxis.

As appealing, at least superficially, as such act utilitarianism or "agapism" may be, it breaks down when one considers that what is "right" is not merely determined by someone's subjective understanding of the "goodness" (or the lovingness) of the outcome. Other, more objective factors seem to matter. Moreover, bringing about a "good" circumstance by thoroughly reprehensible means or performing an accidentally "good" deed (in terms of its outcome) with evil intentions certainly cannot be considered an untarnished "good."

PRINCIPLISM

In their classical work on Medical Ethics Beauchamp and Childress carefully developed a system of medical ethics to which the name "principlism" (sometimes called the "Georgetown Mantra") has been applied.[26] This is not a good term and is, in fact, a misnomer: most ethical theories or systems operate according to some "principles." What is generally understood by that term today is the primacy of the four principles set out by the authors: (1) Autonomy; (2) Beneficence; (3) Non-maleficence and (4) Justice. These "principles" are to be applied in each case and with each problem. On the surface such an approach seems to have merit. And yet it has fallen more and more out of favour and has had to be adapted.

There are several problems with this approach. First of all, like most ethical theories or approaches, instead of constituting important considerations in all cases these four items are treated as "principles" to be applied to a given case or problem. Secondly, it is unclear how any of these stated "principles" are to be defined let alone applied. Aside from the fact that beneficence and non-maleficence are really two sides of the same coin (viz., respect for persons), these so-called principles give no guidance as to their hierarchical nature: what is one to do when autonomy conflicts with beneficence or beneficence with justice?

The reason (aside from the fact that they were developed at Georgetown) the rather unkind term "Georgetown Mantra" has been used identified with these four principles, is that health care professionals often find it convenient to appeal to this form of principlism when approaching concrete problems at the bedside as well when dealing with more general issues in health care. They have become a slogan behind which physicians and other health care professionals can retreat and to which they can then appeal. But when one party insists that their solution (based on the principle of autonomy) is right and another disagrees (and bases theirs on beneficence) there is little left to discuss. There are, furthermore, many more factors in each case or problem and its specific individual or cultural context than is covered by these "principles." Different cultures will not only define but also value these terms differently.

In our view, there is a tendency abroad to forego the agony of decision-making by abandoning patients to such empty "principles." On the way to dealing with a problem we can, conceivably, say "aha: this is a problem of autonomy" and find our answer there. But neither general nor specific problems are quite that easy. These four considerations are critically important in making decisions—but in our view they often conflict and certainly, in and of themselves, cannot be considered to be binding in absolutely every case. When used with discretion and definition they can be helpful but in the view we put forward here they are not straitjackets that tell one what to do.

──────────────── **CARE ETHICS** ────────────────

What has been called the "care ethic" has had a more prominent place in yesterday's than it has in contemporary, at least in contemporary American, ethics literature.[27,28] In many ways it can be seen to be the obverse of Kantian ethics. It is unclear, when one reads the writings of its proponents, whether they are speaking about a theory or a method. What is clear is that the development of this form of ethical thinking (or process) originates in an abhorrence of the cold application of rules to particular cases: what the authors call "justice-based reasoning," which they claim is the type of reasoning used mainly by males and which they oppose to "care-based reasoning," a form of reasoning its proponents claim is generally used by females. The thesis that men and women somehow approach ethical problems differently is not only something one would not expect from feminist philosophers who justifiably and effectively are working for equality between the sexes, but it also cannot be empirically sustained: in general, female physicians or attorneys tend to reason like their male counterparts do, and male nurses tend to approach ethical problems like their female colleagues do.

Justice-based reasoning, it is claimed, operates by applying ethical principles or rules to concrete situations without much bothering about the unique features of each case. On the other hand, care-based reasoning, it is claimed, approaches problems and cases by allowing the feelings (Kant would say "inclinations") of the persons involved to guide the way. Whereas Kant basically distrusts inclinations and wishes to build his ethical structure upon a foundation of rationality and of strong (alas, virtually absolute) rules, care ethicists celebrate feelings and eschew principles and rules. To act well is to involve oneself in a case and then to act in accordance with what one's feelings tell one to do.

Such an "ethic" (whether it is supposedly a theory or a method: albeit, if it is a method it seems to be a method for applying itself) is a theory or method based on emotion and one that largely eschews intellect. Indeed, we would argue that it is an irrational (i.e., anti-intellectual) way of proceeding. Without a framework of theory and of principles and rules to guide one, acting on one's gut feelings, or "letting one's conscience be one's guide," can lead to disastrous results. Caring very much about an issue or a problem in itself can give little guidance: people can care very deeply but in diametrically opposed ways about the same problem or issue. Gassing Jews merely because we are strongly inclined to do so or artificially keeping vegetative patients alive and intervening in their inter-current illnesses because our emotions tell us to do so can be opposed to the actions of others whose inclinations and feelings counsel them in very opposite ways. In medical practice those who care and care very deeply about a given case are by no means apt to reach the same conclusion simply by virtue of caring. The way one cares or what one's conscience tells one to do is a mish-mash of one's personal experiences, assumptions and biases, one's tradition, one's religious views, and one's social setting. While such inclinations and feelings cannot and should not be ignored, they are, as yet, merely knee-jerk reactions of a single individual to the raw data of a shared experience. For this reason, they cannot be allowed to be the sole moral guide to our moral actions. When our actions need to be justified or defended—and they certainly do when they affect the well-being of others—an appeal to the way we feel or felt about a situation is simply insufficient, whether subjectively "true" for us singly, as individuals, or not.[29,30]

There is no question that there are several important features of the care ethic or of something quite similar, an appeal to conscience. In reacting to a sterile application of rules and principles, the proponents of this form of acting make the important point that there is a crucial human, emotive, and contextual side to all moral problems. However, appealing to one's conscience as the sole basis for justifying one's actions is likewise fraught with danger. The consciences of Albert Schweitzer and Adolph Eichmann, one would suppose, were quite different! In virtually jettisoning reason and, instead, celebrating emotion, caring, or conscience, a non-verifiable claim of acting correctly (according to one's conscience, emotion, or way of caring) is made. Conscience, emotion, or ways of caring can serve as an important corrective to blind rule following: they may call one's decision into question and motivate one to re-examine the issue. But by themselves they, in our opinion, will not do.

There is little doubt that blind rule following can lead to disaster. But, because inherently capricious and subjective, conscience, emotion, or particular ways of caring uncontrolled and unmodified by reason may prove to be far more dangerous. Ethical theory spawns the principles and rules with which we choose to govern our behavior. As we derive these principles and think about them, we begin to re-examine our ethical theory. When it comes to dealing with ethical problems in medicine (whether these are the individual patients of the provider–patient relationship or the broader problems of just allocation in society), we must consider context and situation. We must, in other words, apply these principles and rules with "caring." As we do this, some of our predetermined principles and the theories from which they are derived may be called into question and may subtly (or sometimes radically) change. It is a process of learning and growth in which reason and emotion must complement and enrich each other.

A BIOLOGICALLY GROUNDED ETHIC OF SUFFERING

Biology as, in part at least, a basis for morality will, inevitably, be challenged by those who would separate the reasoning process from its biological underpinnings and instead appeal to "pure rationality." The viewpoint that rationality can be separated from its biological underpinnings has, and has had, many adherents. It is a claim that is inherently dualistic in that it would separate "body" and "soul," "brain" and "reason," etc. In suggesting that biology serves as the necessary basis of all reasoning, we are emphatically *not* claiming that the two ("brain" and "reason") are identical but merely that the former is the necessary, even if hardly sufficient, condition for the functioning of the latter. Rationality (i.e., reasoning or thinking) without brain (and, for that matter, specifically without neocortex) is, in the realm of experience, unthinkable.

Data abound to underline this fact. Claiming that a creature may exist somewhere whose substrate for the reasoning process is other than what we call brain does not defeat our thesis; rather, it points out that such a being would then have some other substrate without which its reasoning could, again, not take place. Reasoning without a substrate is as unthinkable for us as digesting without an intestinal tract. Function in the experienced world depends upon, even though it cannot be reduced to, a physical substrate.

Morality, of course, is not grounded in biology in a reductionistic way; i.e., morality cannot be completely reduced to biology. Saying that morality is biologically grounded merely implies that we, as sentient creatures, cannot escape the framework of our biology, which subtends all of our functions, including the functions of thinking and making judgments, things that can occur only within the predetermined framework of our biological possibilities. This is what we mean when we say that it is impossible for morality to escape or transcend biology entirely. That is not to say that we cannot resist biological drives or urges but merely that our ability to resist and, at times, to go counter to such drives is itself expressed in the embrace of biological possibility. Biology is the source of our common experience, moral and otherwise.

In that sense, a biologically grounded ethic can be rationally carved out and can be cautiously employed as a basis for further exploration and progress. Such an ethic is predicated on a "common structure of the mind,"[31] which enables all sentient creatures (be they parakeets, chimpanzees, or humans) to appreciate benefit and harm, and, at the least, to suffer. This theory differs from Kant's in that it does not ground itself merely on rationality but, rather, finds firmer footing in the capacity of all sentient creatures to suffer. If, as we have consistently claimed, ethics is other-directed, then the capacity now or again in the future to be capable of perceiving benefit or harm—at the very least to suffer—can be seen as central. Not to bring harm (or suffering) to entities capable of experiencing harm may be a meager but a sound basis. Of course, "harm" (or "suffering") has to be defined in terms of the entity itself. The capacity to suffer (with suffering defined by the sufferer), then, is central to such an ethic.

Suffering is not quite the same as perceiving a noxious stimulus or having pain. We may say of someone that he or she is experiencing pain but would be amused to have such pain termed "suffering." A teenager having her earlobes pierced is an example. On the other hand, we may suffer without having distinct pain: patients with terminal cancer and with their worst pain obtunded may still be suffering intensely.

The capacity to suffer, then, implies more than the mere ability to perceive or to react to pain. Suffering implies a more sustained perception and one that perforce is integrated into memory and linked to thought. I suffer when I believe that my pain is interminable, when I believe my fate to be hopeless, when I feel myself powerless, when I see the pain of a loved one, or for many other reasons. Suffering is a composite concept. At the very least, to suffer I must have the capacity to remember what has gone before (remember, for example, that my pain was here a little while ago and is here still) and, in the most primitive sense, anticipate the future. To suffer, then, at whatever level, implies a rudimentary ability to sense, to integrate such sensation into however rudimentary a memory, and, beyond this, to have a sense of future at however primitive a level. We would then say that organisms that have, however primitive, a neocortex can, and that those who lack a neocortex cannot, suffer. Making the neocortex the necessary condition of the capacity to suffer is not the same as reducing suffering to the neocortex. It is merely to affirm that in biological systems as we know them (not as we might speculate about them), suffering is inextricably linked to such a substrate.

Grounding an ethic on the capacity to suffer, then, grounds it in a universalizable quality common to all sentient beings.[32] Grounding an ethic on the capacity to experience mere pain forces one into a morass of considerations dealing with the ability to judge such things. When entities experience and when they do not experience pain is difficult to judge: Does an amoeba withdrawing from a sharp object or do worms experience pain? Certainly there is evidence that they react to noxious stimuli, that they withdraw or avoid them. But that is not quite the same as "experiencing" (being aware of) pain, and it is a far cry from suffering. And in that amoebae lack the substrate necessary for suffering our best knowledge today would indicate that they cannot suffer—which does not exclude that what we take as fact today may prove to be error tomorrow.

To experience, or to suffer, organisms must be self-aware and at the very least have the ability at however primitive a level to think. Memory may be definable as the ability to recall however primitively (Kant speaks of this as *re*-cognizing: "knowing again") past events. To experience anything, rather than merely to react reflexively to sudden and at once forgotten stimuli, at the very least, requires such ability. Thought, on the other hand, inevitably linked to memory, may be defined as the ability to integrate external and internal sense experience into memory. Memory and thought—inextricably dependent upon each other—are the necessary conditions for the capacity to suffer. In biological organisms as we know them, memory and thought are necessarily grounded in a neocortex.

——— PERSONHOOD AND MORAL ——— WORTH

In going about their daily tasks, physicians are concerned with the hard questions of moral worth. We here equate "being of moral worth" with "being of moral concern," i.e., that our acting in a way that affects such entities raises some sort of ethical question. What endows entities or objects with moral worth or what makes us concerned about our actions is a fundamental question, one that seeks to find adequate reasons for differentiating between, say, automobiles and college students as objects of moral worth.

Having moral worth (in Kantian language, being "deserving of respect") does not endow objects with absolute rights; it merely says that considerations against arbitrary treatment stand in the way. Concepts of moral worth are fundamental to such diverse issues as cadaver organ donation, abortion, and dementia, to name but a few. For the purpose of these pages, we are concerned with three main types of moral worth: primary, secondary (and its subset of material and symbolic), and prior moral worth. Any of these confer *prima facie* (not absolute) rights against violating the object in question. Having such worth is the necessary condition for being an object of moral consideration. It is the necessary condition because, unless another is somehow actually or potentially benefited or harmed, the question of morality cannot come up.[33,34]

The concept of personhood, traditionally used throughout the ethics literature, has never been well defined. It basically finds its motivation in Kant's statement that all persons must be objects of respect and in defining objects of respect as those entities capable of moral self-legislation (*viz.*, autonomy). But this definition has been less than entirely helpful in illuminating our ethical gropings when confronted with hard decisions. Is personhood to be granted to the human form (the *res extensa*, as Descartes would have it), or does it inhere in the *res cognitans* (the "knowing thing")? Does personhood require continuity—is, for instance, the anaesthetized or the unconscious patient a person? And does personhood need to be actualized—is, for example, the developing fetus a person? What is the standing of the severely mentally defective, the psychotic, the senile, the vegetative, or the brain dead on ventilators? Personhood without an agreed-upon definition of all that personhood does or does not imply has proven to be inadequate. And such a definition has never been agreed upon. When personhood is used in ethical discourse, we tend to forget the problem at hand in quest of a definition. Personhood today unfortunately carries a heavy load of historical definitions and arguments, and its use has become problematic. The question of what endows objects with moral worth and therefore what must, *prima facie*, make us hesitate to deal with them capriciously or merely to satisfy ourselves, remains.

TYPES OF MORAL WORTH

The question of what endows objects with moral worth is one of the fundamental questions of ethics. How and why do we differentiate among stones, flags, amoebas, dogs, and children? What are the features that permit us to deal with one entity almost at will and with another entity only under certain circumstances? Moral worth, or having moral standing, is the ethical feature of entities that we use to discuss this. As we mentioned earlier, "moral worth" is discussed in three basic categories: primary moral worth, secondary moral worth (which can be further subdivided into material and symbolic moral worth), and what I [EHL] have called "prior" moral worth. Another way of putting this is to claim that to have moral standing or worth implies the capacity to have an "interest." Entities capable of having (now or in the future) an interest are entities that have independent moral standing; others are not.[34]

Our ethical concerns are prompted by the benefit or harm that can, directly or indirectly, result from our actions to another. Primary moral worth attaches to an object that in itself is capable of being self-knowingly benefited or harmed: it is an entity that, in other words, is capable of having an interest, or that, at the very least, has the capacity to suffer. The capacity to suffer, at least one necessary condition of primary moral worth, may be actual or potential. Once this stipulation is met, a *prima facie* condition against acting without such an entity's consent or to its apparent benefit exists. Just because I base respect (or moral worth) on an entity's capacity for suffering does not mean that all we need concern ourselves about is suffering. The capacity to suffer is meant to serve as a marker of moral standing or worth: ethically, entities that have such a capacity cannot be acted upon capriciously. Primary worth is always positive: having the actual or potential capacity to suffer makes one of moral concern whether one is a philanthropist, Hitler, or a fetus. The protection of primary worth, however, is *prima facie* and not absolute. At the very least, a condition against being harmed raises serious ethical concerns. A lack of the capacity for self-knowing benefit or harm is what, *inter alia*, differentiates inanimate from animate objects and what divides the sentient from the insentient. Self-knowing presupposes a capacity for awareness and for social interaction.

To have material secondary moral worth, an object must be of material or concrete value to another. It would be silly to believe that a model airplane or an automobile could, in itself, be benefited or harmed. But if the model airplane is dear to an 8-year-old, or the automobile belongs to another person, destroying the airplane or the car has obvious moral overtones. Moral worth, in such cases, is, as it were, conferred by proxy. It is of moral worth because the object in question has material value to another who is him- or herself of primary worth and who, therefore, has moral standing.

Symbolic worth, the other type of secondary moral worth, endows an entity with value neither because it has value to or in itself nor because it has material value to another. Rather, such objects have worth because in the eyes of some they represent important values. Flags, religious symbols, one's reputation and many other objects have symbolic worth. Symbolic value is a frequent concern in medical ethics: it enters into issues of organ donation, brain death, permanent unconsciousness, or the vegetative state as well as ways of thinking about disease. What is symbolic at a given time or to a given individual may be meaningless at other times or to other persons.

Secondary worth (whether material or symbolic) may be positive or negative: material objects as well as symbols may be valued or disvalued, and they may be valued by some and disvalued by others. Entities may simultaneously be endowed with primary, secondary, and symbolic worth. A sick animal may be of primary moral worth since it can be benefited or harmed, may have secondary worth in having a "market" value, and may also have symbolic value in that it stands in someone's mind for a previous owner who loved it. Kant, in a similar vein, speaks of objects as having "dignity" (primary worth), a "market value" or price (secondary worth), or an "affective" (an aesthetic or, perhaps, symbolic worth) value.[6,12]

Having primary moral worth gives objects a *prima facie* hedge against being capriciously harmed. In the clinical situation this may be helpful. When a patient is anencephalic, brain-dead, or permanently comatose or vegetative, primary worth is lost. Such patients are now of symbolic worth to their loved ones and, as representatives of humanity, to the community; and they may be of secondary (material) worth in that they may have either the capacity to serve as organ donors or to consume badly needed resources.

When an entity loses primary worth, professional obligations change. Consider the patient who is now brain-dead or who has lapsed into irreversible coma. Prior to such a time, the physician's obligation was clear: the interests and wishes of the patient were of paramount importance and those of relatives and context were peripheral to the eventual decision. A conflict between competent patient and next of kin was inevitably (or at least inevitably in terms of the ethics of the patient–physician relationship as we understand it) finally resolved in the patient's favor. When, however, patients become permanently brain-dead or lapse into irreversible coma, things change: barring a prior agreement, such patients who now can no longer be knowingly harmed in themselves (who have, in other words, permanently lost the ability to suffer) move from center stage. The wishes of family as well as the desires, feelings, and needs of their context (the needs of the ICU or the hospital, for example) may now legitimately move to center stage.

The presumption against capriciously dealing with or harming entities of primary worth, however, does not go very far. If parakeets, baboons, the mentally retarded, and college students all share this protection, how can one use such a concept in arriving at concrete decisions in specific cases? When to protect one entity another must be dealt with against their will or evident interest, how is one to determine who "trumps" whom? Inevitably, in trying to establish hierarchies of value, for that is the only way such judgments can be made, external standards have to be applied. Arguments that ground themselves on the superior worth of one or another entity by appealing to biological sophistication ("animals are of lesser worth"), intellect, or any other aspect must necessarily appeal to an externally determined standard. Quality-of-life judgments determined by one for another are another example.

Accepting a non-external standard—a standard determined by subjects for themselves—serves only negatively. It may serve when dealing with a life no longer valued by its possessor—say, a man riddled with metastatic cancer who pleads for death. But a non-external standard, a standard that lets each entity determine its own value, cannot serve when it comes to many practical problems. Most, if not all, organisms value their life and their personal welfare above those of all others. This is true not only when it comes to so-called "death and dying" issues but likewise when it comes to many important issues that pit the interests of one against those of another. External standards, when it comes to difficult choices, are inevitable. Many of our judgments in medical ethics perforce will have to grope with this troubling question—a question that can only be answered (and then not for all times and places) in the context of a particular community and a particular time.

The concept of prior worth is one that assumes that there are things that constitute the basis for the existence of all others. Being, nature (ecology), community, and the future are examples of this. Without existing (being), all else is without reality, without the continuance of a healthy ecology, all being is threatened; without community, individual existence stands moot; without the future, troubling about other issues seems empty. Safeguarding such things, therefore, constitutes the necessary conditions for primary and secondary valuing. With today's technical possibilities our choices more critically have the potential for affecting being, ecology, community, and the future itself. Therefore, as Hans Jonas has pointed out, knowledge itself has assumed moral dimensions.[35]

────── LOOKING AT ETHICAL THEORIES ──────

The variety of ethical theories with their advantages and drawbacks again raises the initial troubling question: What is morality?

We can adopt a variety of viewpoints of morality. On the one hand, we can affirm that morality is absolute and that absolute standards of "right" and "wrong" exist. Beyond this, and not quite the same, we may assert that in the human condition such standards are knowable. We can research such a viewpoint from a number of avenues of approach but inevitably will find that our conclusion is based on the presumption that truth exists and that it is, in principle as well as in fact, knowable. Truth, in this view, depends on neither situation nor context. This point of view claims that morality exists as a discoverable truth, an absolute that is not fashioned by men but unchangeable and immutable. "Rights" and "wrongs" are rights and wrongs quite apart from the stage on which their application is played out. Situations may differ, but at most such differences force us to reinterpret old and forever valid principles in a new light.

Beyond this, of course, there are those who embrace such a point of view and who claim that truth is, in fact, known and that only the stubborn recalcitrance of the uninitiated prevents it from being generally accepted. The step from this point of view to a point of view that would result in the use of subterfuge, lies, force, or coercion is not a long one. When health professionals believe that they "know best" and are, therefore, entitled to mislead or coerce patients to pursue a given course of action, that step has been taken. That such a basically fundamentalist point of view when generalized beyond health care threatens peaceful coexistence is obvious.

There are, on the other hand, those who claim that what is and what is not morally acceptable varies with the culture in which we live. This claim rests on the assertion that there are many ways of looking at truths and that such truths are fashioned by men within their own framework of understanding. Depending on our vantage point, there are many coequal visions of reality, a fact the defenders of this doctrine hold to be valid in dealing with the concrete scientific reality of chemistry and physics.[36,37] Such a claim, it would seem, is even more forceful when dealing with morals. As Engelhardt put it so well, "Our construals of reality exist within the embrace of cultural expectations." And our "construals of reality" clearly include our visions of the moral life.[38]

The claim, however, that since our "construals of reality" occur purely within the "embrace of cultural expectations" all visions of reality are necessarily of equal worth and there are no useful standards that we can employ in judging either what we conceive to be material or ethical reality, does not necessarily follow. One can, for example, make the claim that some visions of reality are clearly and demonstrably "wrong," supporting such a claim by empirical observation or by showing that certain visions of reality simply do not work. This is the stronger claim. In rejoinder, it will be said that empirical observation and what works are framed in the very same "embrace." Or one can make the somewhat weaker claim that certain visions, in the context of a given society and historical epoch, seem less valid than others because they confound careful observation or because they simply fail to work when applied to real situations occurring in real societies.

Such a move does not deny that our "visions of reality occur within the embrace of cultural expectations." But while such a move affirms that there are many coexisting realities of similar worth, it also suggests that within the specific context of such "cultural expectations," some realities have little and others much more validity. Such a view neither throws up its hands and grants automatic equal worth nor rigidly enforces one view; rather, it looks upon the problem as one of learning and growth in which realities (both empirical and ethical) are neither rigidly fixed nor entirely subject to *ad hoc* interpretation.

Consider the Babylonian peasant reared in a small community, a community whose integrity is believed to be safeguarded by the annual sacrifice of a selected first-born to Moloch. Should such a peasant, when his first-born is selected for sacrifice, be held blameworthy for sacrificing his child? Or, on the contrary, could a refusal to yield his son be held to be an immoral act endangering the whole village for the sake of his own selfish interests? By what standards are we to judge? By ours, by those of his society, or by an absolute to whose knowledge we pretend?

But hold on! Judging our peasant blameworthy or not is not quite the same thing as judging the act of child sacrifice to be or not to be wrong. The peasant may, in his special context, not be blameworthy (his intentions were good, and he acted according to his own conscience), but the act of infant sacrifice may, on the whole, still be considered wrong. If so, the judgment that infant sacrifice is wrong must accord to some universal principle to which all could subscribe. The peasant truly "knew no better:" for generations, other peasants had done the very same thing and shared the very same belief. He knew no world in which such a practice was not followed.

It is often argued that such a point of view would make it impossible to hold the Nazis, or a particular Nazi, responsible for their multiple atrocities. But such a comparison is not apt. The Nazis indeed did "know better:" their state existed for a mere twelve and a half years, and all those who participated in or tolerated such atrocities were reared in a world in which quite different points of view were espoused. Furthermore, the population of Nazi Germany was not out of contact with the rest of the world. Some objected, many who were, in fact, perpetrators became severely alcoholic or suffered other severe and often permanent psychiatric effects. What could serve as an excuse for a Babylonian peasant could not serve as an excuse for a Nazi storm trooper or for a Ku Klux Klanner who murders Blacks. In today's world there are some universal (or, at the very least, predominant) points of view about such things: even in those states where human rights are crassly violated, at least lip service is paid to another morality.

When, we seek the "good" or seek for an insight as to what "love" is, we tend to go in one of two diametrically opposite directions: Either we affirm some absolute vision of the "good" or "love" to which we appeal, or we claim that no vision of the truth can be appealed to and that standards, therefore, do not exist. Neither the first nor the second claim has much intuitive appeal: an absolutist vision is not readily shared by others (especially by others from different cultures), and a purely situational claim leaves us with no points of reference to act as guideposts on the road to decision making. Pure relativism not only fails to provide us with standards or norms as guideposts by which actions can be judged or rules crafted, but it also makes it (since we lack acceptable norms or standards) impossible to say what does and does not constitute progress.

Utilitarianism (and we shall consider situationism to be a form of act utilitarianism) and deontology may, in effect, necessarily presuppose each other. When Kant speaks of intentionality, it is the intention to bring about a consequence that he is speaking about. One cannot, it seems, have an intention without this: to intend something is to wish to bring about a consequence. On the other hand, when utilitarians or agapists set out to maximize the "good" (or to bring about a "loving" outcome), their vision of what it is to do good, or to be loving, is unavoidably rooted in a prior vision of the good. This prior vision of the good conforms to a logically universalizable principle (the particular vision of "good" or "love" must be "universalizable") as well as to a preexisting social vision (the particular notions of good or love current in a particular society). The nature of such a universalizable principle must, among sentient beings, in turn, conform to the limits imposed by biology. A common biological denominator of such good may offer a firmer grounding.

Theoretical considerations are necessary if we are to act. Even the statement that we shall act without theoretical considerations is itself a theory of how to act. Likewise, principles or rules are inevitable. Even the decision to act without a rule or principle itself serves as a rule or principle. Thinking persons using the building stones of past theories and adapt them to their own needs, experience, and makeup. The relationship among theories, principles, and application can be seen to be an interactive one: We derive our principles or rules of action from our (conscious or subconscious) views of ethics (ethical theory), and in turn we must use such principles when dealing with individual cases or problems by seeing such problems in their particular and peculiar contexts. In other words, we involve ourselves with such problems or cases and begin to "care." While not allowing our feelings to dominate our decisions we allow our feelings about them to help us arrive at a decision. It is a process I [EHL] have referred to elsewhere as compassionate rationality when dealing with problems or issues and rational compassion (not quite the same thing) when dealing with individual problems.[1]

MACROALLOCATION AND PROBLEMS OF JUSTICE

The basis of traditional ethics involves two issues: the relationship of individuals with one another, and the relationship of individuals with their community. In the parlance of medical ethics, the former largely involves issues of micro-allocation in which health care professionals must deal directly with their patients (that is, with identified lives); the latter involves issues of macro-allocation (or unidentified lives; see also Chapter 8). In between these two stands the problem of distributing scarce resources to specific patients: distributing resources needed by all but not sufficient for all to some but not to others (organ distribution or the difficulties of who is to get the last bed in the intensive care unit are examples). (For a more thorough discussion of justice as well as macro-allocation, see Chapter 8.)

Macro-allocation issues—dealing with resource allocations to institutions or to definable groups of people—must be separated into three levels. First, societies allocate their resources in various broad categories: how much for education, how much for welfare, how much for health care, etc. Second, distribution of funds into the constituent enterprises of education, welfare, or health care takes place: those responsible now distribute funds to hospitals, nursing homes, public health, and so forth. Third, individual institutions distribute available resources according to their peculiar needs: decisions of how much to allocate for birthing units, how much to spend for ICUs, and how much to provide for the library of an individual institution are made. Each of these decisions, moreover, by providing funds for one, allocates less to another activity. Macro-allocation differs from micro-allocation in a critical sense: macro-allocation decisions are made for all the individuals within a given group irrespective of the individuals comprising that group. Physicians faced with micro-allocation decisions are faced with judgments made more about individual patients and with the ground rules governing or directly concerning such patients.

It is important to realize that macro-allocation or distribution issues have a quite different emotive impact from issues that involve specific personal relationships. Generally this is spoken about as the difference between "identified" (or known) and "unidentified" or statistical, and therefore, unknown lives. An illustration of this emotive impact is what happens when restrictions are, for whatever reason, applied (transplants of various sorts are an example). A personal appeal by a particular person affected (or by his or her relatives) often results in bypassing such a restriction. It is said that a crying grandmother on television can confound the best-thought-out distribution scheme. We must, however, be aware that unidentified lives (lives not identified by us or by those who decide on allocation or by the public at large) are still very much personal lives, lives that are indeed "identified" by their relatives, friends, and other associates.

In making allocation schemes this inevitable fact means three things: (1) we must be fully cognizant of the fact that our decisions will inevitably influence particular persons; (2) we must be ready to enforce these schemes in an evenhanded manner and not allow those who have the capacity to "make the most noise" to be benefited beyond those unwilling to do so (the crying grandmother, in other words, must be dealt with compassionately but firmly); and (3) we must be willing to make policies with sufficient elasticity that criteria for allowing deviations from the policy would be possible (for example, a policy that precludes a given transplant after a certain age might allow someone of such an age but otherwise in exceptional health to undergo transplantation).

In dealing with ethics, especially ethics on the "macro" level, notions of justice are essential. Although whether or not the idea of justice is applicable at the bedside is questionable,[39] the moral problems that the physician encounters hardly occur in isolation. Notions of justice have been matters of debate since classical times. Ranging from Plato's notion that justice consisted essentially in attending to one's own business to Aristotle's view that justice consisted in giving to everyone what is their due, the evolution of thoughts about justice have undergone changes intimately tied to social systems. The underlying questions, of course, of what *is* one's business or what *is* one's due remain.

John Rawls in his *A Theory of Justice* as well as in his more recent *Political Liberalism* has developed the notion of justice as:

1. Assuring maximal freedom to every member of the community.
2. Assuring those with similar skills and abilities equal access to all offices and positions found in a particular society.
3. Assuring a distribution to benefit maximally the worst off.

He posits a hypothetical "veil of ignorance" behind which all prospective members of a community choose the broad allocation of resources. Members choosing from behind this veil of ignorance do so ignorant of what their own age, sex, or station in life is to be. Prudent choosers are therefore unlikely to disadvantage a group to which they may well belong, and are therefore likely to agree to these principles. A community's vision of the particulars justice entails is, according to Rawls, likely to emerge from such a choice.[40,41] Such a notion, however, still relies largely on single persons who in a sense are asocial beings choosing for themselves and not within the context of a community.

Habermas sees problems of morals as multi-culturally insoluble but those of justice as being prone to communal dialogue ("communicative ethics").[42] Habermas, we think, speaks of "problems of morals" much in the way in which we speak of "personal morality"—that is, ways of acting culturally, religiously, socially or experientially formed and which cannot appeal to the broader framework of common human capacity or inevitable experience. His notion of communicative ethics presupposes a democratic society in which all are equally capable of participating and in which those who are not (the weak, children, the mentally disabled or those who lack other capacities to participate) are amply represented. His rules of justice, then, would evolve out of a broad framework within which personal morality—provided it was within the framework of acceptable justice—would be tolerated. Habermas—different from Rawls (albeit that the "new Rawls"[43] has somewhat modified his prior individualistic concept)—largely sees decisions as being made by single free-standing individuals acting in their own isolated self-interest whereas Habermas sees individual decisions as enmeshed within community.

THE NATURAL LOTTERY

Views of what has been called the "natural lottery" (a term coined by Rawls) intertwine with our notions of justice and our views of community.[45] By the "natural lottery" is meant "chance" or "luck," which supposedly distributes poverty, wealth, beauty, and other endowments as well as health or disease. Emphasis is on the "luck of the draw," which determines our individual fates. The natural lottery determines our being struck by lightning or slipping on a banana peel. All those things not directly attributable to the individual's own doing or clearly caused by another are viewed in this way. There are three basic ways of looking at the lottery:

1. The results of the natural lottery are no one's direct doing, no one's responsibility, and therefore do not confer any obligation on anyone. Plainly speaking, they are simply regrettable, perhaps unfortunate, but certainly not unjust.[40,44]
2. Although no one may be responsible for the results of the natural lottery, the loser has done nothing to deserve being singled out. In that sense, the results are "unjust." Based on beneficence, such a viewpoint may entail an assumption of obligation.
3. Lastly, one can view the "natural lottery" as far from that simple. Being struck by lightning or slipping on a banana peel do not adequately describe the situation that exists when we are born to wealth or have a heart attack.

Looking at being struck by lightning or being born into grinding poverty as both being mere chance somehow fails to ring true. The conditions that create, aim, and hurl lightning are as yet out of human control; the conditions that create, perpetuate, and ignore poverty are not.[45] Moreover, it has been shown over and over again that health and disease are intimately linked with poverty and other social conditions. At the very least, this makes one's state of health substantially different from being struck by lightning. Most happenings result from a combination of factors in which random selection plays a greater or lesser role. In a complex world, health and disease are conditioned by, if not predominantly due to, a social construct. Our viewpoint of obligation, therefore, may change.

VIEWPOINTS OF COMMUNITY

The way in which we view community largely determines our concept of justice, our sense of mutual obligation, and, ultimately, our laws and procedures. There are two basic and contrasting ways of looking at community:

1. On the one hand, we can view community as consisting of members united only by duties of refraining from harm one to another. In

such communities, freedom becomes the necessary condition of morality (a "side constraint" as Nozick would have it) rather than a fundamental value.[46] Freedom is an absolute and cannot be negotiated. Individual freedom can be restricted only to the extent that it directly interferes with another's freedom. The sole, legitimate power of the community is to enforce and defend individual freedom as well as, since this is part and parcel of acting freely, enforcing freely entered contracts. Beyond duties of refraining from harm one to another, persons have the freedom, although not the duty, to help one another. Except when such help is freely and explicitly agreed upon by mutual contract, they have no obligation to respond to their neighbor's weal and woe.[47]

2. On the other hand, community can be seen to have a different structure. Unless they are united by certain ways of behaving towards one another, associations of individuals living together cannot long endure. Refraining from doing harm to each other makes coexistence possible. But that, in this second point of view, is insufficient and not the way we ordinarily think of community. In ordinary parlance, a community demands a commons in which its members work toward their own, as well as their neighbors', good. Freedom, in such a community, may be a fundamental value, but it remains a value of the community and not one of its absolute and necessary conditions. As such, it is subject to negotiation. In such a community, a "minimalist ethic"[1,48,49] is viewed as insufficient; the Kantian perfect duties (in essence the logically necessary duties of refraining from harm one to another) must be leavened by Kant's more optional imperfect duties (duties, because willing their opposite would represent a contradiction of the will).

These two conceptions of community have quite different roots. The first, so-called minimalist point of view considers only an ethic of mutual non-harm as binding: we are obligated not to harm but are not obligated to help one another. More generous forms of ethical thinking (almost all ethical theories have in fact accepted more than merely obligations of non-harm) assign a varying importance to helping one another.

The way we view such obligations explicitly or tacitly depends on the way we understand what has come to be called "social contract." This term is meant to denote the tacit or explicit agreement among those who associate with one another. The models used are not meant to reflect historical reality but are meant to serve as heuristic devices: they have explanatory power. No association without such explicit or tacit agreements can come about.

Those who subscribe to a minimalist ethic are philosophically related to Hobbes, who saw persons in what he calls "the state of nature" (that is, prior to any association) as living as freestanding asocial individuals. Individuals existed first; associations followed. Such persons were apt to attack, harm, or kill each other and because of this constant threat were unable to "get on" with their lives. They lived in terror of one another, and it was this terror and the consequent inability to live undisturbed lives that prompted them to reach an initial agreement for association with one another. By such an agreement, they promised not to harm one another; mutual help formed no part of such an association. To enforce such an agreement, Hobbes envisioned a sovereign whose power, except for taking his subjects' lives, was virtually absolute.[50]

Much of what is called libertarian thinking today is based on such a notion of initial association, though minus the absolute sovereign. To libertarians such as Nozick in social ethics and Engelhardt in medical ethics, personal freedom is an absolute: it is, as Nozick puts it, a "side-constraint" that modifies all else that a society might wish to do.[47] In this view we have only two ethical obligations: to strictly refrain from interfering with the liberty of others and to scrupulously adhere to freely entered contracts. Certain "moral enclaves" may demand helping one another as a condition of membership" but it cannot be a condition of life within a basic society.[51] Such "moral enclaves," which may or may not be religious, are formed by the free association of their members and are free, as a condition of membership, to promulgate and enforce such standards. One could, for example, make helping each other a condition of membership in such moral enclaves, but one could not constrain those who had not explicitly agreed to this. Beneficence, in the libertarian view, may be "nice," but helping each other cannot be a general moral requirement. Beneficence becomes either a requirement of a particular "moral enclave" which we voluntarily join or it is something that we like and takes on an almost aesthetic dimension.

Such a philosophy believes (*pace* Hobbes) that the only common interest persons from diverse backgrounds and cultures share and know about each other is that each wants to pursue his or her interests freely and to live his or her life as unhindered as possible. They are "moral strangers" and therefore incapable of forging what Engelhardt calls a "contentful ethic:" that is, an ethic that, except for its necessary framework of respect for freedom, can have no universally agreed-upon content. Keeping the peace is the prime function of society, and this can only be achieved if we absolutely respect the freedom of all.[52]

A state exists merely to vouchsafe maximum liberty (consistent with the liberty of others) to all its members. Persons are not required to help their neighbors, and states are not entitled to collect taxes (since collecting taxes is an infringement on liberty) except to provide the necessary mechanisms for enforcing liberty and ensuring common defense. Taxes to support social welfare programs or to provide health care, education, or other services cannot be exacted from the members. The ethics of medical behavior is an ethic of the marketplace; medical practice based on entrepreneurialism is not only allowable but also desirable. Health professionals are conceived as "bureaucrats of health" who are (and, according to H.T. Engelhardt, are properly) entrepreneurs and who see themselves as providing their clients (now seen merely as consumers) with all legally available services regardless of the health professional's own personal moral views. Health care professionals will do the best possible job for the lowest possible fee to attract more "customers:" it is, once again, an ethic of "outward performance" rather than one of "inner intention."

Theories of social contract did not start or end with Hobbes. Anyone who has dealt with or who deals with social ethics today inevitably must explicitly or tacitly deal with the notion of social contract. A quite different notion from Hobbes is that of Rousseau: persons prior to association were also freestanding and asocial, but they were not out to harm each other; they were amoral beings. They were, in fact, endowed with what Rousseau calls "a primitive sense of pity" (or compassion), a trait that forces them to view the suffering of others with distaste and seek to come to their help. Pitted against such an impulse was the "sense for self-preservation." When individual choices had to be made, both impulses were at play. Rousseau sees these beings as amoral rather than as basically immoral. To Rousseau, the type of association in which persons find themselves shapes morality. When the level of ethics is deplorable, so, in general, is the society in which it is found. Rousseau too, however, saw the "state of nature" as consisting of asocial beings and, thus, for him the individual also remained ontologically prior to community.[53,34]

What is today called "social Darwinism" is a misnomer and quite unjust to Darwin who stresses the survival value of compassion. Darwin claims—and we believe rightly so—that solidarity in a group (be it a pack of animals, a tribe, a city or the whole human society) has enormous survival value.[55]

The question of such an association can be looked at quite differently. Whether persons preceded community or communities preceded their members can be seen as a chicken-and-egg question. To be comprehensible, community, and the individual (like cause and effect) must always be defined in each other's terms. There is no doubt that individuals are born into some form of human association and that they are born helpless. Infants do not at first know that they are individuals: they cannot dissociate themselves—biologically or psycho-socially—from the rest of their world, and they become self-realizing individual beings only after some months. At birth and for some time thereafter they are completely dependent upon the nurture of others. Far from terror being their first experience, their first experience is ordinarily one of being cared for. Indeed, in the world as it exists, all of us, even when we are full-grown, are at some time in need of the help and beneficence that only others can supply. This fact, it will be remembered, forms one of the arguments Kant uses to buttress his concept of imperfect or optional obligation. Therefore, autonomy is not a freestanding thing: autonomy perforce develops in the embrace of beneficence. The ability to be autonomous and free is enabled and shaped throughout our lives by the community in which we exist. Autonomy without a community that gives it nurture and support is a meaningless concept.

Furthermore, like cause and effect, community and the individual linguistically must be defined on each other's terms. One cannot adequately explain either without invoking the other.

The notion of moral strangers, furthermore, is flawed. Far from being "strangers" who know nothing about each other except that they want to live freely, all sentient beings have (and know of each other that they have) a framework of needs and capacities. This framework is biologically conditioned: it is a fact from which no moral rules can be derived but without which no moral rules can be crafted. As humans we all share in the human condition, and all our activities are limited and shaped by this framework. Unless some such common framework existed it would be peculiar that virtually all religions and all ethical theories end up with quite similar rules: all forbid murder, theft or lying and all counsel helpfulness to one's neighbor. This framework, which I [EHL] call the "existential *a priori*" of ethics, forms the condition within which our lives are led and our ethics are crafted. At the very least this framework consists of the following:

1. A drive for being: under all but pathological circumstance we all strive to exist.
2. Evident biological needs.
3. Social needs.
4. A common sense of very basic logic that allows us to communicate and reason about basic matters: *i.e.,* things like knowing that an object cannot be in two places at the same time.
5. A desire to avoid suffering.
6. The desire to shape our own lives and pursue our own interests.[34]

Note that it is this last human drive that libertarians claim is the only thing we know about each other and the only one that forms the condition of our ethics. But this desire to shape one's own lives and pursue one's own interests is meaningless if the other conditions are not fulfilled adequately. These "*a prioris*" are not truly separate: they are interrelated and, in a sense, must enable one another. Without existence, we have no biological needs; when these are not met, our social needs are moot, our logic soon has no life to support it, and all suffering ends. Living freely presupposes the meeting of these prior conditions. We are neither "moral friends" who can craft a universally valid ethic nor "moral strangers" who can craft no ethic whatsoever and can agree only on being left alone. At the very least we can assume that we are sufficiently morally "acquainted" to begin to craft an ethic and to leave other aspects about which we cannot currently reach consensus to personal choice and, perhaps, to another day.[58]

Considerations and theories of this sort are crucial to our moral function not only as private physicians encountering private patients or as specialized members of a community whose advice is legitimately sought in health matters, but also as members of that community. Moral theories and the moral principles that emerge from them are most useful if they are used as guideposts along the way of moral reasoning. On the other hand, moral theories and the principles that emerge from them can interfere with moral reasoning providing, instead of guideposts, straitjackets. If moral theories are used in such a way, unnecessarily irresolvable conflicts may result. Ethics reduced to principles and applied to problems in a cookie-cutter fashion, without being filtered through our moral sensibilities, makes a mockery of ethics: instead of being a quest, a search, a sometimes agonizing and always stimulating exploration through which learning and growth can occur, ethics is reduced to yet another technical occupation. Under such conditions, ethics loses its soul.

━━━━━━━━━━━━━ **REFERENCES** ━━━━━━━━━━━━━

1. Loewy EH. *Moral Strangers, Moral Acquaintances and Moral Friends: Connectedness and Its Conditions*. Albany, NY: State University of New York Press, 1996.
2. Clouser KD. What is medical ethics? *Ann Intern Med.* 1974;80:657–660.
3. Churchill LR. Bioethical reductionism and our sense of the human. *Man Med.* 1980;5(4):229–249.
4. Churchill LR. Principles and the search for moral certainty. *Soc Sci Med.* 1986; 23(5):461–469.
5. Abelson R, Nielsen K. History of ethics. In: Edwards P, ed. *The Encyclopedia of Philosophy,* Vol III. New York, NY: Macmillan Publishing Company, 1978;81–117.
6. Kant I. Beck LW, trans. *Foundations of the Metaphysics of Morals*. Indianapolis, Ind: Bobbs-Merrill, 1980.
7. Kant I. Infield L, trans. *Duties to Oneself. Lectures on Ethics*. Gloucester, Mass: Peter Smith, 1978;116–125.
8. Toulmin S. The tyranny of principles. *Hastings Center Report.* 1981;11(6):31–39.
9. Bentham J. *The Principles of Morals and Legislation*. New York, NY: Hafner Publishers, 1948.
10. Mill JS. *Utilitarianism*. Indianapolis, Ind: Bobbs-Merrill, 1979.
11. Kant I. Beck LW, trans. *Critique of Practical Reason*. Indianapolis, Ind: Bobbs-Merrill, 1956.
12. Kant I. Gregor M, trans. *The Metaphysics of Morals*. Cambridge, UK: Cambridge University Press, 1991.
13. Rachels J. *The Elements of Moral Philosophy*. New York, NY: Random House, 1986.
14. Ross WD: *The Right and the Good*. Oxford, England; Clarendon Press; 1938
15. Singer P. *Practical Ethics*. Cambridge, UK: Cambridge University Press; 1979
16. Singer P. *Writings on an Ethixal Life*. Hopewell, NJ: Ecco Press; 2000
17. Dewey J. Ethics. In: Boydston, JA, Kolojeski PF, eds. *John Dewey: The Middle Works 1899–1924,* Vol 5. Carbondale, Ill: Southern Illinois University Press, 1978.
18. Dewey J. *Logic: The Theory of Inquiry,* New York, NY: Henry Holt & Co., 1938.
19. MacIntyre A. *After Virtue*. Notre Dame, Ind: University of Notre Dame Press, 1983.
20. Von Wright GH. *The Varieties of Goodness*. New York, NY: Humanities Press, 1965.
21. Reeder JP Jr. Beneficence, supererogation and role duty. In: Shelp EE, ed. *Beneficence and Health Care*. Dordrecht, the Netherlands: D. Reidel, 1982;83–108.
22. Churchill LR. Bioethical reductionism and our sense of the human. *Man Med.* 1980;5(4):229–249.
23. Fletcher J. *Situation Ethics*. Philadelphia, Pa: Westminster Press, 1966.
24. Fletcher J. Situation ethics revisited. *Rel Humanism.* 1982;16:9–13.

25. Frankena WJ. *Ethics*. Englewood Cliffs, NJ: Prentice-Hall, 1973.
26. Beauchamp J, Tom L. (6ᵗʰ Ed.) *Principles of Biomedical Ethics*. New York, NY: Oxford University Press, 2001
27. Gilligan C. *In a Different Voice: Psychological Theory and Women's Development*. Cambridge, Mass: Harvard University Press, 1982.
28. Nodding N. *Caring: A Feminine Approach to Ethics and Moral Education*. Berkeley, Calif: University of California Press, 1984
29. Nelson L. Against caring. *J Clin Ethics*. 1992;3(1):8–15
30. Loewy EH. Care ethics: a concept in search of a framework. *Camb Q*. 1995; 4(1):56–63.
31. Kant I. Smith, NK, trans. *Critique of Pure Reason*. New York, NY: St. Martin's Press, 1965.
32. Loewy EH. *Suffering and the Beneficent Community: Beyond Libertarianism*. Albany, NY: State University of New York Press, 1991.
33. Loewy EH. *Suffering and the Beneficent Community: Beyond Libertarianism*. Albany, NY: SUNY Press; 1992
34. Loewy EH. *Freedom and Community: The Ethics of Interdependence*. Albany, NY: SUNY Press, 1992.
35. Jonas H. *Das Prinzip Verantwortung: Versuch einer Ethik für die technologische Zivilisation*. Frankfurt a/M: Suhrkamp Taschenbuch, 1984.
36. Fleck L; Trenn DJ, Merton RK, eds; Bradley F, Trenn DJ, trans. *Genesis and Development of a Scientific Fact*. Chicago, Ill: University of Chicago Press, 1979.
37. Kuhn T. *The Structure of the Scientific Revolution*. Chicago, Ill: University of Chicago Press, 1970.
38. Engelhardt HT. *The Foundations of Bioethics*. New York, NY: Oxford University Press, 1986.
39. Cassel EJ. Do justice, love mercy: the inappropriateness of the concept of justice applied to bedside decisions. In: Shelp EE, ed. *Justice and Health Care*. Dordrecht, the Netherlands, D. Reidel, 1981;75–82.
40. Rawls J. *A Theory of Justice*. Cambridge, Mass: Harvard University Press, 1971.
41. Rawls J. *A Theory of Justice* (2nd Ed.). New York, NY: Belknap Pr. 1999.
42. Habermas J. *Erläuterung zur Diskursethik*. Frankfurt a/M, Deutschland: Suhrkamp Taschenbuch, 1992.
43. Rawls J. *A Theory of Justice* (2nd Ed.). New York, NY: Belknap Pr, 1999.
44. Engelhardt HT. Health care allocations: responses to the unjust, the unfortunate and the undesireable. In: Shelp EE, ed. *Justice and Health Care*. Dordrecht, the Netherlands; D. Reidel, 1981;121–138.
45. Loewy EH. *Freedom and Community: The Ethics of Interdependence*. Albany, NY: SUNY Press, 1994.
46. Nozick R. *Anarchy, State and Utopia*. New York, NY: Basic Books, 1974.
47. Engelhardt HT. *The Foundations of Bioethics*. New York, NY: Oxford University Press, 1986.
48. Callahan D. Minimalist ethics. *Hastings Center Report*. 1981;11(5):19–25.
49. Loewy EH. *Freedom and Community: The Ethics of Interdependence*. Albany, NY: State University of New York Press, 1989.

50. Hobbes T; Oakshott M, ed. *Leviathan*. New York, NY: Collier Books, 1962.
51. Engelhardt HT. *Bioethics and Secular Humanism: The Search for a Common Morality*. Philadelphia, Pa: Trinity Press International, 1991.
52. Engelhardt HT: Morality for the medical-industrial complex: a code of ethics for the mass marketing of health care. *N Engl J Med*. 1988;319(16):1086–1089.
53. Rousseau JJ: *Du Contrat Social (R. Grmsely, ed.)*. Oxford, UK: Oxford University Press, 1972.
54. Rousseau JJ: *Discours sur l'Origine et les Fondaments de l'inégalité parmis les Hommes*. Paris, France: Gallimard; 1965.
55. Darwin C: *The Descent of Man*. New York, NY: H.M. Caldwell, 1874.

4

Fallibility and the Problem of Blameworthiness in Medicine

In the course of their daily practice, healthcare professionals, just like other people in all walks of life, are confronted with choices. We should not delude ourselves—when we have the capacity to make a choice or the capacity to act or to refrain from acting we have, in fact, chosen or acted. Healthcare professionals, like all other men and women, must accept the fallibility of these choices. The diagnosis made, the treatment determined, the conclusions drawn, the procedure done or how it was done, all may be wrong. In the human condition, error is the risk we take. When shoemakers err and, despite prudence and care, spoil a pair of shoes, they must, if possible, remedy their error or at least learn from it. Regrettably, their error may be irretrievable, and a pair of shoes may be lost. When doctors or other healthcare professionals err and, despite prudence and care, misdiagnose, mistreat, misjudge, or otherwise do something they later recognize as "wrong" or harmful, they too must try to remedy the error and learn from it. And such errors, even more regrettably tragic, may also be irretrievable: the patient may die. In the course of most types of medical practice practitioners must face the fact that their error at one time or another will be responsible for the death of a patient. This fallibility, inherent in medicine as it is in any other human activity, is the price of action in any field.[1] It is only that in medicine the stakes are so high. But no matter how high the stakes, no matter how dreadful the consequences, mortal man is bound to err.

Medicine, it has been said, should not be like other fields: decisions, actions, and consequences are too critical. Physicians and other healthcare professionals, it is often implied, should not be like other men or women. Their material is life, and life is too precious to permit error. But workers, no matter how well trained, no matter how alert, no matter how conscientious, no matter how careful, remain fallible. They may spoil the material with which they must deal, and the nature of the material does not change this basic and certainly disturbing fact.

Physicians as well as other healthcare professionals have come to accept the often heavy burden of their potential fallibility and to deal with its regrettable exemplars in their own way. The honest, intelligent, and strong admit, learn from, and regretfully put the error behind (except, at times, in their dreams); the honest and less strong often cannot and are sometimes crushed by it; the less honest and weak try to manipulate the facts and to structure the evidence until the error of yesterday is seen as no error at all and failure becomes success.

Manipulating facts so as to make what is appear as though it were otherwise, happens in every field of endeavour. Sometimes it is done unconsciously, an act of repression, as it were. Sometimes the attempt is initially a deliberate one that starts consciously and in which eventually we ourselves come to believe in the truth of what we initially pretended. Instead of using our energies in learning from the mistake we acknowledge, we consume our energies in a senseless quest for the blamelessness we seek and, in fact, can never attain.

At this point it is important to consider our attitudes toward "truth" as well as toward solving problems. If we believe that truth is absolute and is "somewhere out there" to be discovered, our attitude toward human fallibility will be quite different than if we believe that truth (whether scientific or ethical "truth") in the human condition is also, in part, crafted. A belief in the absoluteness of truth confronts error either by denying it or by clinging to the idea that our search was somehow flawed. On the other hand, if we believe truth to be created by human effort, we are more likely to take a more tolerant view of fallibility and to use this fallibility as an instrument for growth and learning rather than as a cause for condemnation. In the latter view (which owes a lot to the work of John Dewey), problems are, by their nature, not "soluble:" we can only strive to make an "indeterminate situation more determinate." Such an improvement carries within it new indeterminacies that, as we discover them, will be worked upon so that further improvement can take place. Gaining knowledge, solving problems, or engaging in our daily tasks is an opportunity for continued improvement and continued growth. (See Chapter 3.)

On the whole, physicians have learned to be, if not comfortable with, at least accepting of their technical errors. Often they will acknowledge these only to their inner selves, fearful that honest disclosure will lead to censure or, worse yet, to legal suit. This fear, while generally unjustified, is frequently potent enough to prevent more public disclosure and thereby tends to hamper others from learning from such mistakes. There is, however, more than a fear of censure or suit. Physicians, at least as much as other men and women, fear a loss of prestige and the associated loss of an aura of omnipotence. Such a fear can be as deep-seated as the belief in one's infallibility and the arrogance that nourishes it. Such a fear may originate in a deep inner sense of inferiority, which would find its confirmation in admitting an error—even to oneself. Physicians and other healthcare workers, at least as much as other men and women, partake in a process of delusion that starts with an attempt to delude others and ends by deluding oneself. When, however, physicians fail to acknowledge error even to themselves, self-delusion blocks even self-learning, and errors (no longer recognized as errors) are prone to be repeated.

John Dewey long ago pointed out that analyzing and dealing with moral problems is (methodologically) the same as all other inquiry: We suggest hypotheses and test them in the crucible of praxis. Such "praxis," then, is forever experimental and, despite all of our care, is prone to error. Praxis does not suggest merely acting on our choices, but praxis (or experimenting with the propositions we have put forth) denotes a much broader range of activities: thought experiments or dialogue is as much a part of such "praxis" as is actual concrete action.[2,3]

Knowing that probability is not certainty and that certainty, at best, is really only reasonable and tentative certainty, physicians accept uncertainty and error in technical matters as a reasonable price for action. But when it comes to the moral realm, uncertainty is less easily accepted. None of us expects to be right at all times, but at the very least we want to be perceived as scrupulously virtuous. The fear of blame causes us to seek certainty in the very realm least likely to provide it.[4,5]

Even worse, in making choices about moral matters, we rarely have problems in choosing between the "good" and the "bad." All other things being equal, only fools or psychopaths deliberately or knowingly choose a "bad action" in preference to a "good" one. Our choices are more constricted: we must, in general, sort out one "bad choice" from another more or less "bad" one and then act upon that choice. Such choices rarely leave us with a "good" alternative for action. No matter how hard we try, we are left with a course of action that, when considered by itself, is bad. Blameworthiness, it seems, is difficult to evade.

Actions, objects, or judgments in moral as well as in non-moral matters may be desirable, undesirable, or indifferent to us. We have little trouble choosing between the desirable, on the one hand, and the indifferent or undesirable, on the other. We have no trouble with choosing between having a tooth pulled or seeing *Hamlet*: most people enjoy seeing *Hamlet* and hate having their teeth pulled. It would take a very odd man, indeed, to will pulling a tooth instead of seeing a play (no matter how bad the play or its performance might be!). Our problem is in choosing one from among several attractive objects or courses of action or, on the other hand, one object or course of action from among several unattractive ones. In the old legend, the donkey sitting equidistant between two equally attractive parcels of food starved to death. His was a true dilemma.

We do not generally have severe problems choosing among attractive objects or actions. Deciding whether to go to the theater or to the opera leaves us with choices which are both in themselves attractive and neither, in and of themselves, regrettable. However we choose, we may be sorry not to have picked the other, but we will still be glad to have picked the one. And a normal person is. It would be odd if we let our pleasure at hearing *Don Giovanni* be entirely spoiled by the thought that we missed *Hamlet*. We may say to ourselves or to others: "I am sorry to have missed *Hamlet*" but we do not, therefore, say that seeing *Don Giovanni* was an unpleasant experience. Even if we regret the choice, we regret not having had, as we now see it, a greater pleasure; but we do not therefore conclude that our pleasure, even if lesser, was not a pleasure.

Choosing among unpleasant experiences—say, the choice between having a tooth pulled and having a filling put in—leaves us in much the same way. We may conclude that, on balance, we prefer having a filling, and we go have it done. And we may be glad that we did not choose what we perceive to have been the more painful route. When we come home from the dentist, we may take pleasure in the pain we evaded; we may say that, in comparison, we are glad to have had the smaller rather than the greater pain. But few would now say that having a tooth filled was a pleasure and, on the whole, a nice experience. When we choose between two unpleasant experiences, then, we do not try to pretend that one of them, because we chose the other, is now pleasant or pleasurable. The same holds when we choose one from among several possible pleasant choices: we do not or cannot reasonably claim that what we did not choose was unpleasant but merely, for a variety of reasons that it was or seemed less pleasant.

But why choose? Why not simply do nothing or toss a coin? Few are willing to abrogate their right to choose; few are willing to say that they are unable to make an intelligent choice. Rarely are the cards so evenly stacked that, simply speaking, there is no relevant difference. When we fail to choose, or when we leave the choice up to chance, we deny our freedom to make a choice, refuse to think deeply and critically about relevant differences, or simply are confronted with a situation about which we do not much care and in which thinking deeply or critically is not worth the bother.

Whether we go to *Hamlet* or *Don Giovanni* may, basically, be irrelevant to us. We may not wish to expend the time or effort needed to make the choice. And so we toss a coin, or leave it up to our spouse, glad not to be directly involved. But, however we allow the choice to be ultimately made, refusing to partake is an expression of not caring very much about what happens. It is a form of "copping out." When it comes to choosing between a play and the opera or among various restaurants, the choice may seem too trivial to justify engaging our attention: We truly do not really care which choice is ultimately made. When it comes to making ethical choices, however, leaving the matter up to chance or refusing to choose will not do: it will not do because making such choices by definition affects the weal and woe of another. Refusing to choose or choosing to leave things up to chance is, moreover, very much a choice: in such a case it is a choice not to engage in making often troubling choices about important issues that affect another rather than about merely trivial issues affecting only or predominantly ourselves. It says loudly and clearly that we really do not care.

Ethical choices in medicine confront us with similar considerations. Inevitably perplexing situations fail to have "good" answers, and we are left with alternative courses of action any one of which is bad in and of itself. Not treating pneumonia in a vegetative patient may be the most reasonable and, on balance, the "best" alternative; but that fails to make the non-treatment of pneumonia and, thereby, the hastening of death a "good" and "praiseworthy" thing to do. Such a course of action cannot and should not cause us to go home feeling self-satisfied; it can, and perhaps should, leave us saddened but relieved that, among the terrible alternatives, we have chosen the least bad available. It does not make us praiseworthy for the act (an act intrinsically blameworthy), although it might make us praiseworthy for the agony of choosing. The process of choosing and the choice itself are not the same thing.

Not making choices, evading decisions, and hiding behind immediate technical concerns is tempting and is what is, in fact, frequently done. After all, physicians are trained to watch scientific changes and to intervene in the biological process so as to support homeostatic mechanisms tending to promote life and health. Such work is difficult, demands a great deal of skill, and is often perplexing. It is easy to hide behind technical factors, pretending to oneself not to have time for other matters. And that pretension is often close to the truth. Yet if physicians allow themselves to attend only to the task of balancing electrolytes, adjusting blood gases, and choosing antimicrobials, they will not only evade their prior human responsibility as moral agents, but will also eventually force others (or mere chance) to make such choices for them. Problems in medicine are too critical and too close to the weal and woe of real people to permit the evading of choices.

Refusing to make moral choices does not evade blameworthiness. Rather, it transfers such blameworthiness: physicians now are responsible not for the choice not made but for making the choice not to choose and thus for abrogating and denying their moral agency. When healthcare professionals only pose the "can" question ("can I do something?"—e.g., treat a condition or make a diagnosis), and evade the more troubling "ought" question ("ought I to do what I can do?"—e.g., treat a condition or make a diagnosis), technology ultimately will drive itself. In a day and age when technology (for good or for evil) is ever more powerful such a course of action (since our deciding critically involves another—another who is in our trust) is not an ethically proper choice: it is, in other words, humanly speaking, "bad medicine."

In the human condition, and in the human condition confronted by healthcare workers, moral choices often leave us with the necessity of choosing between acts any one of which, to a greater or lesser extent, is blameworthy in and of itself. This notion, if one is theologically inclined, hearkens back to notions of original sin, in which humans by their very condition inevitably must sin (and, according to Luther, should at least sin "boldly," i.e., forthrightly and without dissimulation). In modern clothes, this Augustinian concept is found in many of our choices in medical ethics. When we evade our blameworthiness by rationalizing or manipulating the facts and circumstances of our action until a claim of praiseworthiness emerges, we run the danger of moral callousness. Doing certain things—killing, for instance—must never be looked at as "good" or praiseworthy; when things that in themselves are clearly "bad" or blameworthy are manipulated and glossed over so that they are now presented or seen as "good" or praiseworthy, moral callousness easily results. Doing such things now is suddenly presented as, and easily becomes, an intrinsically good and praiseworthy act and makes the next such action, even if in different circumstances, all the easier.

On the surface, this realization seems bleak and dark. If, indeed, all humans when confronted with many moral dilemmas are destined to make an inevitably "bad" choice, why choose a field in which the consequences of such choices so often lead to disaster and in which blameworthiness is blameworthiness for such terrible consequences? Why not do something else, where decisions, actions, and consequences are not as stark? Why beat yourself to death? But there is another side to this: most of the daily work that health-care professionals do helps innumerable people and brings deep satisfaction, the price of which may well be an occasional mistake or failure.

There is a difference between the choice made or the action ultimately taken and the process of choosing. Refusing to choose leaves us with a choice made for us by external forces. Choosing between *Hamlet* and *Don Giovanni* or between having a tooth pulled and having it filled is a process internal to us. Like every internal process, it generally reveals more about us than it does about the problem itself.[6] It speaks to our fears, hopes, and values more than it does to the intrinsic merits of the thing we actually choose. Killing, ripped out of the context of the situation and divorced from the moral actors involved in the actual example, becomes an action subject to dispassionate interrogation or examination. But looking at problems in this fashion and making issues or categories out of real problems changes the real problem and makes it an artificial and, therefore, a different one.[7]

Abstracting the practical problem, and making out of it a category to be studied, has practical value only if, after examining this now isolated phenomenon for the sake of greater clarity, we rejoin the issue or category to the actual context and its moral actors and now reexamine the actual problem in that light. The "answers" to the problem, examined out of its context and away from the moral actors that ultimately must act, are part of the material of our ultimate choice. Our actual choosing, however, involves more than this part of the material alone. Our choosing and our ultimate choice involve the context and the moral actors no less than they do the category of our problem.

When we examine a given problem in isolation, away from its context and divorced from the moral actors whose agony of choice must, eventually, be translated into agony of action, we may be blameworthy for a course of action chosen. But we may have no other better choice than this unquestionably blameworthy one. In the human condition, choosing will inevitably confront us with this fact: Our choice, removed from its context and divorced from its moral actors, may be blameworthy in itself. But in choosing among an array of choices all in themselves blameworthy to a variable degree, we may deserve praise for the agony of that choice rather than choosing not to choose or choosing capriciously or carelessly. Healthcare professionals, in going about their daily tasks, must learn to assume and be accepting of their fallibility in technical as well as moral matters. Mistakes will be made, errors committed, and undesired outcomes achieved. Doubts, often nagging doubts, will remain. These not only are the price for action but, used properly, can also serve as prods to and vehicles for learning and for moral growth.

Healthcare professionals must learn to take not only their technical but also their moral fallibility in their stride, to learn from it, and to put it into the perspective of a full and rewarding life. Beating oneself to death, or dwelling on errors as opposed to learning from them and then going on, is a destructive way of dealing with the realization of one's own mortality. Accepting blameworthiness (rather than immaturely rationalizing that which is worthy of blame to be worthy of praise) and making moral choices in the full realization of fallibility and blameworthiness serves to enhance the personal growth of compassionate, thinking persons and, therefore, helps make better persons and, in turn, better healthcare workers.

To be blameworthy for initiating or participating in a bad act does not, although we often think of it that way, make us "bad" or evil human beings. We are judged— by ourselves or by others, if they or we choose to be fair—by the totality and inevitability of our actions and choices, not by actions divorced from their context. A good person is not described by a tabulation of single actions and choices bereft of context but rather, as the Greeks saw it, by their "self-making" or the ability to learn from situations and, in consequence, to change themselves for the better.[8] Our choices say more about us as persons than they do about the problem itself. Whether technical or moral, and the difference is certainly not always clear-cut, our choices are reflective of us as persons and are springboards to moral and intellectual growth. The business of ethics, which cannot be divorced from the business of living or that of practicing medicine, is concerned not merely with single problems, their categorization, or their solution, but also, and with at least as much force, in promoting personal growth and, in a classical sense, the skills and "virtues" of medicine.[8] (See also "virtue ethics" in Chapter 3.)

In recent regulations of the Joint Committee on Hospital Accreditation patients (or, if they lack decisional capacity, their surrogates) must be informed of all medical mistakes.[9] According to statistics such mistakes are not at all rare.[10–12] Many are trivial; most are correctible; but some are fatal. The central question is what one understands as "all mistakes."[11] The "taxonomy" of mistakes badly needs to be clarified.[13] The definition by Wu as "a commission or an omission of with potentially negative consequences for the patient that would have been judged wrong by skilled and knowledgeable peers at the time it occurred, independent of whether there were any negative consequences" seems to be as solid a definition as possible.

Patients most certainly should be informed of mistakes in diagnosis or management—that is part of truth-telling and an issue that is hard to argue against. Hiding such mistakes from patient or family is a violation of truth telling in all spheres of life. We do not tolerate "hit and run" accidents—whether made with the best intentions or not. But hitting the bumper of the car parked in front of you without any visible damage is rather different from denting the same car and is again quite a different matter than hitting a pedestrian and leaving the scene. To say that all mistakes must be shared with the patient can range from the trivial to the sublime. In medicine we need clearer instructions than "all medical mistakes" which may range from the trivial ("I started your i.v. 15 minutes too late"), to the fatal. Usually it falls somewhere in between. Perhaps the worst part of omitting to inform patient, family or members of the treating team is that it supports self-delusion. Physicians frequently appeal to the fear of malpractice suits. The problem (in my [EHL] experience) is not in the mistake itself. Physicians all too often create the impression of infallibility to patient and family. A mistake made by one who claims (implicitly more often than explicitly) infallibility will be judged by family and patient quite more harshly than when such a mistake is made after reasonable attempts at honest disclosure.

Stress is laid on "outcome"—did my mistake lead to a serious outcome or did it fail to lead to a serious, possibly a fatal, outcome? The consequences or lack thereof is, however, hardly the question. A mistake not admitted is a mistake not easily recognized. The outcome is difficult to assess. Terrible mistakes may lead to no serious results and errors which at times may be considered trivial can result in serious setbacks or even prove to be fatal. One of the authors (EHL) who was engaged for about 15 years in the practice of cardiology before he "re-tooled" and began to deal solely with questions of ethics, remembers such mistakes very well. Those that cost the life of the patient and those in which one simply was lucky. I remember a patient who was in a two bed room—two very nice elderly gentlemen. My patient had a heparin canula and was receiving heparin every four hours. The other (by chance the head-nurses father) had sustained a severe upper GI haemorrhage and was not my patient. The nurse was unable to insert the heparin canula so that I offered to do so and did. Before starting I checked the name on the wrist but that had, for some reason, been taken off. When I asked him if "X" was his name, he nodded pleasantly, I inserted the canula and gave 50mg of heparin. It was only then that I saw my patient who had had his curtains closed. I told the patient, his daughter, the attending physician and left a note on the chart. It certainly taught both the nursing staff and me.

Aside from the particular outcome in a particular patient, errors need to be ventilated and discussed. The old "M & M" (mortality and morbidity) conference provided a vehicle for discussing mistakes openly—its aim must not be punitive but rather one of "learning from one's mistakes."

All of us want to minimize error—I purposely do not say abolish it since that is an impossible albeit laudable goal. Errors are by no means necessarily individual error—at times such errors are system related. That, however, does not excuse the particular physician, nurse or other health-care worker. One of the obligations that physician and allied health-care professionals have—and one which will be stressed throughout this book—is to work towards creating a system with sufficient "fail-safe" mechanisms in their office, hospital or other medical setting which will greatly diminish such occurrences. When an error caused by physicians, nurses or other health-care professionals causes or interferes with treatment or healing this should be readily disclosed to the patient.[15] The questions we must clearly address are: (1) what is error in the medical setting? (2) how and by whom should an error be discussed with the family or patient? (3) are we obliged to reveal an error made by a previous physician, especially when they have referred the patient to us?

More difficult than revealing one's own error is revealing to the patient that a prior (or referring) physician has made a severe blunder. Misdiagnosis or mistreatment by another or—most unpleasant—referring physician, nevertheless, must be addressed. The aims properly are safeguarding the patient and teaching the physician. In speaking with the referring or previous physician it may turn out that what appeared like error was in fact and in this case justified—or it may turn out that it was not. When there has been a clear mistake, it seems clear that the patient must be informed—failure to do so could be analogized to knowing who had committed a crime and purposely shielding the criminal. The real question is not informing the patient—for that we are obligated to do—but how to inform the patient and relatives. When physicians see themselves—and are perceived by their patients—as partners instead of implicitly taking on the mantle of infallibility, admitting one's error comes easier and contrary to what many physician's think admitting to an error is not generally perceived by relatives as a statement of incompetence, provided that the physician has treated patient and family as equal human beings right along and has not hesitated to share some of the problems. Patients and their relatives are more than likely to "forgive" such errors in a setting in which communication was and remains open.

One problem closely related to that of admitting error, accepting blame, and being aware of one's fallibility is what we consider to be one of the rarely spoken about but central problems of ethics today: hypocrisy. Not only is hypocrisy, as a form of lying or at least of misleading, evil and dangerous in itself; one of its chief dangers is that it easily leads to a thin veneer of self-delusion when the agent himself begins to believe his own excuses and untruths. In daily life as well as within the healthcare professions, hypocrisy takes many forms and expresses itself in many ways. When we show a degree of concern for someone we in truth do not feel, when we use euphemisms to cover the tracks of what we in fact are doing, or when we show deference to another we really do not respect merely because that other is powerful or wealthy, we are engaging in a form of hypocrisy that, like all hypocrisy, destroys integrity. As we proceed in this book, we shall repeatedly run across examples of this sort of thinking and acting. If any message can be given in a book dealing with ethics, it is this: what one does may be problematic, what one does may be mistaken or even morally wrong, but, at the very least, what one, in fact, does or thinks needs to be freely acknowledged and discussed. The attempt to practice medicine or to go through life in a morally acceptable manner requires at least that much.

REFERENCES

1. Gorovitz S, MacIntyre A. Towards a theory of medical fallibility. *J Med Phil.* 1976;1(1):51–71.
2. Dewey J, *Logic: The Theory of Inquiry.* New York, NY: Henry Holt & Co., 1938.
3. Dewey J. The quest for certainty. In: Boydston JA, ed. *The Later Works of John Dewey,* Vol 4. Carbondale, Ill: Southern Illinois University Press, 1988.
4. Doherty DJ. Ethically permissible. *Arch Intern Med.* 1987;147(8):1381–1384.
5. Loewy EH. The uncertainty of certainty in clinical ethics. *J Med Humanities Bioeth.* 1987;8(1):26–33.
6. Loewy EH. Drunks, livers and values: should social value judgments enter into transplant decisions. *J Clin Gastroenterol.* 1987;9(4):436–441.
7. Churchill LR. Bioethical reductionism and our sense of the human. *Man Med.* 1980;5(4):229–247.
8. Pincoffs E. Quandary ethics. *Mind.* 1971;80:552–571.
9. Johnson BJ, Tzang J. Evaluating a medical error taxonomy. *Proc. AMI Symp,* 2002:71–75.
10. Gabel RA et al. Counting deaths due to medical error. JAMA 2002;288(4):501–507.
11. Leape LL. Reporting of adverse events. NEJM 2002;347(20):1633–1638.
12. Honig P, Phillips J, Woodcock J. How many deaths are due to medication error? JAMA 2000;284(17):2187.
13. Johnson NJ, Zhang J. Evaluating a medical error taxonomy. Proc AMIA Symp 2002, 71–75.
14. American Medical Association Council on Ethical and Judicial Affairs. *Code of Medical Ethics: Current Opinions with Annotations.* Chicago, IL: American Medical Association 1997;sect. 8:12–15.
15. Rosner F, et al. Disclosure and prevention of medical error. Archives Int. Med. 2000;160:2089–2092.

5

The Ongoing Dialectic between Autonomy and Responsibility in a Pluralist World

INTRODUCTION

Autonomy, as usually understood, implies the ability to govern oneself or, in Kantian language, to set one's own rules. One must differentiate between autonomy (or freedom) of the will and autonomy (or freedom) of action. This differentiation is especially important in the medical setting: on the one hand patients (because of dementia, hypoxia, hysteria, drugs, alcohol and many other factors) my lose their freedom of willing (that is, they may be unable to make a rational decision because they cannot grasp the circumstances); on the other hand, they may when ill, hospitalized, and weakened but quite rational often lose the ability to act for themselves. Such patients, although they retain adequate function of willing are incapable of translating their clear will into action. This loss of power and consequently becoming, as it were, a prisoner of the medical system is something especially feared by the patient. Such a loss of ability to act is, of course, variable and may range from the slight to the complete, but it is almost invariably a part of the medical interaction. Even in an outpatient setting the patient's ability to act is constrained by the evident fact that healthcare professionals have more power: the power of greater knowledge as well as the power to provide or to refrain from providing certain services to patients. We will come back to this point throughout this book.

Full autonomy of will or action is an ideal never fully realized or realizable in the human condition. Biological (including genetic) factors impose very real limits on our abilities; environmental factors create conditions to which, whether we like it or not, we must adapt; and matters of cultural background, upbringing, and the social conditions in which we find ourselves severely constrain our willing as well as our acting.[1] Kant, whose name is most closely associated with the concept of autonomy, understood this very well when he stated that only the "divine being" is truly autonomous.[2] But not being entirely autonomous does not mean that man is totally at the mercy of external forces or totally, as Hume would have it, "the slave of the passions."[3] Certainly, the limits of autonomy are set by forces that are, in a sense, external to the will and beyond the control of man (and these may vary from time to time and from situation to situation), but man's freedom to operate within those limits is what we commonly mean when we speak, loosely, of free will or of an "autonomous act."[4] Patients unable to will (that is, patients who are unconscious, confused, unable because of circumstances to grasp the facts presented, to choose among possible options, and to know the probable outcome of such choices, or patients who are unable to give a rational reason for their choice even if it is a reason we would not subscribe to) are said to lack decisional capacity.

Autonomously motivated behavior by, say, person A that is perceived as harmful to person A by another person B may lead person B, if he is capable of so doing, to interfere with A's behavior. Such interference is often defended by an appeal to responsibility. One can be responsible because one is culpable (responsible because one has, positively or negatively, been causally linked to a particular event or circumstance), or one can be responsible for other reasons (responsible, for example, because one has a particular role: as citizen, as human being, or as healthcare professional). Whichever it may be, the feeling of responsibility is a response to an externally or internally imposed or perceived condition. Thus, we may be responsible because of our role (or the way in which we and others conceive that role), because of a promise freely given or a contract freely entered, because of something that we have done (broken a cup or given the wrong treatment, for example), or because of the promptings of an internally felt *noblesse oblige* (as when we see a helpless creature in need of help).

Our view of community and obligation to one another conditions our sense of responsibility and consequently our actions (see Chapter 3). In the sense of causal responsibility (as in breaking a cup or treating a patient in the wrong way) and, in some respects, in the sense of role responsibility, responsibility is determined and judged externally. In role responsibility, however, the delineation is largely a changing social construct determined over time. Communities determine role responsibility as a composite expression of the internally felt responsibility of their individual members. Internally felt responsibility, while to some extent conditioned by extrinsic factors, is also, to some degree, the product of autonomous, rational choice. It is a function of how we view ourselves in relation to others and in relation to community. In that sense, the choices we make and the responsibilities we assume say more about us as moral actors than they do about the problem.[5]

Responsibility for one another, and the feelings of obligation that result, may clash with autonomy. When healthcare professionals feel "responsible" for their patients (as, indeed, they should), when they feel that maintaining, safeguarding, or restoring anatomic and physiological function is the highest good, they are likely to ignore or attempt to ignore what they feel are poor choices on the part of their patient.

Unless physicians and other healthcare professionals accept the fact that supporting or, at times, restoring a patient's autonomy is part of their responsibility, they may choose to treat against a patient's express wishes or attempt to delude patients by withholding or modifying information. On the other hand, physicians, out of respect for an ill-conceived understanding of autonomy, may allow patients, without further efforts, to pursue a course leading to disaster (as it were, abandon patients to their possibly seriously deficient "autonomy"). Medicine is but a microcosm of this daily struggle: when we allow our homeless to wander the streets and freeze to death (mostly because communities lack even reasonably adequate facilities, but sometimes because we accept that some of the homeless allegedly "wish" to do this) without, forcefully if need be, taking responsibility, the same issue is at play. Unthinkingly and unfeelingly abandoning persons to their supposed autonomy is the flip side of paternalism. In many ways it is as grievous an infringement on the physician–patient relationship as is crass paternalism.

——————— PATERNALISM ———————

According to Dworkin, who has given us the standard definition of the term "paternalism," paternalism can be defined as the "interference with a person's liberty of action justified by reasons referring exclusively to the welfare, good, happiness, needs, interests or values of the person coerced."[6] Such a definition is essential but does not, by itself, take us very far. Feinberg took this a step further in first dividing paternalism into a form that seeks to prevent harm and a form that seeks to bring about another's good. Secondly, he distinguished between "weak" and what he called "strong" paternalism. Strong as well as weak paternalism may be motivated by preventing harm or by bringing about good.[7]

In weak paternalism the actor attempts to prevent conduct that is (1) substantially non-voluntary or (2) done without full or adequate knowledge or understanding of the consequences by the person acting; also at times, the actor may temporarily intervene to determine whether an act was truly autonomous or not. An example of the first (protecting patients from non-voluntary harm) might be preventing harm to one under hypnosis or on drugs or even one who is under severe coercion. An example of the second would be giving life-saving therapy to a young child whose parents refuse such treatment; an example of the third might be pushing someone from the path of an oncoming train or treating a patient who has taken an overdose of drugs but whose motives are not clear to us or whose motives are believed to be capricious, not thought out or temporary. Weak paternalism is a form of preventing persons from coming to non-understood harm. It is protecting another from the results of misinformation or non-comprehension. As such, and because such actions are seemingly clearly called for, they are basically morally uninteresting.

Strong paternalism, on the other hand, seeks to prevent harm to (or act for the benefit of) persons by liberty-limiting measures even when their contrary choices were not capricious, were well informed and voluntary (forcing patients who are Jehovah's

Witnesses to be transfused is one of the more frequent examples). It is, as are all paternalistic acts, done (ostensibly, at least) to prevent "harm" or to bring about what is perceived to be the "good" of another—terms which, in this situation, are defined by the actor and not the recipient. In that it is by definition an act that only seeks to prevent harm or to benefit another it is not self-interested but other-directed. Ordering an unnecessary test for a patient so as to benefit the person ordering it is, thus, not paternalism but rather an act in which the patient's good plays but a small role.

Unfortunately and not rarely, "seeking the patient's good" is the reason given for acts that are meant ultimately to benefit the physician, the institution, the family, or the community. Acting for the sake of other interests may not always be something to be condemned: there may well be reasons why the interest of the individual patient ought not to be the sole or even the main motivating factor. Such, for example, may arguably be the case when a test to detect a highly infectious disease needs to be ordered to protect the public and the patient refuses. But pretending that the patient's sole interest is what is important to such a course of action adds hypocrisy to the already suspect act of paternalism.

Paternalism (or parentalism, as it has lately begun to be called) seeks to do one's own good to another instead of facilitating that other's (self-selected) good. It often arises out of a sense of responsibility in which the paternalist's claim to greater knowledge, foresight, wisdom, or experience is the ostensible excuse. The fact that such claims are not always without foundation makes paternalism all the more insidious and therefore dangerous. If paternalism were simply a crass act of one human being callously superimposing values on another, the problem would be easy. Forcing persons to listen to Mozart (or to rock music) because we happen to like Mozart (or rock music) is clearly indefensible. But forcing a panicky patient to undergo emergency treatment to save his life (or forcing a homeless person to seek shelter in a snowstorm) may be quite another matter.

Although clearly presaged by Rousseau, the modern concept of autonomy, as mentioned above originated with Kant. [8] The will is not only subject to the moral law: To be autonomous and, therefore, "worthy of respect," it has to be subject to its own self-legislated, universalizable law. This autonomy comes from within the individual. But autonomy itself does not seem to be enough if one reads Kant clearly: "Nothing in this world ... can possibly be conceived to be good without qualification, except a good will."[8] And so, autonomy can only be an instrumental good, one that depends for its goodness or badness upon the will and the (unfortunately not entirely clear) "goodness" that guides it. Mill, in his entirely different Utilitarian concept of the bases of morality, likewise considers autonomy to be a fundamental fact of the moral life.[9,10] To Mill, autonomy of action must be one of the fundamental principles in order to maximize the good of society. So long as persons' actions do not directly infringe upon their neighbors', such actions *must* be permitted.

In a society in which personal liberty, freedom of thought, and eventually freedom of action was severely restricted, the ideal of autonomy served well. Men were downtrodden by a rapacious state that simultaneously inhibited its citizens' freedom and denied responsibility. Where the state was forced to assume responsibility, it did so hesitatingly, grudgingly, and with humiliating condescension. Charity itself became a tainted word, responsibility for one's fellow man an impoverished concept, and autonomy a dangerous thing. Whether interpreted from the philosophical slant of Kant or that of Mill, autonomy of will and action underwrote the promise of the American, the French, and, later, the Russian revolutions. And it has continued to motivate the overthrow of a regime calling itself Communist but violating most if not all basic Socialist precepts. If capitalist nations do not provide adequately for the basic needs and education of their members, freedom of will and action likewise becomes a sham, and such nations, in turn, court disaster.[11]

If we hold freedom to be an absolute condition of the moral life and look at communities as collections of men united merely by duties of not bringing harm one to another, we will place a supreme value upon autonomy and will find the place of responsibility for one another to be, except under contractually stipulated conditions, quaint.[12,13] If, on the other hand, we hold freedom to have a high communal value but not to be an absolute, and if we look at community as being united by more than the minimal duties of refraining from harm one to another, a different kind of responsibility enters the equation.[14] It seems doubtful that many of us, not knowing what station in life we were to occupy or what our fate is to be, would deliberately choose a community in which beneficence was to have no moral standing.[15] Neither a minimalist ethic (built only on duties of refraining and bereft of the duties of charity, benevolence, and kindness) nor an ethic in which men are coerced to follow another's vision of the good presents the sort of society most rational men would envision for themselves. (See also "Looking at Ethical Theories" in Chapter 3).

In society at large as well as in medical practice, there is an ongoing interplay between autonomy on the one hand and paternalism (couched in terms of responsibility) on the other. Autonomy, seeking to maximize personal moral agency, and responsibility, as an expression of benevolence, both have their places. Neither can become a moral obsession. Autonomy, as a moral obsession, leads to neglect: it is often a moral "cop-out," an excuse for pursuing our own interests mindless of the often very obvious and glaring needs of others. Benevolence as an obsession, on the other hand, too easily eventuates in personal or communal tyranny: it easily serves as an excuse for repressive acts of the crassest kind.

In choosing between alternative courses of action, reason must guide us to choose the least restrictive for all of those most relevantly affected. This interplay has classically been seen as a dialectic in which the goals of one (personal freedom or individual interests) are in conflict with the goals of the other (the interest of community). A *modus vivendi* develops from this interaction. However, such an interaction need not be seen as a "struggle."[16] Rather, such an interaction might be more fruitfully viewed as homeostatic, in which one force balances another in pursuit of a common goal without which the interests of neither can be realized. Individual freedom and autonomy are not possible outside of a nurturing and dynamic community; communal survival without a value for developing individual talents and underwriting individual action is also unlikely.

It has been said that there must be a presumption against paternalistic acts and in favor of autonomy.[17] If individual and communal tyranny is to be prevented, society and medicine share the need for this presumption. Autonomy, never complete, always variable, is an ideal and not a concrete fact. Furthermore, autonomy without a community that allows it to flourish and that provides its limitations and its seedbed is not possible. By itself, and not integrated by a sense of community and responsibility to one another, it leads to a callous and uncaring society. To be effectively expressed, freedom and autonomy must be enunciated, vouchsafed, and actively supported and pursued by community. Autonomy, in other words, develops in the embrace of community and its beneficence; i.e., the community's caring for our good as well as its trying to prevent or ameliorate harm enable solidarity within the community and thus the personal autonomy of those within it.[11,16]

There should, as we have said, be a presumption against paternalistic acts: the burden of proof is (and, if we are to prevent personal or communal tyranny, should be) on the paternalist. Such an initial presumption, however, must be measured against another: the presumption against allowing others to come to harm.[17] The problem, of course, is the strength of one presumption against the other and the meaning of what it is to come to harm. There can, at times, be a stronger justification for paternalism just as, at other times, there can be a stronger justification for allowing persons to come to harm in respect of their autonomy.

Allowing respect for autonomy to result in personal harm to our fellow man, or to our patient, requires justification just as does violating their autonomy. When one or the other presumption must prevail, the initial presumption will usually be to safeguard one another. That, if nothing else, will buy time. In critical emergencies (say, when a patient is in imminent danger), a presumption against allowing another to come to harm seems reasonable and very much in order. When situations are less critical, however, the presumption against paternalistic acts becomes more persuasive. The rational patient bent on suicide may try again; the irrational patient allowed to die is denied that second chance. Once the criteria for autonomous action have been met, the presumption against paternalistic acts should prevail (see also Chapter 6)

———— AUTONOMY, COMPETENCE, AND ———— DECISIONAL CAPACITY

To be autonomous, an action must meet certain minimal criteria: (1) it must be amply informed, (2) it must be the product of sufficient deliberation, (3) it must be free of internal or external coercion, and (4) it must be consistent with an enduring worldview.[18] A judgment as to when such criteria are sufficiently met (and what is considered to be sufficient) is often difficult. However, keeping these criteria carefully in mind at least distinguishes the clearly autonomous decision (say, when a lifelong Jehovah's Witness staunchly refuses a transfusion) from one clearly not so (say, when a hysterical person, confused and in severe pain, refuses a critically needed intervention).

Ample information, in and of itself, may be problematic. Telling patients technical details that are often not understood or that are entirely out of the range of the patient's experience is not truly "informing." Every attempt to enable patients to truly understand in simple, non-technical terms—not only procedures and diagnoses but also options and consequences—must be made. And these must be understood by the patient, a thing which can be determined only by having the patient relate these diagnoses, options and their probable consequences back to the physician: an autonomous decision is predicated on an understanding of the facts, an internalization of external data. More than the giving of data is therefore involved. Healthcare professionals must remember the impact that the emotions of the moment and, often, the lack of experience have, and must try to minimize the gap of understanding between themselves and their patients. (See also the section on consent in Chapter 6.)

Sufficient deliberation implies time—time, as much as possible, away from the acute pressures of critical situations. Time, however, is often just what is sorely lacking. Here healthcare professionals can only try to provide as much time as possible: pressuring patients in order to meet the needs of a busy physician's or nurses' schedule defeats that purpose. Allowing a patient sufficient time to deliberate, discuss the affair with others, make choices, and ask questions consistent with the patient's own values has to be, at times, balanced against the exigencies of a critical situation; but all too often healthcare professionals, by pleading the urgency of a situation, are apt to serve their own convenience.

Coercion can come in many guises. Patients may be coerced by external circumstances to act not according to their own will but according to the will of others. Healthcare professionals can often sense this in the context of situations in the office or hospital. Elderly patients wishing to please their children, husbands wishing to please their wives, and patients wishing to please the medical team fall not rarely into this category. It is the healthcare professional's job to understand and to perceive sensitively such problems and to counter them by speaking to the patient with understanding, with compassion, and, when possible, alone. But coercion, as often as not, is also internal. Panic, fear, pain, and hope can all be coercive and can help obliterate or, at the very least, impair autonomy. And so, regrettably, can economic factors in our society.

Familiarity with the patient's worldview is, perhaps, of the greatest help. Courses of action that accord with such a prior worldview are called "authentic." An "authentic" action is one those who know me well would expect me to make. Choices made under the influence of fear, pain, or depression, for example, may nevertheless be considered autonomous if they are consistent with a previous enduring and well-understood worldview. Choices which would puzzle those nearest us need not be dismissed—though they do need to be carefully re-evaluated to ascertain that patients truly understand, truly believe that what they are being told is the case (for if they do not believe it consultation is called for) and that they are not acting under undue coercion.

Knowledge of a patient's prior worldview is best obtained in the context of an ongoing and enduring professional–patient relationship. Ideally, sympathy and understanding for each other's worldviews have slowly developed over time. Unfortunately, such relationships are more the exception than the rule today. More often, physicians and patients are virtual strangers, and often physicians, nurses, and other healthcare professionals must deal with acutely ill patients about whom they know little or nothing beyond the immediate medical situation. At best, impressions can be gleaned from many inevitably biased sources, which may or may not have the patient's "best interests" in mind: interests which ought to be defined, within the limits of reason, on the patient's and not on the source's own terms. An equally biased professional must then distill such impressions. The final picture often resembles more what we would like it to be than what it actually is. One of the pitfalls of ethics consultation—and to a lesser extent of ethics committees—is just that: Often, without conscious intent, consultants (and, at times, committees) may manipulate the situation to serve their own (and not their patients') values and worldviews.

Deciding whether a patient's action is acceptably autonomous is not an easy task. Individual decisions will depend upon specific circumstances, on prior knowledge, and on consultation with the patient's family and colleagues and among members of the healthcare team. At times, psychologically schooled personnel may give invaluable assistance for an understanding of the dynamics informing patients' decisions. But, when all is said and done, the final decision—to accept or not to accept a patient's choice—will have to be made by the physician. Consultants (of any kind) can be helpful in reaching a decision (and most certainly are responsible for the care they take in giving their advice) but ultimately the physician will have to be and remain responsible for that choice.

In general, healthcare professionals have tended to override the wishes of depressed patients, especially when such wishes would limit treatment. This canon of our teaching and action may well be wrong. It is often still used as an excuse for crass paternalistic acts. Depression, first of all, is not all of one cloth: endogenous depression, psychiatrists will tell one, differs markedly from situational depression. When a patient who finds out that he/she has metastatic cancer, whose wife or husband of many years has just died, whose business has failed, and whose house has burned down is "depressed," health professionals often take this as sufficient reason to deny them the right to choose. Being depressed under such circumstances is hardly psychiatrically abnormal: imagine if such a person were not!

Recent literature has cast considerable doubt on the role of depression in a patient's choice. According to some papers, choices are not greatly changed. This leaves us in a quandary: to abandon a long-held belief and embrace what is now suggested, or to decline to modify one's behavior. A middle course may, for the time being, be appropriate: if a depressed but not psychotic patient's wishes seem to accord with his prior worldview (a thing not always easy to determine but something with which family and friends can be of great help), more credence might be given to him than if it had not.[18-21]

Although we have repeatedly used the term "competence" (and although the ethics literature tends to do the same thing), judging "competence," strictly speaking, is something only a judge can do: the presumption, legally speaking, is that all adult patients, no matter what their age or condition, are "competent." It is only when a court of law rules them "incompetent" that, strictly speaking, such a term should be used. Legal competence does not denote decision-making ability, just as legally determined "incompetent" patients do not necessarily lack the capacity to make healthcare decisions for themselves in a clinical setting. Establishing competence is a legal prerogative. Determining "decision-making capacity" is the physician's task: a task that, when there is question, will hopefully be done in concert with the rest of the immediate healthcare team and with the help of psychologists and ethicists.

Since we assume ourselves to have decision-making capacity, we rarely question such capacity when patients make choices consistent with our own. When, on the other hand, the patient's choice conflicts with our own, questions about the validity of the choice tend to come up. The more blatant the conflict, and the more we disagree with the road chosen, the more we tend to question the patient's ability to choose ...and that is natural. Conflicts of this sort may arise from a lack of factual material, a different understanding of such facts, or a profound difference of worldviews. (See also Chapter 6.) Problems of judging competence or autonomy arise because competence and autonomy are not unbounded.

Patients may, for example, be incompetent to handle their own finances but be quite competent to order dinner or to determine which theater to go to. Competence to make one's own will and competence to determine one's own course of therapy are not necessarily related matters. Determining that patients lack decision-making to consent to or to refuse treatment must be adjudicated on a one-to-one basis depending upon individual circumstances. We cannot presume that lack of such capacity is an all-or-nothing phenomenon or that it does not change with time and circumstance. Denying competence to choose a course of treatment may, on the one hand, deny the patient's individual dignity when the capacity to choose such treatment is present; on the other hand, affirming that a frightened, ill-informed, or otherwise incapable patient is capable of refusing treatment may deny the real beneficent responsibility that lies at the core of medicine. Abandoning patients to their autonomy is all too easily done today. Respecting autonomy in the competent person presupposes beneficence: when persons are competent to choose, even when the choice is not one we ourselves would make, respecting their choice is a beneficent act. It allows their will to be done in circumstances directly affecting them. Respecting autonomy in those lacking decisional capacity, however, is a hollow mockery that denotes callow non-caring rather than beneficence. Allowing an uninformed coerced, or confused will to be done makes a mockery both of autonomy and of beneficence.

Using the criteria for autonomy cited above may help sort out specific cases. Healthcare professionals have the obligation to provide patients with all pertinent information necessary for informed choice concerning their case, and they are obligated to provide it in a manner understandable to the patient. Further, healthcare professionals should make sure that such information is truly comprehended. Comprehension means more than merely the ability to parrot facts. True understanding, in addition to an essential cognitive part, includes understanding on an emotional as well as, where possible, an experiential plane. It must include some understanding by the healthcare professional of what the diagnosis or condition means to patients: not just what it is scientifically, but what it connotes to and for patients: how it will be seen to impact on their daily lives and what it means emotionally for them, given their personal worldviews.

Physicians often assume that their patients understand far more than is actually the case. A little time spent asking some gentle but penetrating questions may be most enlightening and helpful. In addition, and where possible, time for deliberation must be provided, and snap judgments guarded against. Seeing that patients are as free from coercion as possible during this time and gently probing patients' prior worldviews likewise are obligations of the healthcare team. Forcing patients to make hurried choices with inadequate information, sometimes presented in an unnecessarily threatening manner, violates basic respect for autonomy and, incidentally, is destructive to the professional relationship. It is also—even when not recognized—destructive to the health-professional's character and sense of self-worth.

At best, problems remain. Patients may make apparently autonomous decisions and then, when the situation is upon them, change their minds. This confronts the healthcare team with agonizing problems. In general, but by no means always, we tend to honor the more recent rather than the more distant choice. Such decisions are predicated on the assumption that both decisions were autonomous, that both were competent choices.[22]

When patients change their minds in circumstances when reasonable autonomy appears present (when, in other words, information and time for deliberation are reasonable, coercion is held to a minimum, and the change is not entirely at variance with a previously enduring worldview), respect for the more recent over the more remote decision will generally be granted, though there are times we might want to re-confirm that the patient is able to offer reasons for his or her change of mind. When, however, the choice appears to be the result of ignorance, fear, or panic, matters may stand differently. It is at such times that difficult choices will have to be made.

TRUTH-TELLING

Truth-telling is intimately linked with problems of autonomy. Persons who hold autonomy to be an absolute principle will, under all circumstances, tell their patients the absolute and unvarnished truth. They even may do so quite brutally and with little or no visible compassion—which often is merely a veneer for the pain felt by the health-care professional…though sometimes is not. Paternalists, on the other hand, are apt to judge what is, in their view, to the patient's benefit to know and what not, and act accordingly. The fact that paternalists often misjudge their patient's good,[23,24] substituting conjectures and personal values instead, has become well known. In practice, a presumption for telling patients the truth can be overcome only by extremely weighty evidence. In law this is called "clinical privilege"—that is, physicians for good reason feel that telling the truth would cause irreparable and grievous harm. When challenged it is rarely upheld by the courts.

But truth-telling, like other principles, acts as a guideline to moral behavior and not as an absolute. Blindly following principles (e.g., always telling the truth) can become an end in itself instead of a means to a moral end. Following principles in this way substitutes ironclad rules for moral deliberation and severely limits moral agency and its necessary choices. Although there is a heavy presumption for truth-telling, ethics reduced to pat principles applied to preconceived problems without being filtered through our moral sensibilities is a technical and not a moral activity.[23]

Not telling the truth (never a praiseworthy and invariably a blameworthy act) may, in rare situations, nevertheless be the best choice to make from a range of poor options. Patients at times may not wish to have the truth told to them and may ask to be spared certain knowledge. And while this is rare and while it certainly imposes a heavy burden on the physician, the patient's desire not to know can be as autonomous a decision as the opposite. Such a decision by the patient should be carefully recorded in the records and where possible countersigned by the patient or by a witness. Rarely there may be other reasons for being less than candid with an occasional patient. The decision to be less than candid must, at all times, be a weighty one, not one made for the sake of expedience or convenience, or out of cowardice. On some occasions, it may, all things being equal, be the only humane option open to the physician. When truth-telling succeeds only in removing all hope from dying patients, discharging an absolute moral duty exacts a heavy price. Ethics not tempered by compassion and understanding becomes like "random cords on a piano,"[24] and loses its intrinsic value.

One other issue must be briefly mentioned: truth-telling, especially when it comes to healthcare, is very much a culturally modified issue. The basic ethical principle is that a patient with decisional capacity ultimately has to be allowed to decide his destiny in his own way. The way this principle is played out, however, is quite different in different cultures and carries different cultural expectations. In rural areas around the Mediterranean, for example, patients themselves are generally not informed of critical diagnoses, and the decision is often made by the husband for the wife or by the families concerned.[25,26]

There are many other cultural differences depending upon the particular culture studied. This may confront healthcare professionals who must deal with patients from other cultures with difficulties. Insisting, cultural customs not withstanding, that the blunt truth must be told to patients is a form of ethical imperialism that insists that what is done in "our" culture is necessarily the "right thing to do." When physicians are confronted with such situations, they are well advised to consult the patient's wishes. Saying to the patient that you have been told that in their particular culture information is generally given to husband or family and asking them if this is what they wish done preserves the basic principle: any patient who has decisional capacity continues to decide. In this case it is a decision not to be informed but it is a decision properly made by the one entitled to make it and not by the healthcare professional acting through his or her own idiosyncratic vision of what it is to "act rightly." One should likewise avoid cultural stereotyping: remember that not all persons who belong to a given culture or religion conform to it. Just as patients brought up in a Mediterranean culture may, in fact, wish to be informed, patients brought up in the United States may not want to be. It behooves physicians and other healthcare professionals to understand and respect the patient, the culture, and the patient's wishes—at all times leaving the door open so that patients can, if they so desire, change their minds.

—— PLURALISM AND HEALTHCARE ——
PROFESSIONALS

We live in a pluralist world in which vastly different forms of belief and worldview must coexist or extinguish each other. Worldviews differ, often radically, and can at times not be reconciled. Attempting to coerce each other into behaving in certain ways is not only impractical; in a deeper sense it is immoral. Coercing each other seems immoral because, by relying on explicit or tacit power, it is inevitably a violation by the stronger of the personal freedom and moral agency of the weaker and thus ends by defining "might" as "right" and "is" as "ought." By that type of analysis, one would assume that guidelines and norms do not exist or that ultimately all ethical decision becomes merely a matter of culture or of personal whim and caprice. Between these two extreme beliefs, the one a variation of "do as I say (or do) because I say (or do) so," and the other an "anything goes" approach, there is a middle ground that would make some, but very few, rights and wrongs normative except as they are normative in and for a given context. Although particulars differ and although these differences may, in the way that particular decisions play themselves out, be starkly different from one another, the basic framework of existential *a prioris* that unites all sentient beings remains the same. (See Chapter 3.)

Basically, ethical considerations arise because our actions impinge on others and because their impinging on others matters to us. Were this not so, ethics would make no sense. We think, on balance, that to do right is, at minimum, not to harm another (or to harm another as little as possible under existing circumstances); to do wrong is to bring unnecessary, or needlessly severe, harm to another. But in all ethical considerations, the harm or benefit done, or potentially done, to another is at stake. This statement, despite its emphasis on consequences, should not be interpreted as being purely a matter of utility. It can be equally well defended in more deontological language: the aim of all imperatives, maxims, and principles cannot be easily or entirely separated from the consequences they would eventually bring about. My wish that the maxim that guides my actions under a particular set of circumstances might become a "universal law of nature" is motivated precisely by a set of circumstances that such actions can be seen to bring about. The way our actions affect others is, at the very least, a critical factor in the moral equation.

This consideration for others, then, is a basic norm: not to bring needless harm to another. Such a norm is rooted in the realization that man's common structure of the mind allows us, among other things, to share the ability to rejoice and to suffer. Starkly different things may bring about rejoicing or suffering, in differing societies and among different individuals. But the capacity for joy and pain is a universal of sentient beings: a shared capacity and a shared value that may serve as a starting point in the quest for peaceful agreement. It is a reference point—a norm—against which to judge our actions as moral or not.

The norm (or principle or ethical obligation) of refraining when possible from doing harm to another who can perceive such harm (or to his or her actual or symbolic possessions) is a purely negative one. As we have discussed in Chapter 3, such an obligation is a necessary but insufficient building block of the moral life. It can permit a somewhat eased coexistence, but it cannot suffice to provide communal solidarity or suffice to describe the moral life. Beyond such a minimalist ethics, and where possible, we are obligated to help one another. Individuals can derive this obligation from the fact that they themselves were and inevitably will be again in need of another's help or they can derive this obligation from the obligation to consider a realm of ends when formulating their own goals. Communities must accept this obligation if they wish to be held together by the solidarity necessary for their survival, a solidarity that cannot come about if only obligations of mutual non-harm and, therefore, a minimalist ethic are accepted. (See Chapter 3.) Thus beneficent action or "doing another good" becomes another obligation of the moral life. In this book it is the extent but not the existence of this moral obligation that is at issue.

Differing civilizations, and differing enclaves within the same civilization, define the "good" in different ways. It is this lack of uniformity that underlies the often radically different judgments made by patients and their physicians; this is what causes us "not to understand" the Jehovah's Witness or comprehend divergent attitudes toward abortion. Nevertheless, no matter how described, all sentient beings strive for their particular and self-selected good and are, on their own terms, benefited by its realization and harmed by its removal or interdiction.

The universal of harm and benefit, the universal of the capacity to suffer and rejoice, are not trivial considerations. While self-evident, they are often ignored. We are only too ready to inflict suffering for the sake of doing others our own (instead of their own) "good" and to rationalize doing so by an appeal to a "universal" standard that, when carefully examined, is merely our own. The Crusades, the religious wars, the attempt of various fundamentalists today, the behavior of Communist and capitalist alike, all provide ample evidence of man's incessant desire to convert—forcefully if necessary—the world to one particular and peculiar belief. On a practical level, the world has become too small and the weapons have become too powerful to permit intolerance of this sort. Medicine, embodied in this world, must likewise examine its own standards, norms, and behaviors if it is effectively to accomplish its mission in such a world.

The pluralist society in which we live, then, is in need of reconciliation and understanding. Not only is this true in comparing, say, Uganda to Sweden; in a fluid world, it is just as true within national entities such as the United States or the EU. Many social, economic, cultural, and moral issues must be addressed if this reconciliation, understanding, and ability to live and work together is to be effected. Extravagant social and economic differences (not only between regions of the world but within the very borders of what we consider national entities, including our own) must be leveled at least to a tolerable degree; cultural exchanges enabling understanding and facilitating cultural diffusion must occur. If human life is to survive, ethical systems must be reconciled sufficiently to permit mutual toleration and function toward a common goal.[11]

Medicine exists in a community. Prevalent attitudes and prevalent moral senses within a given community are inevitably and at least to some degree shared. A contentious and intolerant society, unwilling to allow others to pursue a different vision of the "good" and bent on enforcing one view of life as "correct," is unlikely to produce professionals otherwise inclined. A society entirely without moral standards, on the other hand, will tolerate amoral healthcare professionals who see themselves merely as neutral or uninvolved biomedical facilitators of their patients' (or, even more ominously, their state's or their HMO's?) wishes. In the first instance, crass paternalism running roughshod over other values and views will result, and patients will be left in physicians' hands to enjoy (or suffer) the "good" purely as defined by the professional. In such an intolerant society, the relationship between healthcare professionals and their patients is seen as intensely personal, dominated by the healthcare professional's (usually the physician's) personality and totally unequal. The enterprise of medicine now has an evangelical flavor.

In the second instance, physicians abrogate their moral agency and become their patients' technical agents to be bound merely by strict contract devoid of beneficence or a sense of mutual obligation. Here the relationship is one in which patients bring their complaints and desires to buy a "cure" from the now entirely technically defined healthcare professional. In this model, healthcare professionals in the role of healthcare professionals assume the character of civil servants, bureaucrats who operate under bureaucratic rules and during working hours leave their notions of right and wrong at home.[13]

If one subscribes to such a worldview, moral agency is replaced by bureaucratic (or institutional and unchallenged) rules, and professionals become vending machines dispensing their bureaucratically stipulated wares to all comers provided only that payment is made. All too easily, as bureaucratic rules change, healthcare professionals can become agents of the state ready to execute, help torture, or exterminate others.

If we subscribe to the bureaucratic model, the problem of conflicting moral agency is "resolved" by being abolished. A conflict between moral agents cannot occur so long as the patient's request for services does not fall outside arbitrarily established and legally stipulated norms. The traditional vision of the patient–physician relationship is replaced by one in which physicians and other healthcare professionals are seen merely as competent technicians whose technical competency and an unquestioning adherence to explicit contract and bureaucratic rules define their moral duty. It is a point of view to which this work does not subscribe. In examining the clash of moral agents, the legitimate moral agency of all actors is presumed.

Problems of healthcare ethics and the answers we choose to accept are frequently underwritten by deeply held and as frequently unexamined moral and metaphysical beliefs. Such beliefs involve deeply divisive and fundamental issues, highlight moral systems, and are intimately associated with emotive, aesthetic, and religious considerations. For that reason, medical ethics provides a suitable paradigm to examine the more basic problem: the coexistence of diverse ethical beliefs in a world whose diverse cultures have become of necessity more interrelated. (See also Chapters 3, 6 and 11.)

Moral agents are sentient beings capable of making moral choices. In Kantian terms, the moral agent is capable of making a moral choice, and moral agency is the action taken by a moral agent in the moral sphere.[8] Moral agency will be defined as the assumption of moral responsibility for one's acts. It denotes deliberate choice made in the light of moral belief and, consequently, entails responsibility and accountability for choice and action. A decision to act, if it can be said to be moral, cannot be made on technical grounds alone, but involves a careful consideration of alternative options and of the moral issues involved. Sentient beings are, by definition, capable of moral choice, and exerting moral agency is their primary ethical duty. When physicians or other healthcare professionals, as professionals and as sentient beings, refuse to partake in the agony of decision-making and leave decisions to authority or, perhaps even worse, to whim, chance, or the luck of the draw, they have violated this first of all moral precepts without which all others stand moot. All concerned (healthcare professionals as well as patients) are moral agents in their own right, with none, therefore, entitled to run roughshod over the other's beliefs or convictions. If one subscribes to the bureaucratic model, on the other hand, a healthcare professional's moral agency, like his overcoat, is hung up and suspended for the duration of his professional function.

Human beings in exerting moral agency will conflict, often sharply, in what moral sense to follow. While practical decisions among persons of goodwill are often—but not by any means invariably—similar one to another, the principles to which such decisions are appealed often differ greatly.[27] Professionals involved with the care of patients and with making decisions about them must deal with patients of kaleidoscopically differing backgrounds and beliefs; moreover, these professionals have among themselves greatly differing backgrounds and worldviews. It is not surprising that conflicts and misunderstandings occur.

Healthcare professionals are faced with a variety of ethical dilemmas in medical practice that must be resolved or adjudicated before deliberate action can take place. These dilemmas are of two kinds. The first is the universal internal human dilemma in which agents confront themselves, their beliefs, and their own clashing contexts with often differing claims. Healthcare professionals, like all individuals, are a composite of often-conflicting forces that must be internally reconciled: The outcome of this internal dilemma creates our worldview and determines moral attitudes held toward broad categories of problems. Moreover, this shaping of a moral view is an ongoing process: as we live, think, act, accumulate more knowledge or information and gain experience, our worldviews (unless we are confirmed and hardened absolutists) will evolve and change.

Libertarians would ignore this internal conflict when it comes to a healthcare professional's function: somehow health care professionals are to set aside the outcome of this internal struggle and resolution during "business hours."

The second of these dilemmas is the external dilemma in which healthcare professionals and their patients must reconcile differing points of view with each other and with the family, with other team members, and with the demands of community and law. Physicians and their colleagues must justify the conclusions and the process internally and externally and must try to produce consensus and understanding. Finally, they must act and must assume responsibility for that action.[28] In adjudicating either the internal or the external dilemma of conflicting beliefs, points of view, and contexts, a search for shared values is essential. Such "shared values" are values that exist prior to any of the conflicting specific beliefs.[29]

Disagreements between healthcare professionals and patients can be disagreements about the desired ends to be gained, about the means utilized to gain an agreed-upon end, or about the moral issues involved at any point in the process. Conflicts about ends may arise when the ends have not been examined. This usually happens when certain key assumptions are taken for granted: the assumption, for example, that the end to be pursued under all circumstances is the patient's life and health. Our own personal "ends" or goals are the product of our internal conflict and, although changing over time, generally present a continuity and thus an evolving authenticity peculiar to our own worldview. Our goals or ends differ accordingly and are apt to evolve over our lifetime. Judging one as "better" or "more true" than another (except for themselves) is not within a health professional's province.

Disputes about means may be technical (the patient and physician may disagree about the best means of delivering a baby) or moral, and they may involve a hierarchy of values concerning desirable ends: The Jehovah's Witness, for example, may desperately want to live but be unwilling to take blood and, within his or her belief system, jeopardize salvation. When the dispute is technical, its resolution is quite different from when it is moral. In a technical dispute, the patient simply does not believe what the healthcare professional has said: he or she may simply not accept the diagnosis, prognosis, or treatment. What is called for here are consultation, explanation, and perhaps even referral to another institution or healthcare provider.

When the dispute is moral, as when a Jehovah's Witness refuses blood, it is really a dispute about goals: saving life versus saving one's chance for going to heaven. Unfortunately, the moral end is often hidden. Jehovah's Witnesses not rarely will attempt to dispute with healthcare professionals about the utility of giving blood (claiming that transfusions are simply not needed to save life, for example) rather than directly stating their deeply held moral conviction that accepting blood constitutes a sin. Healthcare professionals must attempt to give all reasonable information and data supporting their point of view to the patient but are ill advised to engage in a technical argument. Instead, they should indicate that, while they do not agree with the patient's moral point of view, they are willing either to accept it or to refer the patient elsewhere.

Physicians may disagree with their patients about the need for intervention either in an emergency or non-emergency, life-saving or non-life-saving situations. In emergency situations in which the patient's life is at stake (say, a patient refusing needed surgery to stanch hemorrhage), the presumption will be to act so as to safeguard life. Allowing the patient to refuse entails the conviction that the patient's refusal is a truly autonomously derived one and not the product of panic or fear. The "reasonable person" doctrine (which holds that in emergency circumstances, where either consent cannot be obtained or refusal seems confused by the patient's state, the physician should proceed to do what a "reasonable person" would want under such circumstances) is the doctrine usually applied to handle such cases. This doctrine, of course, can be dangerous, for it presupposes that majority views are reasonable and other views are "unreasonable." Under emergency circumstances, however, physicians must either proceed to save the life of their patient or forego action. Using the "reasonable person" doctrine under such circumstances merely affirms the obvious wisdom in pursuing the more likely rather than the less likely course of action. If nothing more, it buys necessary time so that a deliberate autonomous choice, instead of death, may take place.

In non-emergency situations (whether the situation is one with the patient's life ultimately at stake or not), more deliberate effort to ascertain the patient's state of mind and worldview is possible. An example of this might be the patient with a small colon cancer who refuses surgery or a leukemia victim who refuses chemotherapy. When the patient's decision seems to be autonomous (i.e., when it meets the criteria of sufficient information, sufficient time for deliberation, lack of coercion, and consistency with prior worldview), the patient's will is the ultimate deciding factor. Physicians as well as other healthcare professionals, when they themselves subscribe to a far different moral set of beliefs from their patients, must have the option of relegating further care to another professional. Such sharply divergent beliefs are often but not invariably religious in nature. What individual physicians will do in individual cases depends on their personality, their own worldview, the specific relationship between the physician and the patient, and the context in which the problem occurs. But dealing roughshod with their patients' wishes or belittling their patient's choices (e.g., calling the patient a "sinner"—outrageous, yes, but we've seen it happen!) is not an acceptable moral option.

We have emphasized the importance of differentiating between consent to and refusal of a life-saving intervention.[30] (See also Chapter 6.) In general, healthcare professionals must use due care in accepting either, but "due care," depending on which one we are talking about, should, if we are to discharge our moral obligation of safeguarding and caring for the patient, each require the application of a different set of criteria. When a reasonably competent patient readily assents to a life-saving transfusion or to surgery for a small colon cancer, one certainly needs to make sure that he understands the clinical situation and the options. But one would not, and would not morally, feel obliged to dwell on it much beyond this. When, however, a seemingly competent patient refuses such an intervention (and thereby virtually signs his own death warrant or, at least, accepts a far more dismal clinical prognosis), healthcare professionals, while not entitled to coerce, should carefully go over the entire territory of explanation. Healthcare professionals must make certain that patients believe what they are told, understand the problem and the possible options, are not under severe coercion, and have thought the problem through. Health professionals must take the time to discuss and help patients come to terms with how their decision and their worldview fit together. It is not rare that patients are so paralyzed by fears of surgery (which are often irrational) that they will, without truly understanding, refuse a procedure to which they might otherwise agree. Here bringing patients together with others who have had similar problems, giving them literature, asking them to discuss their problems with those close to them, and many other things can be done to help an eventual decision. Beyond this, healthcare professionals may try to advise and persuade, but they are not morally entitled to coerce or to lie to their patients.

Patients may desire to be treated in ways that are morally repugnant to the health care professional. Examples of this, of course, deal with abortion, birth control, and many more subjects. Here the problem usually involves the morality of the means; in abortion and birth control, the end—not being pregnant, for example—is not in dispute. Healthcare professionals generally find this quite disturbing, though those who subscribe to the bureaucratic model will encounter no problems…their moral feelings are left entirely at home.

When, however, healthcare professionals see themselves as persons extended through time with their worldviews, opinions, and idiosyncrasies intact, obvious problems come to the fore. Such problems cannot be solved in a morally acceptable manner by coercing or lying to patients; nor, if the healthcare professional's moral integrity is to be respected, can they be solved in a morally acceptable manner by expecting healthcare professionals to become "bureaucrats of healthcare." All too often, one or the other is done. Lying to patients or coercing them to prevent an abortion is not exactly a rare event—and it is one defended by many who are doctrinaire religious. A former federal ruling (now, luckily, no longer in effect) forbade healthcare workers to discuss or even to mention the option of abortion to pregnant women seeking help in an even partially federally funded clinic. At the current writing we once again stand in danger of having such a rule implemented or introduced explicitly or by some subterfuge (such as eliminating funding). Such a rule, in that it denies patients a true picture of all legally available options, is, at the very least, disingenuous. It not only forces patients along a path that they might not themselves choose to go, but likewise reduces healthcare professionals to mere bureaucrats and rule followers.

In a free society, healthcare professionals are neither compelled to subjugate their personal moral views nor entitled to impose such views on others. Problems of this kind can only be resolved by frank and compassionate discussion that enables patients to make their own choices and to reach their own conclusions. Resolutions can often be found when compassionate and caring persons who have respect for each other and for each other's points of view search together for a basis of shared values. When healthcare professionals and patients continue to be unable to resolve their differences and continue to differ in these circumstances, patients should be referred to several competent practitioners who may be more in tune with the patient's moral views. Even referral may be repugnant to some who may feel that it is aiding and abetting an act they consider immoral—but the alternative of lying to patients or keeping them hostage to one's own beliefs is hardly a better option.

Healthcare professionals who fail to offer patients un-disparaged referral act in a crassly paternalistic fashion. As their patients' medical advisers, they are bound to offer their reasons for their own beliefs, but they are not entitled to prevent patients from following their own moral dictates. Healthcare professionals who proselytize or argue with their patients (as distinct from merely offering their reasons for their own views) abuse their implicit power and take unfair advantage of a professional relationship in an attempt to change their patients' minds and convert them to their own idiosyncratic moral point of view. On the other hand, healthcare professionals who simply comply with their patients' wishes and perform procedures or do other things that they themselves find morally repugnant abrogate their own moral agency, are apt to lose their integrity, and, by abrogating their moral agency, can be argued to be acting immorally.

We have considered only the clash of moral values and agency between healthcare professionals and patients. Other clashes are not without importance. These include disagreements with rules of the institution or with current laws, differences among the various moral views of other team members (nursing, social work, etc.), problems with family members, etc. Physicians and other healthcare professionals in their role as teachers must strive to explain, adjudicate, and, when possible, convince; they cannot simply override or curtly dismiss other concerned views. When physicians or other healthcare professionals fail to consult with each other and fail to respectfully listen to each other's moral view and to consider it, they fail to show the basic respect one person owes to another. The larger the team and the more fluid it is, the more differences in worldviews and values are apt to occur: Teams that have long and with mutual respect worked together generally have many fewer problems. But in the final analysis, physicians and, in their own sphere, other healthcare professionals ultimately remain, so long as nothing contrary to their own moral sense is demanded, their patient's agents and must seek and defend each patient's vision of the good.

——— SELLING OUT TO AUTONOMY ———

Lately there has been a tendency to give patients their diagnoses, therapeutic choices (including non-treatment) and prognoses and to ask them to make their choice. For the patient it is similar to going to a restaurant and being given a menu—except that the lack of knowledge about each item is apt to be far greater and the stakes often infinitely higher. When patients consult healthcare professionals they do so for a number of reasons, among which the preeminent are to gain information about their state of health and to seek advice. Even when we go to our favorite restaurant we are apt to ask our waiter what he or she recommends. When we seek advice from our physician that, too, is exactly what we expect. We may not like the choice and may, indeed, make another—albeit in the medical setting we are far less apt to do so.

Being told that two courses of action have similar outcomes in the literature can be fairly opaque to us; it may fail to give us the information we really seek when what we really want to know is what our physician thinks is most appropriate in our particular case and, thus, should be done. There is more than statistical information in such a choice—hopefully our physician knows the overall context in which our problem occurs. Furthermore he or she has most likely had more experience with one rather than another modality of treatment and, therefore, feels more comfortable with (and is probably far safer providing) its application.

The practice of handing patients a menu of possible options should be resisted. While it may, at first blush, simplify the life of the healthcare professional it is destructive of the physician–patient relationship—one regrettably and inevitably based on a difference in power in which the weaker has turned to the stronger for advice and help. Being coldly handed a set of options without compassionate advice and truthful reason for that advice truly abandons patients to their autonomy.

———————————————— **REFERENCES** ————————————————

1. Morison RS. The limits of autonomy. *Hastings Center Report*. 1984;14(6):43–49.
2. Kant I. Beck LW, trans. *Foundations of the Metaphysics of Morals*. New York, NY: Macmillan, 1986.
3. Hume D. *A Treatise of Human Nature*, Book II, Part III, Sect. III. Oxford, UK: Clarendon Press, 1968.
4. Dennet DC. *Elbow Room: The Varieties of Free Will Worth Having*. Cambridge, Mass: MIT Press, 1985.
5. Churchill LR. Bioethical reductionism and our sense of the human. *Man Med*. 1980;5(4):229–249.
6. Dworkin G. Paternalism. *Monist*. 1972;56:64–84.
7. Feinberg J. Legal paternalism. *Can J Phil*. 1971;1:105–124
8. Kant I. Beck LW, trans. *Foundations of the Metaphysics of Morals*. Indianapolis, Ind: Bobbs-Merrill, 1980.
9. Mill JS. *Utilitarianism*. Indianapolis, Ind: Bobbs-Merrill, 1975.
10. Mill JS. *On Liberty*. New York, NY: W.W. Norton, 1975.
11. Loewy EH. *Moral Strangers, Moral Acquaintances and Moral Friends: Connectedness and Its Conditions*. Albany, NY: State University of New York Press, 1996.
12. Nozick R. *Anarchy, State and Utopia*. New York, NY: Basic Books, 1972.
13. Engelhardt HT. *The Foundations of Bioethics*. New York, NY: Oxford University Press, 1986.
14. Loewy EH. Communities, obligations and health care. *Soc Sci Med*. 1987;25(7):783–791.
15. Rawls J. *A Theory of Justice*. Cambridge, Mass: Harvard University Press, 1971.
16. Loewy EH. *Freedom and Community: The Ethics of Interdependence*. Albany, NY: State University of New York Press, 1993.
17. Bassford HA. The justification of medical paternalism. *Soc Sci Med*. 1982;16:731–739.
18. Lee MA, Ganzini L. Depression in the elderly: effect on patient attitudes toward life-sustaining therapy. *J Am Geriatr Soc*. 1992;40:983–988.
19. Loewy EH. Of depression, anecdote and prejudice: a confession. *J Am Geriatr Soc*. 1992;40:1068–1074.
20. Ernlé WD, Young JC, Johnson R. Does depression invalidate competence? Consultants', ethical, psychiatric and legal considerations. *Cambr Q*. 1993;2(4):505–516.
21. Kübler-Ross E. *On Death and Dying*. New York, NY: Macmillan, 1975.
22. Loewy EH. Changing one's mind: is Odysseus to be believed? *J Gen Intern Med*. 1987;3(1):54–58.
23. Churchill LR, Simán JJ. Principles and the search for moral certainty. *Soc Sci Med*. 1986;23(5):461–468.
24. Cassel EJ. Life as a work of art. *Hastings Center Report*. 1984;14(5):35–37.
25. Dalla-Vorgie P, Katsauyanni K, Garvanis TN, et al. Attitudes of Mediterranean

populations to the truth-telling issue. *J Med Ethics*. 1992;18:67–74.

26. Gordon DR. Culture, cancer and community in Italy. *Anthropol Med*. 1991;7: 137–156.

27. Toulmin S. The tyranny of principles. *Hastings Center Report*. 1981;11(6):31–39.

28. Loewy EH. Patient, family, physician: agreement, disagreement, and resolution. *Fam Med*. 1986;18(6):375–378.

29. Gorovitz S. Resolving moral conflict. In: Gorovitz S, ed. *Doctor's Dilemmas: Moral Conflict and Medical Care*. New York, NY: Oxford University Press, 1982.

30. Drane JF. The many faces of competency. *Hastings Center Report*. 1985;15(2):7–21.

Patients, Society and Healthcare Professionals

6

INTRODUCTION

In former days, patients related to their physicians on a direct one-to-one basis in the context of their home and relatives. Few other caregivers or institutions were involved. In medieval times, the clergy began to intrude into the relationship and, as we have seen, to make demands apart from the direct problem of illness. The Catholic Church often assumed a controlling role. As the medical profession developed and took on new tasks, as more modern hospitals emerged, and as severely ill patients and, eventually, the patients less severely ill, began to be institutionalized, the relationship between patient and healthcare professionals took place in a different, and increasingly more impersonal, setting. Further, other caregivers began to develop their own particular expertise, to assume critical roles, and to justifiably demand recognition of their own profession, skills, and moral agency.

Hospitals play a critical role in communities and, even when privately owned, serve public functions. Such institutions invariably serve the general public and inevitably are not entirely separable from the public coffer. Therefore, hospitals not only establish internal rules but are also governed by a set of external rules through which the community attempts to control their integrity, their quality and last but not least their expenditures. Additionally, third-party payers (a mish-mash of insurance carriers, industry, various levels of government, HMOs, MCO's etc.) have started to play a critical role in American medicine. These third-party payers not only control the hospitals ("he who pays the piper calls the tune") but are also beginning to assume a significant role in controlling the function of private physicians in their own offices. More and more insurance companies play a central role in determining "medical necessity," length of stay, and questions of whether procedures should be done in the hospital or in an outpatient setting. A large number of healthcare professionals (whether physicians, psychologists, nurses, technicians, social workers, or others), moreover, are employed by hospitals or HMOs as well as by industry or other institutions that, among other things, provide healthcare. Moreover the institution of "managed care" spread to the point that it controls an ever-expanding section of what has (so unpleasantly we think) been called the "healthcare market." Under these circumstances, physicians and other healthcare providers can hardly be conceived as free agents whose obligations are merely to their patients and who are somehow free (if they ever were) to ignore external demands. As time goes on physicians view themselves and are increasingly viewed by the public as "employees" of institutions instead of as trusted advisors seeking solely the patients "good," however conceived.

Many patients, furthermore, are not ill: they see physicians to have their health evaluated or certified, to be examined for employment, or to meet a number of other requirements not directly associated with illness. Healthcare professionals serve in the military, in industry, and in jails. Their tasks extend beyond dealing with "illness" or "disease." While plastic surgeons restore what accident has shattered, they also pander to the vanity of those who do not like their appearance or who want to appear younger, slimmer or to have smaller or larger breasts. Although pregnancy is hardly an illness, physicians prescribe birth-control medication and devices and ligate tubes and sperm ducts. In the past 50 years, the relationship between and among physicians, nurses, other healthcare professionals and patients has become complicated beyond imagination. And yet there remains and very likely always will remain a basic and unavoidably deeply private relationship. Birth, death, and illness or the threat of illness bring out primitive fears, hopes, and drives; they generally cause people to seek out help, ideally from people they feel they can trust. Reflecting this fact and combined with a slowly evolving traditional vision of such relationships, the relationship between and among healthcare professionals and their patients remains inevitably based on trust, fear, and hope. Lately and, in our view, unfortunately this has often undergone a radical and not beneficial change.

Our vision of the relationship between and among healthcare professionals and patients is central to the way in which we perceive the obligation of physicians and other healthcare professionals *vis-à-vis* individual patients—and the way in which we bestow them with our trust. It has developed slowly over the ages, changing and adapting to the development of basic and clinical science as well as to its social context. Except in its crudest outlines, the relationship is not one codified in law. It has been affirmed by social contract and has been largely accepted as the ground of the physician's function. Without this tacit understanding—reinforced here and there by law—disruption would occur. Enforcement of social contract is generally through social mechanisms involving praise, censure, and, at times, even stigmatization.

ROOTS OF HEALTH CARE RELATIONSHIPS

The relationship between and among healthcare professionals has at least three roots: (1) a root of social contract relying upon a mutual perception of interpersonal obligations as well as upon profession; (2) a root developing out of the historical tradition of society and profession; and (3) a personal root that gains its strength from the unique relationship produced by an interaction of the various personalities: patients as well as the differing personalities of members of the healthcare team. These three roots nourish a relationship that has found expression in three basic models, first described by Szasz and Hollender.[1] These essentially behavioral models (activity–passivity, guidance–cooperation, and mutual participation) coexist with and flesh out the more attitudinal paternalistic, scientific, entrepreneurial, and, lately, interactive models.

The root of social contract is expressed, on the one hand, in the immense privileges and power given to physicians and, on the other hand, in the high and often unrealistic expectations communities have of physicians. Physicians enjoy high status, special prerogatives, unusual rights, and ample material reward. Physicians strip strangers, administer poisons, inflict wounds, and hold legal power over determining life and death. To a lesser degree, other healthcare professionals likewise have special power and privilege not granted to laypersons. The community has vested this trust in healthcare professionals because of the tacit and communally accepted promise held out by their profession. It is this very relationship which today stands in jeopardy.

Communal expectations are expressed in the community's view of professionals in general and the medical professional in particular. Skills, whether technical or intellectual, are merely the instruments and not the essence of a profession. To be a professional implies a willingness to use requisite skills in a manner consonant with the moral ends implied in the contract. It is to declare oneself freely willing to assume an obligation: in medicine, the obligation to "perform a good act of healing in the face of the fact of illness."[2] Technical competence is the necessary condition without which the act is fraudulent, but it is insufficient to describe the professionalism implicit in the social contract that binds doctors to communities and thus to their patients. The act of healing implies moral choice and moral sensitivity.[3] Beyond a willingness to perform the technical act is the willingness to participate in and to guide the choice. Social contract, then, binds all healthcare professionals to technical competence as well as to moral discretion, and it enjoins them to use both.

Social contracts evolve through the ages. In our vision of the physician–patient relationship today, healthcare professionals are expected to attend to restorative as well as hopeless illness; that same contract in ancient as well as medieval times enjoined no such expectation.[4] The physician–patient relationship, cemented by specific contract (the contract as understood to exist between doctors and patients) as well as by a larger social contract (the communal contract, which promises fidelity to such contracts) evolves with communal notions. It endures and evolves through time. Today as rarely before and especially in the United States it seems in jeopardy. A contract between other healthcare professionals and their patients (albeit its historical roots are much shallower, they nevertheless necessarily intertwine with those of the medical profession itself) likewise exists. Beyond this, relationships and problems with such relationships among healthcare professionals affect the way patients will be treated. Such relationships and the problems they may entail cannot, therefore, be ignored. A collegial relationship and a team spirit are critically important to successful medical practice and are therefore ethically relevant.

Within the larger society as well as within the profession, the relationship between and among healthcare professionals and their patients has rested on historical tradition. Patients seeking out healers have done so with the fundamental assumption that such healers will, above all else, be dedicated to their personal "good," no matter what. Disagreements about the nature of the "good," about who legitimately defines it, and about the means necessary for its attainment may exist. But as long as such relationships have endured, the central fact was never in doubt: healthcare professionals, if they are to fulfill the social contract, must be dedicated to their patients' good. Trust validates this assumption. Although from time to time healthcare professionals have unscrupulously broken the implicit contract, have sought their own good, or have become the willing tools of the state[5] (by participating in torture, executions, or experiments on non-volunteers or by practicing "acute remunerative medicine"), the immediate or at least eventual negative communal response and the healthcare professional's evident personally felt need for justification and defense speak for themselves. Many aspects of medical care have changed; many definitions and ways of defining the "good" have come and gone; but the central fact has remained: healthcare professionals must seek their patients' good (however defined within a given personal contract and within the community) and never their patients' harm. The patient remains central; in the vision of the contract between healthcare professionals and patients current in the Western world today, the interests of the family, of the institution, or of the state vary in importance but are all, ultimately, peripheral. If we are to preserve what people treasure and have treasured in their healers, today's danger of medicine for institutional profit will also become past history.

Intertwining with the other roots is a root formed out of the specific personality of healthcare professionals and patients. Each relationship, therefore, is unique and changing over time. At times, relationships may be deep and pervasive (an old-time patient who has become a friend[6]); at other times, almost nonexistent (the unknown patient brought to the hospital in an unconscious state). Deep, pervasive relationships are informed by the enduring world-view of both parties expressed over time: "They have gotten to know each other." Their relationship is underwritten more by personal than by communal understanding. When, however, no such previous relationship exists, the communal contract, and the symbolism inherent in the contract, comes to the fore. Healthcare professionals doing their duty as they see it are informed in their vision of the contract by the social forces in which they are enmeshed.

——————— MODELS AND THEIR USES ———————

The behavioral models posited by Szasz and Hollender assume that physicians or other health care workers are primarily responsible to individual patients.[7] A model of activity–passivity in which treatment takes place "irrespective of the patient's contribution and regardless of the outcome" is best adapted to the unconscious, critically ill, or irrational patient who has not executed advance directives or given any other guidance to decision making. In the guidance–cooperation model, in which the physician is invested with great power by virtue of the patient's internal coercion by fear or pain, the physician is attuned to situations of serious illness in which patients, while conscious, are reduced in their capacity for making reasonable and informed choices. Here the patient's power and consequent autonomy are reduced (but not lost), and the physician's authority is enhanced. It is the situation in which most hospitalized and severely ill patients find themselves. It is here where abandoning patients to their autonomy" assumes dangerous proportions. In the mutual participation model, the physician and the patient, cooperating for an end satisfying to both, are seen as mutually interdependent and gifted (albeit in different ways) in power. It is model that readily suggests itself in various chronic states of which diabetes, hypertension, or coronary disease may be examples.

Many physicians and to some extent other health care professionals have felt most at ease when their power was great but when patient participation (and therefore "consent") was possible. Physicians have been willing to assume an entirely active role but have, traditionally, felt unequal to the task. The charge of paternalism in a setting where the activity–passivity model is most appropriate does not ring true. It is generally, and barring the existence of advance directives or competent surrogates, a paternalism of necessity—a form of "soft" paternalism" (see Chapter 5). Since, by definition, there is no one else to define the "good" or choose the means, physicians, ultimately and hopefully with the aid of the family and in concert with other members of the team, must do so. It is the guidance–cooperation setting, in which illness has distorted and unbalanced power that lends itself most readily to paternalism and in which, ultimately, a certain amount of cautious paternalism is even sometimes appropriate and occasionally inevitable. Here the patient seeks firm guidance and often willingly (and not always wisely) surrenders all decisions into the hands of the physician. Mutual participation, where patient and clinician are never really entirely equal, finds the patient most ready to disagree both with the definition of the "good" sought and the means used. Power is never really entirely equal: the physician still controls pad and pen. But power may go the other way: patients control the purse strings and, in a sense, the physician's reputation. A balance exists, and it is here that skillful and humane interaction and negotiation are most useful and necessary.

Cassel has pointed out that sick persons lose their sense of control over themselves and their world. They are not merely normal persons with the "knapsack of illness" strapped to their backs. Often they may lose their adulthood and revert to a more childish form of existence: in a sense, we might say that they exhibit autonomy-surrendering behavior At this time, their attribution of unrealistic power and competence to the physician is maximal. This is characteristically seen in the guidance–cooperation model. Consent here is often token consent: power has been yielded to the physician, and autonomy is virtually lost. Supporting, fostering, and restoring this autonomy is an ethically important task of medicine.[8,9]

Physicians and other health care professionals bring technical expertise, experience with similar problems, judgment in analogous situations, and, hopefully, integrity to the relationship. Patients bring their needs, their hopes, and their implicit promise of "payment," whether that "payment" is directly from the patient, through an insurance carrier, through the government, or merely by a psychological mechanism of gratefulness and enhanced reputation. Thus patients who, for a variety of reasons, do not pay in a material way still bring enrichment to professionals: they provide them with an opportunity to pay a small bit of their great debt to society, enable them to feel self-esteem, and promote their feelings of humanity as well as sometimes providing a much desired professional challenge. The accomplishment of such a task is a social responsibility. Beyond licensure, and despite malpractice suits, the law has little practical relevance upon most individual situations in which professionals and patients find themselves.

There are other ways and other models by which the physician–patient relationship can be described. Simplistically, of course, the relationship of healer to patient can be viewed simply as that of the "healer" attending a "sick" patient. Such a concept evades several important factors: it fails to address the concept "healer" as well as that of "sickness," and, above all, it takes the situation in which the healing act occurs outside its inevitable social and communal matrix.

"Healer" is a broad concept that has evolved from a unitary conception in which religious and medical functions were united in the same person.[10] The term "healer" has had various connotations throughout history and therefore has evoked different expectations. The shamans of the prehistoric world and those surviving today were not unique in uniting the religious and the "medical" function: from the Asclepiads of the Greeks (who coexisted and at times exchanged patients with their Hippocratic colleagues) to medieval times, when many if not most physicians were priests, to the faith-healer of today, the function of the "healer" has often been combined with more priestly functions. To physicians, the concept of what "healers" are today may be quite firm and obvious, but these concepts by no means invariably match historical precedents or, at times, even come close to the varying conceptions of healer by today's lay public. Furthermore, the various cultural backgrounds with which modern-day health professionals need to deal have quite different views of healers, healing, and the function of healing and of healers. Expectations are closely tied to such concepts.

Furthermore, what has been considered a crime, a disease, a sin, a sign of holiness, or even nothing in particular has varied throughout history with the particular subject in question, sometimes fluctuating among all of these. Consider only the fact that masturbation in the 19th century was considered a disease, a disease associated with demonstrable pathological findings and, furthermore, a disease that not only had a surgical cure but also was listed as a cause of death on death certificates.[11] Or consider the ways in which epilepsy, leprosy, and homosexuality have been variously categorized. The category in which we place such things (whether we hold them to be crime, disease, sin, or an irrelevant condition) makes a radical difference in the way we deal with them. While the World Health Organization definition of health as a "state of complete physical, mental and social well-being and not merely the absence of disease and infirmity"[12] may hold out an unattainable goal, it does serve as a point of reference. Nevertheless, it likewise, requires a social definition of disease and infirmity.

The social matrix within which the healing act occurs and within which doctors and other health professionals, as well as patients, function conditions both the way we look at such matters and how we look at ourselves. If health and disease are in fact social constructs, and if what we consider to be a "healer" is conditioned by societal viewpoints of health and disease, then the physician–patient relationship is also very much molded by the society within which it occurs. Concepts do not have a straight linear arrangement with one another. Rather, concepts are web-like, interacting with each other and, ultimately, with their social context.[13]

The "sick role" is defined by one's particular culture and history.[14] It determines how we as patients behave and what, as professionals, our expectation of patients will be. As has been shown, this can vary widely in differing cultures.[15–17] In consequence, the physician–patient relationship will be different in differing cultures. Models that seem appropriate to one culture and to one historical epoch cannot simply be transplanted and expected to flourish.

In the way health care is structured in the Western world today (and this likewise seems true for the rest of the world), other professionals have partly or completely taken over or complemented the physician's function. Midwives (used throughout history but only lately staging a comeback in the Western world), nurse practitioners, physician assistants, nurses, and others have increasingly and deservedly (as their training has increased and as credentialing has become stricter) gained credibility. This variety of skills and functions can greatly enhance patient care or it can result either in a Tower of Babel or in a power play between various groups. A Tower of Babel or a power struggle is eventually not only unseemly and detrimental to all concerned but, above all, can be most injurious to the patient seeking help. If these various professionals are to function as more than mere technicians (if, in other words, they are to be truly professionals), they must be trained and socialized into their respective professions and willing to play their part in assuming moral agency. In one sense, then, moral agency emerges out of consensus among the members of a team and membership in a profession, and is no longer the province of merely a single person.

A slightly different version of the models outlined in the preceding pages can be historically identified. Fundamental to almost all has been the centrality of acting for the patient's benefit, however defined.[18] It is the bringing about of a "proper act of healing in the face of the fact of illness" that has been and remains the proper concern of medicine.[2] It is the definition of what is proper, and who decides, that has given rise to many of the problems today.

In this version, three models have also been identified. The first of these models, which with considerable variation has lasted until recently, has been the paternalist model. In this model, physicians decide both their patient's ends or goals and the means necessary for their attainment. In such a model, physicians simply decide legitimate goals in specific situations: for example, whether the goal of health by giving a transfusion (the means to reach that goal) is to be preferred over the patient's goal of not jeopardizing salvation or, perhaps, whether a "do not resuscitate" (DNR) order serves the patient's interests (as defined by the doctor) best. In other words, the physician chooses both ends and means: the presumption is that "father knows best." Not only does "father" choose the means since, obviously, "father" has much more knowledge and experience with such matters; "father" also feels entitled to choose the goals despite the fact that goals of health, salvation, or beauty are highly individual matters.

The scientific model, ushered in by Francis Bacon (late 16th and early 17th centuries), evolved until by the mid-19th century the patient more and more became an object to be scientifically studied and acted upon. Rather than seeing themselves primarily as serving their patients, physicians progressively began to see themselves as scientists serving science: meticulous attention to technical details, unfortunately at times associated with callous neglect of the human dimension and often with very real neglect of the problems of pain and suffering, followed. The relationship between healer and patient sometimes became one in which the centrality of the patient was lost: instead of science being a means to serving the patient's ends, the roles were reversed, and patients became a means of serving science. This, of course, was especially true in experimental settings, but it penetrated into, and was reflected in the clinical arena. That is not to downplay the importance and, indeed, the central nature of science to medicine: it is beyond dispute that the last 200 years have seen an ever greater possibility to prevent, ameliorate, or even cure disease. But it is to sound a note of caution: prevention, amelioration, and cure are not ends in themselves but must, above all, truly serve the patient's (self-selected) interests.

As the economics of the marketplace changed and as the Western world progressively became more individualistic and less communitarian, an economic or entrepreneurial model developed. Increasingly, medicine was seen not so much as a profession as a business, and patients became consumers. Notions of the "health care industry," of "packaging," and of obtaining as much of the "health care dollar" as possible emerged. Physicians as entrepreneurs, or as workers in an entrepreneurial enterprise, were enmeshed in mutual competition. An ever-growing conflict of interests inevitably followed. (see discussion of managed care below). At best a contractual and at worst an adversarial relationship with a marked increase in litigation not surprisingly followed.

Clearly the paternalist model is not suitable in the modern world. Patients are not, if they ever were, children whose good we medical professionals must decide; rather, they are sophisticated beings whose capacity to pick and choose is limited only by their lack of specific medical knowledge and experience as well as by the facts of their illness, which to a greater or lesser extent limits their autonomy.[19] Restoring the patient's necessarily more or less limited autonomy, and affirming as much of the autonomy that remains as possible, are acknowledged to be among the central functions of proper medical practice today. But restoring or affirming autonomy is not equivalent to abandoning patients to their autonomy (see also Chapter 5) and certainly is quite different from seeing one's patients as customers and one's colleagues as rivals.

Science and a competent understanding of science are essential to proper medical practice. Developing and maintaining competence is the necessary but insufficient obligation of medical function. But crucial as science is to the practice of medicine, it forms a proper tool and not a proper goal in dealing with patients in the clinical setting. Unless physicians are abruptly to break with a tradition that mandates the centrality of the patient, the patient–physician relationship cannot properly be reduced to this model. Patients come to physicians to be healed, and the patient's, not abstract science's, good must first of all be served.

The entrepreneurial model would break even more with a tradition that sees the physician's function as primarily aimed at the patient's good and would reduce the relationship between healer and patient to one of coldly exchanging goods and, therefore, inevitably to one of competition. Explicit and, at times, well-written contracts, and not human relationships, form the basis of entrepreneurialism. It is, furthermore, a contract in which physicians invariably have the "upper hand:" they have far greater knowledge, have far more extensive experience, and, at least as importantly, are not ill, frightened, or in pain.

New models, which are now beginning to emerge, are sorely needed. An emerging interactive model promises not only to be more adaptive to the realities of modern life but also to be more respectful of the evolving tradition.[20] This model derives the physician's status from his or her (undoubtedly) greater expertise in medical matters and sees the physician's role as one of enhancing the patient's ability to choose. Physicians and patients in such a model are partners in health care with mutual respect and concern for one another. In this model, the commitment of physician to patient is grounded in a prior commitment to the patient's good and to the necessity of having that good enunciated by the patient. Any relationship in such a model takes place in a social matrix and is shaped by it.[21]

A more recent way of looking at these relationships uses many of the ideas of Dewey and Habermas. This way of looking at relationships has been recently elaborated by one of the authors (RSL). It sees a need for a perspective "rich enough to account for the dynamic interplay between the psychological and social components of relationships." In the medical setting it sees the weakness of one partner as complementing the strength of another as well as the converse. In such a relationship (which may be called a "consensus model"), relationships can neither be understood from only one perspective nor viewed from some ideal "God's-eye" view. It is a dynamic model in which "only in the communal activity of reflectively comparing—examining, challenging, defending, testing—multiple perspectives can problems be resolved and persons and their relations discriminated."[22]

Models must not hamstring one. More often than not they overlap and, furthermore, they differ from health-profession to health-professional and from time to time. Models serve as convenient examples, but they cannot fully describe relationships between people. Professionals not only are different from one another but are also themselves changing and evolving—and so are patients. Relationships, as has been mentioned, contain a personal root shaped not only by a particular patient and a particular professional; such relationships are not comprehensible unless one understands the history of the relationship. Furthermore, relationships at any one moment are dependent upon circumstances outside the relationship that impinge on each of the actors. My relationship with a particular student (or patient) at a particular time not only is informed by a social understanding of how students and professors (or patients and doctors) are expected to interact with one another in a given culture and community, but also depends on more personal factors. The way our day has gone, for example, inevitably shapes the way in which we interact with others who, in their turn, shape our further relations. These changes are subtle, but they nevertheless constitute the realities of life as well as, inevitably, those of medical practice.

RANKING THE PATIENT'S GOOD

When patients and physicians interact, their interaction is inevitably underpinned by an often tacit and loose understanding of fundamental assumptions. Here a ranking of "goods" is helpful. Pellegrino and Thomasma have most helpfully proposed that such a ranking can proceed at four levels: (1) the patient's ultimate good or "good of last resort;" (2) the good of the patient as a human person; (3) the patient's particular good; and (4) the biomedical good.[21] It is this ranking, within a social framework, that enables physicians and patients to negotiate goals and treatment plans.

The patient's ultimate good (or "good of last resort," as it is aptly labeled) is the highest good that the patient has autonomously chosen for him or herself. Obviously such choosing is not entirely "autonomous:" history, culture, and community frame the array of choices open to the patient; nor is it always evident—to self or others—unless or until consciously reflected upon. Yet, it is inevitably within such a framework that choices called "autonomous" are made. Such a "good of last resort" may be a religious vision, a secularly enunciated belief, or any other that is appealed to when the "chips are down." Paternalism, by reason of respect for persons, is inappropriate in the choice of such goods.

The good of the patient as a human person involves the choices—whether tacit or explicit—made by the patient concerning his or her vision of him- or herself as a human being. This good is the personal freedom to make choices, and it implies that health care professionals, in honoring such choices, must do everything possible to enhance, and not to interfere with, the patient's competence. Insofar as possible, patients must therefore be supplied with complete information as well as with a complete set of options. Drugging patients to ensure compliance, or deliberately giving them less than complete information concerning their options, is, *under almost all circumstances*, at least ethically suspect and in most instances ethically inadmissible. The recent ruling (and currently not in force) that health care professionals who work in clinics funded by federal dollars are not allowed to discuss or even bring up the option of abortion to pregnant women is an example of such an ethically inadmissible violation of this "good of the patient as a human person."

The particular good a patient may choose emerges from these other considerations. Here the patient chooses whether, in view of the previous considerations, a procedure is or is not worthwhile. A patient may, for example, choose to take or not to take a greater or lesser risk (say, a woman who must decide on various options when breast cancer is diagnosed). Here serious conflict between physicians and patients may occur (as when a patient makes a choice morally intolerable to the physician), and compassionate referral to another health care professional is sometimes the only option. Compassionate negotiation should, as far as that is possible, precede such a last-resort move.

The biomedical good is the *prima facie* good of the physician–patient interaction. After all, when patients come to see physicians, they come primarily with that good in mind. Only when a higher good interferes will the biomedical good be neglected or postponed. It is often here that negotiation is at its most fruitful. Within the context of the patient–physician relationship, patients cannot be forced to pursue the biomedical good if they believe it violates a higher value; on the other hand, it is here that patients cannot simply be abandoned to their autonomy (see Chapter 5).

——— HEALTH CARE PROFESSIONALS ——— AS THEIR PATIENTS' FRIEND

There have been those who claim that physicians and perhaps other health care professionals "ought" to be their patients' friends. Such a claim uses the term "ought" sloppily and is generally made without defining the term "friend."

In our culture, the term "friend" is often haphazardly used.[6] You hear persons speak of a "friend" when they mean a colleague or someone they happened to meet at a party and barely know. If we use the term "friend" in that sense, then the patient as the doctor's friend becomes an equally possible and meaningless proposition: if everyone we know and do not for some reason actively dislike is our "friend," then most patients fill that bill. If, on the other hand, we mean by the term "friend" someone with whom we share many interests and with whom we are by tacit as well as by explicit relationships deeply connected and who, so to speak, forms a significant part of ourselves, then the idea that physicians should be "friends" to their patients makes no sense.

The term "ought" can be used predictively (since you love Haydn and Beethoven, you ought to like Mozart) or prescriptively (you "ought" not to steal or "ought" not cheat). In ethics, mixing up these two meanings of the word can have disastrous consequences. When it comes to matters of feeling, taste, or affinity, the prescriptive use of the word is out of place: one cannot, no matter how hard one tries, force oneself to like a particular thing—one simply likes the thing or one doesn't. Kant's suspicion of inclinations in part rests on this evident fact: one can prescribe possible actions, but one cannot prescribe possible feelings. One can, prescriptively and plausibly, require that I not kill my enemy, but one cannot plausibly prescribe my feelings toward him. Likewise, you can, within the limits of reason, prescribe how health care professionals ought to act toward their patients, but prescribing that they feel friendship (or anything else) for them is to misunderstand both feelings and the act of prescribing.

The relationship of health care professionals and their patients, whatever else it requires, requires a peculiar mixture of detachment and involvement.[6] Health care professionals must do many things that are distasteful, disagreeable, painful, or dangerous to their patients. Many patients must be hurt so as to be helped. Furthermore, health care professionals often must do things that are aesthetically distasteful to them. Rationality has to control emotion and has to modify what Rousseau has called "the primitive sense of pity," or compassion. When rationality alone controls what we do, our actions are likely to be cold, dispassionate, and often unnecessarily cruel. A person controlled by reason alone may well act rather cavalierly toward pain relief or be largely influenced by the patient's usefulness to their careers or pocketbooks. (See also Chapter 8.)

Rational compassion (it differs from compassionate rationality, about which we shall have more to say in the chapter dealing with distribution) is one's sense of compassion tempered by reason: it allows one (or counsels one) to do things that compassion would, at first blush, forbid one to do. Compassion alone easily leads to sentimentality, and sentimentality can lead to doing some ultimately destructive things. Rationality without compassion is cold and sterile. Neither by itself is sufficient, and both are necessary when it comes to dealing with ethical problems, especially with ethical problems in health care. When we are directly involved with the fate of another (when we become "close to them"), we become more sensitive to their immediate suffering. When this happens in the health care setting, professionals can easily be led to abandon rationality. In that it often has disastrous long-term consequences (we neglect to do something very disagreeable or painful that would have been necessary to save someone's life or do something that, while it ameliorates the immediate problem, jeopardizes the long-term goal), such excessive "compassion" is ill conceived. Ultimately giving in to such excessive compassion and allowing it to swamp or dominate reason panders to our own emotions rather than acting for the good of the person for whom we are responsible: acting so as to produce long-term benefit but short-term pain is felt to be too painful for us ourselves. Patients are ill served when the persons to whom they have entrusted their lives and ultimate welfare act as if paralyzed by their own emotions.[6]

It is possible that a relationship that starts out as an ordinary relationship between a patient and a health care provider slowly and over time evolves into a true friendship. Patients and health care professionals may find that they share a large number of interests and may begin to experience a relationship transcending the professional. At that time—depending on the health care worker's capacity to control his or her own emotions and act rationally—health care workers and their charges will have to re-examine themselves and decide whether to continue the professional relationship. Such a choice depends on a number of factors, among which the personality of the physician or other health care worker is critically important. Here, as in all else, one of the first requirements of ethical action is honesty.[6]

CULTURAL AND LINGUISTIC PROBLEMS

A relationship among sentient beings requires them to communicate in some fashion. By its nature, a relationship is bilateral: we cannot, in a more than symbolic sense, have a relationship with a person who lacks awareness or understanding or with whom we cannot, though not necessarily by language, communicate. Of course, we often say that we have a "relationship" with an object that has become dear to us, but it is a "relationship" only between our imagination and us. We can and often do say that we have a relationship with a person whom we knew and who has died or who is in profound coma but, again, it is a relationship that is within us. In addition, relationships necessitate reciprocity: both have to communicate with, listen to, and attempt to understand each other.

While a good part of communication can be and certainly is non-linguistic, some ability to communicate explicitly in language among health care providers and patients is generally necessary. One can "listen" with a good deal of understanding to persons (or, for that matter, to animals) whose language we do not understand: We can usually tell if they are happy, unhappy, or in pain, and we can even begin to communicate ideas and concepts. But in the medical setting, where an accurate understanding by all parties involved is essential, some common ground of language at the very least is helpful. But beyond this: all the common ground of language still does not help if cultural or personal differences distort meaning. The concept "pain" or "illness," for example, has different meanings in different cultures and within the same culture to different people. Cultural differences, furthermore, can cause health care workers to make terrible "*faux pas*:" tuberculosis or cancer, in some cultural settings, has a meaning similar to venereal disease in the more usual Anglo-European setting. Some cultural settings expect and appreciate "touching;" in others it may be highly patronizing or even insulting. Eye contact is taken as a sign of mutual honest dealing in some cultures, but as demeaning in other cultures. Physicians who today must deal with patients from a large number of diverse cultures should seek to familiarize themselves with some of these cultural idiosyncrasies as well as attempt to understand their own. It is also important to encourage patients to discuss anything that might seem puzzling or confusing to them about their treatment or the health care team's behavior towards them.

Many patients do not share the same language with their physician or may be inhibited by an extremely limited vocabulary or a heavy accent. Unbelievably, but all too often, health care professionals respond to this with anger, and all too often having an accent is equated with an inability to handle the language properly or to understand, or even with stupidity or illiteracy. (One of the authors [EHL] could write volumes about this point!) What happens is understandable but hardly excusable: trying to communicate with someone whose understanding of language is limited or trying to understand someone with a heavy accent takes effort and time and is, therefore, too often resented.

Very often translators have to be used. Health care professionals must be aware that communicating through a translator, while at times unavoidable, has inherent dangers: the understanding of the translator with regard to what is being said may be limited or at any rate will be funneled through their own understanding; not rarely, translators have their own agenda or their own ideas of what the patient can or ought to be told. These considerations, important as they are when it comes to communicating the symptoms of the patient to the physician, the questions of the physician to the patient, or the physician's diagnosis to the patient, assume even greater importance when it comes to ethical questions. Health-care professionals are wise to speak with the translator and find out as much as they can about the translator's cultural attitude towards disease in general and the patient's disease in particular. Mutual comprehension will be more readily achieved and translation will go much smoother.

A patient's values filtered through the translator's understanding and fidelity have

be accepted with a great deal of skepticism. When a member of the family translates, it is sometimes well to have other family members present and to watch the
interaction and dynamics. One of us [EHL] has heard translators translating when he
understood both languages equally well and has, at times, been astonished at the lack
of correspondence between what was said and what was translated as having been
said—a phenomenon also rather notorious in diplomatic circles.

The problem is different when it comes to patients who cannot speak at all. Such
patients are often wrongly assumed to lack decisional capacity and are ignored when
it comes to making decisions. The assumption that patients who cannot speak are
necessarily incapable of making decisions is, obviously, untrue. In respecting such
patients as persons, physicians are obligated to do all that is possible to establish their
decision-making capacity and, if the patient is capable of making decisions, to communicate with them directly. Here the involvement of a speech therapist may be crucial. There are few patients (thank heavens!) who are truly "locked in": that is, who
have maintained cognitive functions but are entirely incapable of communicating (see
Chapter 11 on end-of-life issues). In general, but often only with great effort, a good
speech therapist can help establish communication by pointing to letters on a board
or by other means. But thus does much more than just establish communication—it
gives the patient some sense of being once again the captain of his fate instead if an
object to be tossed about at someone's will.

CONFIDENTIALITY

One of the enduring cornerstones of medical practice is the confidentiality of information obtained in the context of medical practice. This stricture—not to reveal information about patients to anyone and under (almost) any circumstances—has
endured through recorded time. Patients expecting to be helped must be truthful with
their physician (or with their attorney). This, in turn, according to some, necessitates
a strict (almost absolute) obligation of confidentiality. It is codified in law and has
come to be seen as an (almost) absolute condition of proper practice. There are those
who argue that, for a variety of both utilitarian (encourages full disclosure) and deontological (absolute respect for persons) reasons, it is near-absolute.[23] Confidentiality, as Rawls says, is an agreement bound by the principle of fairness.[24] But that does
not make it absolute.

Problems of confidentiality, even though at first blush we may think of confidentiality as near-absolute, are frequent: with whom to share hospital or office records,
to whom to reveal diagnoses, how to safeguard information at a time when multiple
health professionals are engaged with and must share information about common
patients or are members of the same group or HMO, and how to deal with information of possible communal impact (infectious disease, the revelation of crime, dangerous forms of insanity, etc.). Confidentiality, according to some, has become a "decrepit
concept:" one that, because of the necessity of multiple persons having access to
patient's records, is non-dischargeable in the context of proper medical practice.[25]

That does not mean that abuses of confidentiality by inadvertence or idle prattle (which Siegler, probably quite correctly, argues are those breaches patients fear most) can be condoned. It does mean that, rather than being declared "absolute," the concept needs thoughtful and compassionate attention in the light of current medical realities. Patients must be made aware of the complexity and changing nature of this concept; physicians must exercise caution and prudence when making chart entries. Claiming to be adhering to what, in many instances, is a non-dischargeable obligation adds hypocrisy to the violation.

Saying that confidentiality is an absolute and must never be breached is untrue and impractical and, at times, may cause disaster (as when a man kills his wife after telling his psychiatrist that he might do so). Laws, almost universally accepted as legitimate, that safeguard the community and enable public health, insist that we report infectious disease; others force us to assign a cause of death on the death certificate. Surely confidentiality can be neither absolute nor can it fail to be tempered by common sense. Sharing information with other health care professionals is a necessary condition of medical practice; without it, cure often cannot be effected. Most patients assume this as an implicit norm. Today with information increasingly available on often interconnected and shared computers and with the possibility of having such data instantly available globally, a new era of access to what may be quite personal and potentially destructive information seems more than likely. On the other hand, the benefits to patients who have suddenly become ill in a setting where nothing about them is known and who may be unable to give any information can be significant and, therefore, must be weighed against this threat. A life may well depend upon such information. It may well be that a solution lies in having it known publicly that information considered by health professionals as essential will be recorded and shared under normal circumstances but allowing that the patient may—knowing the risk of doing so—refuse and not have his information recorded.

Providing information to insurance carriers is also usually done with the patient's explicit or implicit consent. This consent, especially when it comes to health insurance, is increasingly often a matter of coercion—refusal to consent may result in denial of benefits for conditions judged to be pre-existent. In addition, problems may arise when the health care professionals are privy to potentially embarrassing information or to information with legal implications for the patient. Although under such conditions the physician's first obligation is to safeguard the patient's trust, that obligation is neither absolute nor universal.[26]

The obligation to keep confidentiality can be argued on utilitarian and deontological grounds. There are, however, times when obligations clash. Such clashes involve a conflict between confidentiality and the rights of the community (public health issues), threatened third parties (as when patients threaten to do harm to others or in cases in which their condition threatens others) and, at times, conflicts between preserving confidentiality with the patient and not allowing that patient to come to harm (as, for example, when patient threatens suicide).[26]

The rights of the community and the physician's communal obligations may clash with the patient's wishes (see also Chapters 5 and 6). The reporting of communicable diseases is an example. Physicians here have a clear legal obligation and, sometimes, an ethical quandary. Patients, when they seek a physician's help in the context of our community, are generally aware that certain conditions must be reported. This has become a particular problem with HIV infection: on the one hand we make the public claim that HIV infection is "not different from any other disease;" on the other we—and especially the public—continue to treat it differently. This disingenuous state of affairs is not aided by not making HIV infections a reportable transmittable disease like any other but ultimately only by public education. Leprosy some years ago was treated in a rather similar fashion. The requirement of reporting transcends any legal obligation: it forms part of the social contract in which the patient and physician, both as citizens, are enmeshed. Reporting such diseases, then, may be against a patient's expressed will, but has nevertheless been tacitly agreed upon by the community. Patients themselves have benefited, and expect to benefit, from such laws. They are part of the community in which such tacit agreements occur. However, physicians have an obligation beyond the mere reporting: they have the duty to see to it, as best they can, that rules of confidentiality effectively extend beyond their offices or beyond the hospital's walls so that information still does not become easily accessible or a matter of public record.

Threatened third parties are another often troublesome, issue. There is at least a difference in degree between the person who does not wish the nature of his or her illness revealed to their mate (even though the illness may be venereal or, perhaps, fatal), the patient who refuses to make a will, and the person who seriously threatens to murder another. Among other considerations, the question of sanity—and therefore ultimately of true autonomy—looms larger when a patient seriously threatens physical harm to another. Contracts, agreements, or covenants entered into with the questionably sane, and, therefore, questionably autonomous, cannot have the same force as contracts, agreements, and covenants under better circumstances. Here clinical judgment (judgment that determines the seriousness of the threat and the mental state of the patient) and discretion are of the essence. Ultimately, physicians must make agonizing choices in the full realization that a violation of covenant (be it with the patient or with the community) is inevitable. A given choice in a particular circumstance, while in itself blameworthy, may be the lesser evil.

When patients are infected with a communicable disease and refuse to inform those to whom they could communicate the illness, many of us would argue that while the ethical course of action is unclear, much supports the view that an innocent victim must be shielded. If, for example, the patient is infected with HIV and refuses to inform his or her partner, the physician is confronted with two distinctly unsavory courses of action: keep confidentiality and put an innocent person at risk, or violate confidentiality and protect that person. In such a situation there is no clear course of action and often not even legal guidance. At times, if persuasion fails, a form of coercion (also not a good thing!) may not be ethically inappropriate. Informing the patient that there is no choice but to tell the threatened other and to indicate that one is ready to help convey the information may make it easier for the patient to accept this course of action.

In a similar vein, there are occasional patients who have made no provisions for their dependent families and who wish to hide their fatal illness from them. There may be minor children involved, and the patient's unwillingness to be candid may seriously affect their future. Here again the issue is far from clear-cut: the threat to the family is not physical. But the consequences of such a patient's unwillingness to provide accurate information and make reasonable plans may have disastrous consequences for those left behind. Often a patient's unwillingness to share a bad prognosis with his or her family is not so much a desire to shield them from pain as it is an unwillingness to come to terms with his or her own condition. At times, counseling a religious patient to speak with and to involve their clergy-person may help in resolving such a problem. But at any rate, the physician's obligation is not simply "not to break confidentiality" but to prod, advise, and sometimes even cajole the patient to do what is necessary to safeguard others.

The question of confidentiality when patients threaten harm to themselves is a more troubling issue. A retired person with operable cancer of the colon who refuses surgery and insists that his spouse not be told, or a non-psychiatrically ill person determined to commit suicide would serve as examples. When physicians intervene in such circumstances, intervention is strictly paternalistic (see Chapter 5). Physicians and patients see the "good" differently, and, if physicians break confidence, they are now imposing their vision of "good" on the patient. Provided the patient is sane—not severely depressed—and meets the criteria of autonomy (and that, after all, is a largely clinical judgment), a breach of confidence would be hard to defend within the context of our current vision of the physician–patient relationship. But this does not relieve health care professionals from the obligation of repeatedly and intensively talking with such patients and attempting to influence their course of action. And that takes time.

Individual cases must always be adjudicated in the light of the obligation envisioned and the context in which the obligation occurs. Physicians have a *prima facie* obligation to preserve confidentiality, but such *prima facie* duties can be overridden for weighty reasons. When confidentiality is breached, no matter what the weight of the argument, physicians are blameworthy (see Chapter 3). Breaching a confidence and violating a trust are not now, nor can they ever be, "good" things. But they can, on the grounds of harm and benefit to others, be a better, and often far better, alternative than passively allowing great harm to occur.

With the development of information technology, e-mail and computer storage of information we are facing a new set of problems. The ability to store a patients past medical history in a (theoretically) readily accessible computer from anywhere in the world has obvious and at times life-saving potential. The "flip-side" of that coin is that information readily available to a far away physician may, if conditions are right, be likewise readily accessible to insurance companies, prospective employers or a host of other people the patient would rather not have know about his medical condition. Likewise there are obvious advantages to communicating by e-mail with patients and here again the question of confidentiality is not one that is prone to a permanent solution.

There are obvious safeguards that can be employed but—like "star-wars" defense—every safeguard can eventually be penetrated. As is the case with all ethical problems: by their very nature—do not have easy or "good" solution. They are something that must be worked upon consistently and with an attempt to deal with the ethical problems as best we can. In this essay we mainly wish to warn about some of the ethical pitfalls that a new way of relating has raised. If it has led those involved in this sort of communication to begin thinking and discussing some of this pitfalls among themselves and with their patients it will have served its purpose. It would be ethically proper to inform patients of such possible "leaks," to assure them that as far as possible one will safeguard against them and to give the patient—once he or she are fully aware of the advantages and possibly life-saving advantages of such communication—the option to decline to have their particular information handled in that way.

CONFLICTS OF INTEREST

Physicians and other health care professionals are confronted with a variety of problems unique to the setting of their practice. Practice in a private solo practice, in HMOs of various types, as the employee of various organizations (hospital, insurance company, industry), or within an organ of the state (the armed forces, prisons, etc.) all influence and distort the classical tradition. This problem has increased and continues to increase as managed care proliferates (see section on managed care).

It is often very difficult for health care professionals seeking their patients good within the context of these various constraints and often-conflicting obligations and loyalties to remain honorable. All these contexts offer the carrot of greater pay or advancement in the organization itself for "proper" behavior—proper, of course, as defined by the organization. Furthermore, rewards and punishments are rarely blatant, and they are usually given a veneer of probity: they tend to be more analogous to seduction rather than rape. Such slick hypocrisy is an especially dangerous feature. It remains for physicians to examine themselves and their function in the light of social contract and to reach conclusions with honesty and integrity.

Health care professionals other than physicians are involved to a lesser degree. Although they have traditionally "worked for someone" (a condition many physicians increasingly find themselves in also) and in some respects their opportunity to be coerced or seduced was even greater, their power of decision making was believed to be far less great. However, as the power of other health care professionals has increased and as some have gone into private practice, they too run the same risks.

Blatant infringements of the relationship between health care professionals and patients occur and have occurred in the service of the state. The yearly reports of Amnesty International do not leave physicians or other health care professionals free of blame. As far back as the first century, Scribonius Largus spoke of the duty toward all patients—regardless of war or peace—that united physicians. Our current vision of the relationship between health care professionals and their patients would seem not to endorse practices in which health care professionals become the allies of others bent on their patient's harm or destruction. Such gross examples as uninformed and un-consented to experimentation,[5,27] torture, or interrogation (or assisting, aiding, or abetting torture or interrogation)[28-30] or participation in capital punishment in any form with the tools of medicine are so flagrantly wrong that they ought not to require extensive condemnation. Nevertheless, despite pronouncements, guidelines, and rules, these outrages continue throughout most of the world and regrettably also within the borders of the United States. And even though medical organizations have spoken out against physician participation in and with capital punishment, they have failed to consider sanctions or even to define participation.

Because torture and experimentation on non-consenting subjects are so evidently wrong, they are often less of a problem than the role of health care professionals in the day-to-day working of jails, prisons, or, at times, the armed forces. It is here that utmost vigilance is called for. Further, the problems of an entrepreneurial system that rewards physicians and increasingly other health care professionals for performing or for not performing tests, for hospitalizing or for not hospitalizing patients, etc., presents similar dangers couched in different terms. Institutional pressures on physicians, pressures that again favor minimalism at almost all costs, are likewise considerable.

Health care professionals frequently experience ever-growing conflicts of interests today. Since physicians and increasingly other health care professionals in many respects control their own incomes and often generate their own work—by ordering or not ordering tests from which they profit, by having financial interests in diagnostic or surgical centers and sometimes by financial inducements given by companies whose employees they are or whose products they use, and ultimately by determining the frequency of patient visits—the conflict of interests is formidable.

Moreover, physicians are often put under considerable pressure by their institutions to order more tests or to do more procedures: expensive equipment (equipment often purchased so as more effectively to "compete" with another health care institution) must be amortized. On the other hand, physicians working within an HMO or MCO are generally expected to limit the use of tests or procedures so as to save money for the HMO or MCO—and often they are rewarded by an end-of-year bonus when they do so...and thereby "punished" when they don't. Here, the medical care of the patient, which ethically must be a health care professional's first priority, conflicts with the desire to maximize income or, increasingly, simply to make a decent living. Whether by ordering fewer tests or procedures than needed, by ordering more of such tests, procedures, or return visits than needed, or by ordering marginally needed tests, physicians are failing in their obligations to their patients. It would seem that this is an intolerable situation.

It is difficult to come to terms with these problems. No pat solutions are available. The danger is not so much in the deliberate choice to do more or less than needed (although that danger too is real); the danger is that subconscious self-interest will influence decisions and enter into what should be decisions made on the basis of another's (and not one's own) good. That self-interest should play a role in our decision-making is not ethically wrong; since decisions we make inevitably affect us in some way, it is inevitable. But when self-interest conflicts with a professional duty to one's patients, it should be recognized and admitted for what it is and steps taken, singly and/or collectively (whether by one's professional organizations or by broader, social consensus), to correct—or at least to minimize—it's harm to patients.

Conflicts of interest may have two effects: they may incline us to choose in a way that would profit us or they may cause us to lean over backward and deliberately choose the opposite. For example, when physicians have to decide whether to perform a test or do a procedure that would be profitable for them, they may deliberately choose to maximize their own income or they may be so afraid of choosing in their self-interest that they end up delaying far longer than they otherwise might have. Neither course of action serves the patient's interests. When building a health care system, such temptations should be avoided. Some temptations will, however, in virtually any conceivable system, probably always exist. Sensitivity, honesty, and a careful understanding of the roles and obligations of physicians in society can go far toward helping to find reasonable solutions.

Ethics in general and health care ethics in particular do not exist outside a social nexus that, to a large extent, determines not only many of the rules but also how such rules will be played out. It is not possible to practice ethically in an unethical system, just as it is not possible to create a truly just institution in a basically unjust society. As institutions (such as health care systems) change, ethical problems and answers are likewise apt to change. Health care professionals who wish to practice in an ethically acceptable manner should therefore involve themselves in shaping the nature of the institution within which they practice and ultimately need concern themselves about the social nexus in which such institutions and their practice are embedded. In other words, just as patients cannot be reduced to their biomedical situation, so health care professionals cannot be reduced to biomedical technicians: patients are human beings whose complex interests include, but also extend beyond having their biomedical situations remedied and health care professionals are human beings whose obligations include, but extend well beyond self-interest or simple technical expertise.

───── PROFESSIONALISM AND RISKS ─────

Persons choosing their life's work must make choices that reflect their own deeply embedded values and personality. In choosing their life's work, they assume certain known risks: firemen may get burned, policemen shot, and health professionals infected. Risks, however, are also, to some extent at least, imponderable and may appear during the course of a life's work: fire-fighting equipment and the nature of blazes change, criminals adopt new methods and weapons, and the nature of a given infection evolves. Every occupation has its advantages and drawbacks, its risks and benefits. Medicine is no exception.

Physicians and other health professionals are exposed to risks throughout their professional lives. These risks are rarely explicitly spelled out; nor can they be. Members of social structures, when first coming together, have established communities with far differing notions of what communities are all about (see also Chapter 3). No matter what our notions of the specifics of this contract may be, its existence cannot be much in doubt. The alternative, no social contract and no understanding or agreement of mutual obligation, cannot be called a community. Community is not merely a collection of individuals held together only by explicit undertakings. Such explicit compacts, such affirmations of mutual responsibility, cannot come about without the tacit undertakings and expectations that enabled them in the very beginning.[31]

Historically, health professionals, when confronted with infectious disease, have had to fear contagion. Fear (here defined as a sensation or feeling of anxiety caused by the realization, perception, or expectation of impotency in the face of perceived or expected danger or evil) subsumes qualities of dread and awe and further has other emotive and aesthetic elements.[32] Counterpoised against such fears are the presumed duties of the profession: not only the obligations assumed by moral agents in recognition of the moral law as distilled through the vision of specific social contract by particular societies, but likewise the more specific obligations inherent in being a professional of a particular type. Courage (the "disposition to voluntarily act, perhaps fearfully, in dangerous circumstances," its essence being the "mastery of fear for the preservation of a perceived good against dangers") gives the edge to doing what one perceives to be the right thing despite one's fears.[33] What health professionals perceive to be "the right thing," however, derives from their understanding of social contract applied, in this instance, to the way in which the implicit covenant with the community is envisioned. And such a vision is historically grounded.

Health professionals throughout history have assumed obligations to treat patients despite personal risks. Presuming that health professionals were aware of the possibility of contagion (and that therefore they were quite mindful that they could contract the disease in epidemics), epidemic disease can serve as a paradigm for such an examination. Although the knowledge of what causes infection was still far in the future, there is sound evidence that it was soon clear that some disease could be spread by personal contact. Thucydides, in describing the plague of Athens (5th century BCE), mentions the disproportionate number of physicians who died there, and Hippocrates carefully instructed physicians in methods of avoiding infection. By the time of the Justinian Plague (540–590 CE), there is no question that knowledge of contagion (albeit hardly of its mechanisms) was firmly entrenched. Laypersons as well as professionals were obviously quite aware of the risks.

Many factors enter into our clinical or personal decisions to take, or not to take, risks. Some of these factors are technical: "What kind of risk am I taking?" "How much risk is there?"—to name but two. The answers here are crucial to our ethical deliberations. If undertaking a given course would result in certain death, a different set of considerations pertains than if the risk is moderate or small. Even in the first instance, there is a critical difference between the heroism that gives a life to save another and an action that gives a life with no hope of saving another. Giving one's life to save another may, under most circumstances, be a supererogatory act; doing so with no hope of saving a life in turn for one's own may surely be even more problematic. Under most circumstances, neither can be simply viewed as a clear-cut and absolute moral obligation that must, under all circumstances, be discharged.

For physicians and other health professionals, there is, furthermore, a consideration at least as important as the saving of life. There is a great deal of difference whether, beyond saving a life, significant comfort can be given. The obligation of health professionals clearly does not end with the saving of life. Historically the obligation to give comfort is far more enduring than is the obligation to save life, and this ancient obligation is presumed today. Health professionals must consider both the saving of life and the amelioration of suffering. When a disease is hopeless, ameliorating suffering moves into the foreground of professional obligation. As long as patients have not irretrievably lost consciousness, health professionals are obligated to provide what comforts they can. Such an obligation is grounded in the shared historical vision of the patient–physician relationship.

Both physicians and communities have historically profited from their vision of the social contract. Health professionals gain a tremendous amount from their side of the bargain. Physicians, and to a lesser extent other health care professionals, have been blessed with immense privilege, prerogatives, and power as well as with considerable material reward; communities have profited from their healer's skill and from the security entailed in the knowledge that the contract will be honored in times of need. Like all contracts, social contract implies mutuality and bilateral agreement.

What about the HIV-infected physician or other health care professional? Do such persons have an obligation to inform others about their condition, or is seeking such information a violation of that person's privacy? Is mandatory testing for health care professionals a reasonable incursion on their private liberty, or is it not? As with all problems, one must start one's inquiry by gathering "facts"—at least the best facts that are available. At this writing, no single case of physician-to-patient transmission has occurred. One dentist who in the course of his work appears to have infected several patients apparently did so under particularly peculiar circumstances that make it likely that he, whether deliberately or not, failed to sterilize his instruments properly.

Although laypersons seem very concerned about possible HIV infection in their physicians, such concern does not appear to be based on factual evidence but on rumor, fear, and hysteria. Although some feel that physicians, especially those who do invasive procedures, have an obligation to inform their patients of their status, a persuasive argument in the face of overwhelming data that such transmission rarely if ever occurs is difficult to make. Laypersons who are afraid of such transmission are, it seems, laboring under a false assumption: since data will not substantiate this belief, it is a form of prejudice and one that can be ruinous for the person against whom this prejudice is directed. Rumors, fear, hysteria, and prejudice are not properly addressed by restricting another's freedom of action, but are properly addressed by the education of those who are misinformed. Furthermore, even though a theoretical risk can certainly be argued, patients are not ordinarily informed about many aspects of their physician's private lives that may constitute a risk to them. If a surgeon sleeps badly the night before surgery or if he or she is overworked, worried, or otherwise troubled, the risk to the patient is a very actual one. And yet we do not think about forcing surgeons to reveal their lack of sleep, the fact that they had been to a party, or their domestic or financial worries to their patients.

If physicians were forced to reveal their HIV status to their institutions, licensing boards, or patients, their ability to have a successful and satisfying practice would be severely limited. The minimal risk their patients might face (and all of us every day assume small risks when we go about our business) is out of balance with the destruction such information would cause to their lives.

CONSENT

The way we understand and use the notion of consent in dealing with patients is intimately connected to our understanding of the professional–patient relationship.

We are often told that "consent" is a new idea, one not considered in former times, and this is undoubtedly true of formal "consent." Certainly what is called "informed consent" is a quite recent development. Yet, when patients came to Hippocratic or Asclepiad physicians they came expecting a certain approach, and they were enmeshed in a social relationship that they understood in certain ways. They came first of all to be helped, and second with a social concept of a particular physician–patient relationship. Furthermore, surgeons could not "cut for stone," or physicians administer potions, unless the patient submitted himself or herself to such a procedure willingly. Thus, although not formal and far from as informed as we require today, consent was, nevertheless, a consent of sorts. Then as now, consent is not merely an explicit agreement between two or more individuals but has to be understood as enmeshed in a particular cultural and communal matrix.

In our culture today as never before, we take for granted the necessity of obtaining informed consent. Often this is merely to protect ourselves from legal repercussions; properly (if by "properly" one means being mindful of the richness of the relationship between and among health care professionals and their patients and the consequent obligations), it is in order to make patients willing partners in a joint enterprise, an enterprise in which patients cannot truly be partners unless they understand its "facts," meanings, and dimensions. Consent to do a thing to another is necessary if we are to respect one another. The act of profession requires us to pursue our patient's "good;" respect for others requires that we define that good on that patient's terms. A respect for autonomy presupposes a sense of beneficence. Since my "good," under ordinary circumstances, is properly a "good" defined by me, acting beneficently is to have a regard for and to respect that good. Caring enough for another's welfare to respect their autonomy, ultimately, is a beneficent thing to do.

Consent, as Ramsey has so eloquently stated, can be understood as a "statement of fidelity between the man (or woman) who performs medical procedures and the man (or woman) upon whom they are performed." At its best, consent is grounded in a "canon of loyalty," which requires more than merely sterile assent.[34] Consent implies a fiduciary relationship that assumes that the patient's good is to be done and assumes that patients consent because they fully (or as fully as possible) understand not only what it is that is to be done (the means) but also the ultimate goal (or end) of doing it. The procedure (the means) and the ultimate goal (the end) are necessarily interrelated and interactive; choosing one, in a sense, determines or endorses the other. When fully informed patients consent to a procedure or treatment, they agree both with the goal and with the means toward its achievement. When such patients refuse, they refuse because they disagree with the goal to be achieved, find the means (the suggested therapy) either inappropriate or intolerable, or, at times, have failed to understand the issue. Rather than merely going through motions, physicians, by accepting either consent or refusal as valid must be reasonably certain that patients have, as fully as possible, understood the implications of their chosen course of action.

When patients consent to what we want them to do, we do not question such consent unduly. Persons who agree with our course of action are obviously eminently well informed, sane, and intelligent! When, however, patients disagree with us, we are prone to question the extent of their information, their sanity, or their intelligence. Patients who agree with our recommendations generally share our goals and are willing to conform to our means, and most do. And yet consent too glibly given should be subject to at least some questioning. Patients may not have understood fully (or, at least, as fully as in their particular circumstances they really could), may be frightened into assent (just as others may be frightened into dissenting), or may be unaware that they are, even if they simply fail to act, committing themselves to a course of action. Therefore, it behooves health care professionals to maintain a degree of skepticism for consent too readily given. Health care professionals should reexamine their patients' depth of understanding either when patients dissent or when consent comes too readily.

Accepting a patient's refusal for lifesaving or critical procedures requires a different level of justification from accepting a patient's consent[35–37] (see also Chapter 5). In general, more is at stake. That does not mean that consent should be *pro forma*: physicians must be sure that the consent they obtain is truly informed (the patient has understood diagnosis, prognosis, treatment, and alternatives, and has understood them on his or her own terms) and that the patient is thinking clearly. It does mean that when patients refuse, a much more intensive dialogue needs to take place. Here physicians must seek by all possible means to ascertain that patients (1) understand the facts, (2) believe them, and (3) have reasonable decisionmaking capacity. For example, a patient told that he or she needs an operation or a blood transfusion must not only know that "fact," but must also believe that this opinion (to serve the goal they together with health care providers have set forth) is correct. When patients disbelieve what to their health care providers are "facts" a reasonable amount of consultation and perhaps persuasion without coercion is called for. It is critical, furthermore, to ascertain what a certain diagnosis means on the patient's terms: what it means to the patient and in that particular patient's life. Refusal can easily be based not so much on a misunderstanding of "facts" as on a misinterpretation of meaning. (See also the chapter on autonomy and the section in that chapter that deals with decision-making.)

When we become ill, our autonomy is, of necessity, diminished.[38] This is true even when—much as we may not want to admit this—physicians or other health professionals inevitably also become patients. Laypersons, in addition, have a variable but inevitable lack of cognitive knowledge and a lack of experience with similar cases, which, among other things, distorts their emotive understanding of problems. Further, as if this were not enough, patients—health professionals and others alike—are at least concerned if not, in fact, frightened, and their ability to think clearly and dispassionately is therefore hampered by the very disease that has made them patients in the first place.

In the clinical situation, power is inevitably unevenly divided. Besides the cognitive, experiential, and biological factors we spoke about, physicians have the power and, not to be underestimated, the mystique of the medical setting at their side. In addition, healers are invariably seen by those they heal (or pretend to heal) as endowed with more than their actual power. And this attribution of power may not be entirely unimportant in the healing process. Therefore, physicians and others who concern themselves with obtaining consent must try, as best they can, to promote as much understanding and offer as much true choice as is possible. The relative weakness of patients does not make them "more deserving," but it introduces a special obligation. Discharging such an obligation requires a great deal of compassion, tact, patience, and understanding.

Decisions must be understood within the patient's peculiar social milieu and background beliefs. A disagreement about goals or about the relative value of competing goals cannot usually be solved by dispute: the Jehovah's Witness who refuses blood is not apt to be persuaded by lectures on the safety of transfusion or about the threat that such refusal has for life. A misunderstanding about means (as long as these means are not precluded because they are seen as distorting the goals as in the blood transfusion example) is more apt to yield to negotiation, persuasion, or consultation. Persons who fear that a given treatment or procedure would have a dreadful cost in terms of pain or suffering or who, on the other hand, unreasonably fear some outlandish misadventure may be persuaded by reasoning or by supplying additional facts. Often it is helpful to confront them with patients who have undergone the same or a similar procedure or, if that is appropriate, to introduce persons who are not yet acute ICU patients to the ICU setting.

Unfortunately, the way consent is obtained in clinical practice is often a caricature of both informing and consenting. Patients are generally told their diagnosis (though euphemisms such as "growth" for cancer and so forth are still all too commonly used), but two things are often missing: (1) telling patients such a diagnosis on their own terms and in their own language and then making sure that such information is really comprehended and truly understood; and (2) inquiring what such a diagnosis means in the patient's terms and in the patient's life. Even when patients understand what gallstones are, for example, such a diagnosis may be a far different thing in the life of someone who rarely leaves town and lives near a hospital from what it might be for one who travels extensively and sometimes to remote areas.

Consent, likewise, is often waffled. If, on the day of or the day before surgery a nurse or junior resident is assigned the task, confronts the already admitted patient with a piece of paper and asks the patient to sign, it is quite likely that neither the patient nor the health care professional has had time to read—let alone think or inquire about—the various options and complications listed.

Of course, situations change the degree of information and the nature of consent. In an emergency the presumption that a reasonable person would want to optimize their chances of survival is an ethically sound one. But most surgical procedures and most medical treatments are not emergencies. They are done in a perfunctory manner largely at the convenience of the institution or the professional involved. If more than lip service is to be paid to the notion of respect, patients deserve full information and the time to deliberate and to talk things over and to ask further questions. Above all, health care professionals should seek to understand what such a diagnosis and such a proposed treatment mean in the context of a patient's life.

Truly "full" information is, of course, not possible. The far greater knowledge and experience of health care professionals as well as the illness or anxiety about illness on the part of the patient makes this impossible. But most reasonably intelligent patients can be informed adequately enough that they can truly understand, truly decide, and truly participate. Health care professionals must reveal more than diagnosis, options, and prognosis: they should (unless one believes that they are no more than "bureaucrats of health") also guide and advise patients. Patients seek out their physician not only to receive a diagnosis and be treated; they hopefully turn to healthcare professionals with sufficient trust in their wisdom to seek counsel from them. When patients are informed about alternatives, physicians arguably should also give their advice and the reasons for it. After all, even when we go to our favorite restaurant we trust that the waiter will not only hand us a menu but might also advise us on what seems and what does not seem to be particularly "good" that day!

EXPERIMENTATION

Experimentation in medicine offers a troublesome dimension. Little formal attention was paid to this issue until after World War II, when the outrages committed by Nazi doctors revolted the civilized world.[5,39-41] Yet such outrages were hardly limited to Nazi Germany. From 1932 to 1972 (covering the period preceding the Nazi experiments and extending well beyond their condemnation at Nürnberg by the civilized world— including the United States), the United States conducted a systematic study in which over 400 black Alabama sharecroppers were studied to determine the effects of untreated syphilis in a day when syphilis was highly treatable (and when, incidentally, the effects of untreated syphilis were well known).[27] The Tuskegee Study, funded by the United States Public Health Service, has assumed its rightful place among man's medical atrocities alongside the Nazi and other such experiences. And it is hardly an isolated instance. Racism aside (though, granted, it is difficult in either the Nazi or the Tuskegee experience to put it aside), the dedication of researchers to science has often resulted in experimentation and innovation involving non-consenting human subjects.

Other examples of this sort have continued to emerge: the United States armed services have tested wind currents and the possibility of bacterial warfare by liberating an organism first over San Francisco and then over the Midwest river valleys, first resulting in a flu-like syndrome for many and later in a continued problem with such organisms for patients in ICU settings. Likewise, the armed services experimented with LSD in servicemen and with radioactive materials in patients who had no idea that they were being subject to such procedures. Lately experiments with radioactive substances—entirely unknown to the individuals exposed—have been conducted under government auspices even in civilian hospitals. Prisoners (sometimes with *pro forma* consent and sometimes without even that) have not rarely been used for such purposes. Unfortunately, such things continue to this day and, beyond an occasional flurry of concern, little has been done to effect real change. Such experiments require the participation of physicians and other health care professionals and, unfortunately, usually the *knowing* participation of such persons.

It is, of course, not altogether easy to say what does and what does not constitute experimentation, let alone innovation. Physicians and surgeons, in trying out a slightly different technique or by prescribing a drug under slightly other than stereotyped circumstances, may be accused of experimentation or, at least, innovation. Some such maneuvers are part and parcel of everyday practice, and these are not what we have in mind. Furthermore, innovation has often given the impetus for further development and often forms the grounds for later disciplined experimentation. As in most things, minor innovation (trying an instrument in a somewhat different way during a surgical procedure or administering a somewhat different dosage of a drug when this seems wise to do) is more or less readily separated from true experimentation (trying a new and untried procedure or substance when we have little idea of what might or might not happen). As always, the problem is often in the gray zone.

There is a logical difference between experimentation and therapeusis.[42] The goal of experimentation is the creation of new information; if the individual patient is served, that is a bonus, but it is not the main goal. In therapeusis the main goal is helping an individual patient; producing new knowledge or better understanding is desirable but is not the purpose. Simply put, treatments that lack a "track record" and that are done to gather new knowledge can be considered experimental; those that have a "track record" and are used to benefit a particular patient are "therapeutic." Ethically there is a logical conflict between the obligation experimenters have toward their experiment (which is to produce "truth" or new knowledge) and the obligation experimenters have toward the subjects of their experimentation (which is to hold the patients' good above all else). Some have suggested that, just as in transplantation, two different teams (one whose main purpose was to safeguard the integrity of the protocol and the other whose main purpose was safeguarding the patient) might at times be used.

Experimentation, as ordinarily conceived, involves an adventure into what is at least partially unknown. Those who join in the adventure should, at the very least, be fully aware of what is known about the journey and what is not or cannot be known. As far as possible (and as far as they are known), risks need to be spelled out. But that is not sufficient: patients must also understand that by the very nature of the experiment some risks may not be anticipated or anticipatable. Patient consent should be freely given. (This raises obvious problems with the use of prisoners, medical students, and others who may be prone to more than the usual coercive pressures.)

There are further difficulties. Research can be done on healthy persons either to gain physiological information, to test new drugs, or even to produce illness. Subjects cannot directly benefit from such interventions (albeit they may feel pleased to have contributed to science or medicine). Such experiments may carry little or considerable risk as well as be associated with much or little inconvenience to the subject. On the other hand, research can be done on patients with a particular disease to learn more about that disease and, perhaps, to help them.

But those are far from being the only variations: research can be done on those capable of consent or on those in whom consent is impossible (children, the demented, the insane, the unconscious, those who come to the emergency room in critical condition, etc.). At first blush, experimenting on someone who cannot consent seems ethically illicit, and allowing someone else to consent to something that cannot help the person being acted upon seems problematic. Unless, however, experimentation proceeds with such patients, progress in learning to help such patients grinds to a halt. Ultimately one must use patients with Alzheimer's disease to learn more about Alzheimer's disease, the insane to learn more about insanity, children to better understand child physiology, and the victims of sudden severe illness to learn more about their conditions.

In general, experimentation has come to operate under a set of guidelines constructed to safeguard subjects and meant to ensure that research follows reasonable and ethical standards. Research, whenever possible, should first be done on inanimate models (tissue cultures, computer models, or plants) before being done in animals and should be done on lower animals before higher. When treating disease in humans, acceptable research should not deny a group of patients suffering from a dangerous illness treatment of their disease in order to establish "no-treatment" controls: new treatment for a disease would have to compare current to new treatment rather than current to no treatment. When a treatment for an untreatable disease shows a clear and statistically significant benefit when contrasted to no treatment, the experiment must be stopped, and the new treatment offered to all suffering from the disease (for example, when AZT was tested against AIDS and its benefits became clear, the experiment was stopped, and AZT therapy was offered). Unfortunately, pressure by the contingent from the United States (presumably motivated by the interest of powerful pharmaceutical companies) has managed to delay this requirement, delaying its inclusion into the newest changes in the original Nürnberg code on experimentation. These periodic changes, called Helsinki agreements, are made by international consent and form international standards. Guidelines for conducting research on prisoners, children, and the mentally infirm—on vulnerable subjects, in other words—have evolved but continue to require revision.[43–45]

The requirements for conducting ethical research are also quite different when phase I, II or III studies are involved. Phase I studies involve the first use of a new modality in a human subject—even when great care is taken, little is known of dosage or effect, the chance of benefit is minimal and the risk not inconsiderable. Phase II studies establish this information and phase III studies further advance the process to clinical trials—often a multi-center study being involved—which considerably increases the chance of patient benefit. It is peculiar that most patients are most eager to participate even in phase I studies hoping that they might be in the "treatment" instead of in the "control" group—this despite the fact that given the very nature of an experiment their chance of doing better in the established rather than in the new type of treatment remains unknown. Were it not so, it would no longer be experimental but established therapeutic practice.

It should—but does not always—go without saying that once clear statistical evidence of one group showing a distinct advantage over another can be demonstrated, the experiment must be terminated and all patients switched to the most effective form of treatment. Likewise it should go without saying that statistical criteria for such an event must be determined by an expert in advance and adhered to: a mere impression will not do and, in fact, is liable to do more harm than good.

Research, to be acceptable, must be approved by an institutional review board (IRB). Such IRBs, if they are to function properly, must not only deal with ethical standards or concern themselves about informed consent but should also at the very beginning make sure that the experiment is scientifically sound, that it stands a reasonable chance of producing the information sought, and that the experimenter is well qualified. Unless such preconditions are met, no experiment can be considered to be ethically sound. All institutions conducting human research are expected to have and to utilize such IRBs. This is a laudable, yet hardly foolproof, step forward. IRBs are composed of people. At their best, people are not entirely impervious to political pressures or unmindful of the fact that their colleague whose research they must approve or not approve today will tomorrow approve or not approve their own. Furthermore, passage by an IRB does little to solve the quandary in which researchers find themselves when they must look out both for a particular patient's welfare and for the welfare of their experiment. Inevitably, at times, they are caught between two mutually exclusive, or at least mutually somewhat contrary, goals.

Inevitably, the relationship between health care professionals and their patients (or, in the case of experimentation, subjects) is distorted or, at the very least, strained by research protocols. And yet, if medicine is to advance, research is vitally necessary. Sometimes it is possible to diffuse the problem (for example, by having two different persons responsible for the treatment, one mainly concerned with the patient's welfare and the other conducting the experiment, with the former having veto power over the latter). Often the best that can be done is to be vigilant, to be aware as much as possible of one's own motivation, to be mindful of the problem, and to be honest in one's dealings with the subjects of the research. This having been said, it is curious that so many IRBs in the United States today lack ethicists on their boards!

——— HEALTH CARE PROFESSIONALS ——— AS CITIZENS

Men and women must fulfill various roles in life. Each of these roles has constraints, duties, and obligations peculiar to it. Nevertheless, and fundamentally, all people are members of a community—"citizens" in the sense of being members bound by social contract with one another and, therefore, sharing in a different but more universal set of constraints, duties, and obligations. Beyond this, communities themselves can be seen as corporate individuals united in a larger world community: the members of these diverse communities share their common humanity, with all that this entails, with each other as well as sharing a necessary interest in the peace, prosperity, and stability of their planet.

A fundamental set of constraints, duties, and obligations common to all persons as sharing in common humanity underwrites the constraints, duties, and obligations peculiar to a citizen's specific role. While clashes between these various interlocking roles are inevitable (the duty as a citizen to report criminals conflicts with the physician's obligation for confidentiality to the patient who is a housebreaker, for example), it is nevertheless essential to remember that role constraints, duties, and obligations, developed in a communal setting, are sustained by communal values and are therefore informed by communal strictures and expectations. In the last analysis, even the obligation to confidentiality, which prompts the physician not to report a patient who is a housebreaker to the police, is the product of communal values, strictures, and expectations. It is an obligation because the larger community promotes and sanctions such an obligation.

Roles are, in part at least, defined by these constraints, duties, and obligations as well as by the rights they entail. Although professionals and many other occupational groups to a large degree define themselves and their roles, the expectations a given society has of the members of such professional and occupational groups within it are the necessary setting for such self-definition. Roles and the constraints, duties, obligations, and rights they entail are ever-changing and dynamic constructs whose conception and definition at any particular moment in time and within any particular society reflect an interplay between their own tradition, their view of themselves, and the expectations societies have.

Expectations do not necessarily (or at times at all) determine what is and what is not moral.[46] We may expect an acquaintance to lend us money, for example, but his not doing so, although perhaps irksome and even unkind, is not immoral. No moral duties are entailed by one-sided expectations. But in an evolutionary sense, in the sense in which roles in society emerged and were affirmed, expectations for one another are important in determining the morality of an act. When such expectations are in fact fulfilled—when, let us say, firemen are expected to enter burning buildings and do so, or physicians are expected to take risks of infection and take such risks—expectations are confirmed by practice, underwritten by values, and, at times, affirmed by legal (or at least social) strictures and sanctions. A functional precedent is set. Communal expectations, legitimized by consistent performance, thus form part of the matrix of considerations that determines the morality of an action.

Our viewpoint of community determines our viewpoint toward obligations (see also Chapter 3). Are communities to be viewed as collections of individuals held together merely by duties of refraining from harm to one another? If, in such communities, freedom is the absolute condition of morality, and not a value to be adjusted mindful of other values, communal obligations will be limited to securing absolute liberty (short of harming each other) for all. The obligations of physicians would then be purely those stipulated by freely entered upon contract.[47] If, on the other hand, we consider refraining from doing harm to each other to be the necessary but insufficient condition of community, and if, furthermore, we concede to freedom the standing of "value" to be cautiously traded and bartered for other goods, then the duties and obligations of community as well as those of its component institutions (including medicine) will emerge in a different light.[48]

It is easy to demand that "physicians meet their social responsibilities" in the practice of medicine. No one will seriously doubt that if physicians are to discharge their obligations adequately, more than merely strictly technical or "medical" function is entailed. The World Health Organization, in its statement on health, defines health as "a state of complete physical, mental and social well being and not merely the absence of disease and infirmity."[49] This definition, while holding out an unattainable ideal goal, nevertheless serves as a point of reference. If, in the context of this definition, physicians as well as other health care professionals are obliged to care for the health of their patients, social responsibilities cannot be evaded. On the other hand, if we subscribe to the narrower definition of health as being merely the absence of disease, then the obligation of health care professionals, too, is narrower. But even if we merely accept this narrower definition, we may argue that at the very least the unavoidable public health, occupational, and social aspects of many diseases will involve the physician in social concerns.

The recognition that diseases have partly social causes is not new. The manner of life conducive to health that Hippocrates wrote about includes rules for self-care and diet affordable only by the wealthy leisure class.[50] Detailed instructions for those of other classes tacitly make the point that, even without mentioning the insights into public health, physicians in ancient times were well aware of the social implications of medicine. The descriptions of the different diseases afflicting various social classes—gladiators, slaves working in mines, sailors, etc.—make the same point.

That different occupational groups suffered from different diseases and that, therefore and at least to that extent, disease is a social construct, was systematized by Ramazzini in the 17th century.[51] In the last century, social activists in medicine pointed out the intimate association between health and social conditions and, hence, the physician's necessary function as social architect. Virchow was not alone in his sentiment that the physician should be the "natural attorney for the poor."[52] Among Central European physicians, socialism, stemming from social concern for patients, was not rare. To become, as Lowinger asks physicians to become, "healers of social as well as individual pathology"[53] is a fine sentiment, but it is also a tall order. Nevertheless, physicians as well as other health care professionals, if they are truly to discharge their obligations must, at least to some degree, involve themselves with social issues. This is an analytic statement if one accepts that health care professionals must at the very least be concerned with their patient's health and accepts that social factors invariably at least modulate and, at times, directly cause disease.

Organizations such as the AMA or ANA have taken a laudable role in opposing smoking, working for helmet laws and against permitting public boxing, and have even, of late, shown an interest in helping to formulate a more equitable health care system. But such professional organizations have largely failed to speak out against hunger, poverty, poor education, or ghetto-ization and all that these entail. That is regrettable, for it is beyond dispute that there is an internationally valid connection between income level and incidence of (almost every) disease. Likewise, poor education, poverty, hunger, racism, and ghetto-ization are intimately connected with poverty as well as, independently, with the incidence of poor health. If health care professionals are truly serious about their responsibility not only to treat disease once it is established but also to decrease, as far as that is possible, its incidence, their responsibility to address social issues (and to try to see to it that their organizations address such issues) seems clear. But when one chastises organizations like the AMA or ANA one should, first of all, look homeward. The American Society for Bioethics and Humanities (formed a few years ago by joining together of the Society for Health and Human Values, the Society for Bioethics Consultation and the American Association of Bioethics as well as the Association for Practical and Professional Ethics (all of them dedicated to ethics within health-care) have (and in our viewpoint to their utter shame) consistently up to the time this is being written refused to take a stand—not only on poverty but even on access to health care. At least other medical societies have generally taken a stand on these issues!

Physicians and other health care professionals sub-serve roles other than primary patient care. After all, they teach, work as public health officials, play their part in industry, and work for insurance companies and in a host of other settings each of which entails different obligations to different persons or to the corporation itself. The relationship between a single health care professional and his or her particular patient has always existed in a context of society, family, and concerned others. But today this relationship has become more complex. As central as the physician's obligation to his or her patient may be, it cannot any longer (if it ever was) be the sole criterion of action. Nevertheless, the responsibility of most physicians and of many other health care professionals, directly or indirectly, is involved with the cure of disease. Thus, physicians and their colleagues in the other health care professions have the primary obligation to show "due care and personal concern for their patients."[54] In the view of Jonsen and Jameton, other concerns are not primary and are not to be met at the expense of direct patient–physician obligations. However, such a view would, one would think, depend upon the specific role that a physician has in society.

One can, as Childress points out, start with the social and political responsibilities of all citizens one to another and to the community and derive activities of physicians expressive of these or one can start in the opposite direction and examine those special roles of physicians that give rise to communal or social responsibilities.[55] In the former view, the obligations as citizen are primary and are modified by the special expertise, experience, and role duties of physicians; in the latter view, obligations to specific patients are central, and communal obligations are a spin-off.

If one starts with the social presumption and derives the physician's duty from those of the citizen (specialized and, at times, modified by technical expertise in the field of medicine), one will conclude that physicians are, *inter alia*, obligated to strive for justice in health care. If, on the other hand, one starts with the physician's charge of maintaining the personal health of the patient, physicians, because of their special knowledge and expertise, will be obliged to attend to public health matters within their purview. Except as citizens, however, they would at first blush seem to have no special obligation to strive for justice in health care. But even here, if one (1) accepts that all members of a just community have the obligation to work for and maintain just institutions; (2) affirms that health care professionals are part of the greater community; (3) accepts that to varying degrees all health care professionals are technically expert in matters dealing with health and disease; and (4) acknowledges that the kind of health care institutions and the availability of health care ultimately affects all citizens, one will perforce conclude that health care professionals have an obligation to work for justice in the availability of medical care for all patients. Whichever direction the argument takes, whether we start with the duty of physicians as well as of other health care professionals to patients or with the duties of health care professionals as citizens, physicians as well as other health care professionals, to varying degrees and with varying force, have obligations to be concerned with the social parameters of disease and with social justice.

Furthermore, responsibility in the contemporary world—with its awesome power of technology to be used for good or evil—has changed. We cannot evade the responsibility that comes with this change—the responsibility to use technology wisely, not only for the sake of our patients but also for the sake of the future. In a sense, we need to be able to foretell the future, to re-enunciate norms and standards as substitutes for the norms and standards left behind by technology. If we fail to do this, the future is bleak. And yet, as Hans Jonas pointed out, while the rapidity of change in today's world has brought about a state of affairs in which our capacity to predict the future and foretell the consequences of our courses of action is increased in one sense, it has significantly decreased in another. The better our technological capacity, the more easily should we be able to predict; but the better and more complicated our technological capacity and, consequently, the more rapid the changes such capacities bring about, the less can we truly predict the consequences of a particular action, let alone the consequences of a large number of actions. Only great care and forethought can preserve our future.[56] The obligations of health care professionals to their community, depending upon the way we derive them, entail at least a few obvious duties: duties to discharge their professional obligations with competence and fidelity, duties to serve as advisers in health matters, duties to participate in disaster and other public service, to name but a few. Like all such duties of positive action, discretion will have to guide individual performance under specific circumstances, but medicine and the allied professions, as organized groups, are obliged to see that their roles are properly fulfilled. Organized medicine and the other organizations that speak for health care professionals have the broader responsibility of seeing that communal obligations are met and to see to it, if need be by sanctions, that individual practitioners fulfill their individual obligations.

Health care professionals, if they are to concern themselves with matters that threaten the health of their patients rather than only with alleviating established disease, are obliged to concern themselves, at least in the context of their particular practice, with issues of prevention. In a wider sense, however, they are obligated not only to speak out for such things as sanitary conditions, clean water, safe food, rational immunization programs, and smoking policies or seat belt laws but likewise to concern themselves with far wider issues. This obligation emerges from the physician's citizenship obligation refined through the peculiar technical expertise and knowledge that physicians, as a result of their training, are expected to possess. Physicians are thus obligated to concern themselves with issues of hunger, inadequate housing, poverty, hopelessness, crime, and other social conditions that inevitably threaten health. Physicians therefore, inevitably and to a greater or lesser extent, cannot evade the obligation to be social architects (or, at least, advisers to social architects) and, in Virchow's words, "attorneys for the poor."[52]

Beyond such issues are issues of pollution, war, peace, and overpopulation. Few things in the modern world threaten the health of our patients as much as such issues. The ravages visited upon our environment threaten far more than merely the economic well being of parts of this Earth. They have been shown to be intimately associated with a large variety of diseases. Overpopulation not only threatens the food supply for all but also greatly aggravates the social conditions that produce poverty, ignorance, illiteracy, and crime. It has become a critical world problem: one that many of us feel is one of the fundamental (and generally overlooked or glossed over) ethical issues of the day. Artificial birth control—according to at least some—is morally problematic. While we most certainly do not share this view and even if one were to see artificial birth control as an evil, such an evil would clearly be outweighed by the picture of starving and abandoned children.

Reliable estimates show that only an immediate adherence by the entire world population to a basically vegetarian diet could manage to feed the Earth's population today and that if population growth continues at the present rate even such a course of action will not feed the entire world population in a few years. Famine, disease, and ultimately chaos and war must follow. None of us can allow this to happen. Health care professionals have the expertise to play an active role and, therefore, an obligation to participate in discussions and to take actions eventually aimed at ameliorating and finally eliminating the problem of overpopulation.

Beyond this, although less acutely today than a few years ago, the threat of war on today's terms is perhaps the ultimate threat to public health. The consequences of war, not to speak of nuclear war, are consequences that physicians cannot, if they are to meet their obligations, help but work against. Organizations such as the International Physicians for the Prevention of Nuclear War and Physicians for Social Responsibility have made an admirable beginning, and the AMA in its *Journal* has likewise spoken out against the insanity that impels nations to dance on the edge of this volcano. Health care professionals will have to rethink their obligations: when asked, as they were a few years ago, to help "prepare" for nuclear or other war, they are put before a difficult ethical choice. On the one hand, physicians and other health care professionals may feel impelled to help in such preparations in order to try to ameliorate (if that is possible) the effects of such a holocaust. On the other hand, by helping in such preparations, health care professionals not only give tacit approval but also help to lull the public into the belief that preparing for nuclear war is, in fact, a viable alternative.

The public health role of the physician and other health care professionals extends beyond the reporting of disease and compliance with public health laws. Physicians, since ancient times, have been obligated to be teachers (the term "doctor," after all, is derived from teacher). As such, their obligation extends beyond the technical application of specific treatment to specific disease. Physicians must teach other health professionals (and, in turn, be receptive to their teaching), and all health care professionals must teach their patients how to live healthy lives. To the extent that health is threatened by social conditions, health care professionals are obliged to speak out. Physicians, while primarily obliged by their occupation to deal with immediate matters of health and disease, cannot in good faith ignore dangerous social conditions.

When values, obligations, and loyalties conflict, the specifics are all-important, and no ready, pat solutions are at hand. Safeguarding and caring for patients has a *prima facie* claim on the physician's actions and choices. But safeguarding and caring for patients takes place in a context that, furthermore, has played a dominant role in fashioning our conception of what it is to safeguard and care. It can therefore not be ignored. Important as principles and rules may be, specific problems will demand specific choices and actions made by thoughtful, responsible, and compassionate people, not the blind application of predetermined rules.

——— PHYSICIANS AND OTHER ——— HEALTH CARE PROFESSIONALS

Making patients well (or at least making them better) is generally not a task that merely involves a single patient with a single physician. Other health care professionals are almost inevitably involved, and the family, furthermore, plays a critical role. Even when a patient visits a physician's office for a relatively minor reason, the physician's office staff and more than likely some laboratory workers or pharmacists will explicitly or tacitly be involved. Relationships among these actors are critical if ethical medicine is to be practiced.

Ambiguous relationships not only lead to conflicts of obligation but may also lead to complete fragmentation of loyalties. Physicians and other health care professionals share in such dilemmas and, in their everyday practice, must strive to come to terms with them. Dilemmas occur not only when health care professionals must deal with problems involving themselves and particular patients. Internal dilemmas both among members of the same professional group (among physicians, for example) and among such professional groups (between physicians and nurses, for example) are frequent and detrimental to the mission of the professional. Consultants who find their recommendations simply ignored while the now shared patient worsens or the nurse who finds himself or herself torn among duties to the patient, to the physician, to colleagues, or to the institution are but two examples. Let us remember, as was mentioned earlier, that many of these conflicts are mirrored in the internal conflict moral agents have within themselves when they must adjudicate between the demands and interests of often very different beliefs and roles they themselves incorporate.

If serving patients' best interests as well as serving the communities' best interests when it comes to health care decisions is part of all health professionals' duties, then maintaining a collegial relationship among all who are concerned with health care is an ethical obligation. Patient care today requires teamwork; it can no longer be done in isolation. Unless members of a team respect each other's diverse and complementary expertise, listen to one another, show solidarity, and work smoothly together, their common mission suffers. Physicians, nurses, EMTs, social workers, and many others who form this team each have a somewhat different and yet overlapping role and different and yet overlapping expertise. They can and should learn from each other. When ethical problems in patient care are complex, the different worldviews, roles, skills, and expertise of each of these members of the team may greatly contribute to finding a tolerable solution.

In the hospital or nursing home a team approach is increasingly important. Different members of the team have different strengths and weaknesses, have undergone different forms of training, and are capable of helping each other deal with the problem at hand. They are truly colleagues whose smooth interaction is critical to the success of their combined mission. Games of rivalry or one-upmanship are ill conceived and ultimately detrimental. While ultimately someone must have the power to make a specific decision and while that person is ultimately responsible, deciding what to do is often a team effort in which a fruitful dialogue can result in a consensus. Since all feel that they have contributed, all will feel more eager to cooperate and work with each other toward the commonly decided goal by the agreed-upon means.

MANAGED CARE

A variation of this marketplace model can be seen in the proliferation of managed care. It is obvious that all care is and always has been "managed." From Hammurabi's time (with perhaps the exception of Greece) some sort of licensure and control over the practice of medicine was maintained. Medical school, residency, licensing, requirements for continued medical education or re-certification all are means of managing care. What is new in "managing" is not the emphasis to ensure the quality of the medical care offered, but the emphasis to hold down costs and maximize profits. Holding down costs in and of itself is undeniably important; much rationalization can and need be done. But holding down costs and maximizing profit are two quite different matters.

Managed care as it exists today can be "for profit" or "non-profit." To what extent "not for profit" managed care is, in fact, "non-profit" in the competitive marketplace in which it exists today is another matter. Theoretically a universalized "not for profit" system could easily form the backbone of a national health-care system (see chapter on macro-distribution). As it exists today it is a competitive system in which the professional and, therefore, the ethical elbow room of health-care professionals has been sharply curtailed, in which physicians and other health-care professionals spend countless hours hassling and being hassled by clerks and other non-professional employees so as to obtain what they consider essential for their patients' diagnoses, safety, and comfort. Often this results in either undue delay or in patients not receiving essential services either because they are denied or because patients (often unfamiliar with the system and discouraged by interminable delay before they can speak to a "real person") simply give up. This, known as the "hassle factor" is said to be a well-calculated strategy by organizations out to maximize the profit not of health-care providers or healthcare institutions but of stockholders and highly paid executive officers.

Health-care professionals feel "locked in" by these institutional demands. Often they are torn between their ancient obligation of doing the best for their patients and their perceived obligation to their direct employers. The patient is increasingly seen as the "customer" of a given health-care organization and the health-care professional as its employee or representative. Trust, so critical to the patient–physician relationship, tends—to the extreme regret of both—to be attenuated and finally lost. Indeed, even when as is often the case physicians go all out of their way to secure proper treatment—the very suspicion by the patient disrupts the relationship. Further, time wise, it is impossible for each physician four or five times a day to hassle with persons who often do not even understand the basics of disease but, rather, are obliged to rely solely on a predetermined checklist of approved or disapproved items. Likewise, job satisfaction, so very important to the health-care professional and so critical to his or her proper function tends to decrease and be lost—early burnout easily results.

Beyond this a system of ever-increasing co-payments has made a sad joke of the concept of being "fully insured." A chronic and severe illness these days can easily consume between one and two hundred dollars a month, an amount a well-paid person can but a lesser-paid person (who may have to chose between shoes for the children and medical care for themselves) cannot afford. As will be mentioned in the chapter on macro-distribution, Medicare has carried on the fiction of "adequate" health-care coverage for the elderly for many years.

Under current laws (laws which are apt to change and which have an interpretation of ERISA at their basis) managed care organizations are not legally liable for malpractice: hospitals are. This, of course, is to say the least disingenuous. When organizations permit or forbid diagnostic modalities, therapeutic procedures or drugs (and even try to direct what dosage can be described) they can hardly be said to be doing other than practicing medicine! Any unlicensed person trying to make such decision would most likely be charged with practicing medicine without a license.

If healthcare professionals are to practice truly ethical medicine they must have the elbowroom to do so and, therefore, must take an active part in securing that elbowroom. In this they have traditionally failed. This elbowroom is institutional but ultimately societal—it is difficult, if not indeed nearly impossible, to practice ethical medicine in the context of an institution not predicated on ethical principles; likewise, it is nigh onto impossible to establish a just institution in a society that pays only lip service to justice. Therefore, the obligation of the individual health-care professional extends beyond the care of the individual patient. By their efforts and through their organizations healthcare professionals arguably have an obligation to establish an atmosphere in which ethical practice is possible.

There are, of course, those who defend today's version of managed care: in general they do not include health-care professionals or patients but those who derive profit. Managed care has done nothing to ease the main problem of the uninsured (indeed it has made it worse—there are now an ever increasing number of uninsured as well as many more under-insured) and it has not (even though it initially did) lowered costs. What it appears to have done is to have restricted access by those who need it, made the practice of ethical medicine infinitely more difficult, decreased quality, increased the problem faced by teaching hospitals and consequently interfered with good teaching. It has managed to increase profits for those who have invested their money in such institutions. Often these are insurance companies—it is well known that these control vast amounts of money and contribute generously to both of the major political parties. It is not, at least in its present competitive and for-profit form, a system that recommends itself or which, in our view, can last.

CAPITATION

In general parlance capitation refers either to a health professional's being directly salaried or being paid a contractual amount by an employer. It may be by giving a stipulated amount "per head" accepted as "patient"—that is, each patient, no matter how frequently or infrequently seen, is paid for by a lump sum. Until recently such has been the case for many health care professionals as well as for physicians employed by government or some charitable organizations, but for physicians in practice in the United States capitation has only recently become an increasingly frequent method of reimbursement

Objections—of an ethical as well as practical nature—have been raised against this practice. Some of these objections are unsubstantiated claims: for example, that patients "will abuse the system" or that "physicians will lose their initiative to provide good service." Other objections have been of a more serious nature; for example, it has been shown that the incidence of coronary bypass surgery, cataract surgery and other procedures falls sharply when capitation is introduced. If that is the case—and unfortunately several studies have shown it to be the case—then the implications are indeed ominous: it indicates that either too many unnecessary procedures were done before (the motive, rather obviously, being greed) or that now too few necessary procedures are being performed (the motive here also being greed, but the greed of third party others who then deny any wrong-doing.

The objections raised that patients will abuse the system have not proven to be true. Even in a fee for service system (as well as managed care or other HMO arrangements) some patients will be persistent visitors—either because they are hopelessly neurotic, really do have symptoms which, however, defy current explanations or are simply lonely and find a visit to the physician an important social occasion. None of these constitute true abuse but may, in fact, require referral or social intervention. And some—in any system—are unavoidable. In the final analysis however, few of us equate a visit to the physician with a pleasant afternoon in the park!

That a physician's or other health care professional's initiative will diminish with capitation payments may, on a very few occasion, be true. Few people enter the health care professions unless they are, at least to a good part, motivated either by scientific curiosity, a strong desire to help others or by both. Unfortunately as time goes on, as unnecessary fatigue ensues, and as institutional strictures supervene, this motivation often diminishes or is lost. Be that as it may, intolerable circumstances of employment happen in any system—doing away with capitation is not the solution: making it fair, equitable and just and giving health care professionals more of a say may well attenuate, if not indeed, eliminate the problem.

After all is said and done, capitation is merely another instance of a conflict of interest in which physicians are, in our view, obliged and well served not only to attend to individual patients but to take some joint obligation for helping to create a fair system for all.

——————— HEALTH PROFESSIONALS ——————— AND STRIKES

Often it is suggested that it is ethically improper for health professionals to resort to strikes.[57] Such a claim has been made especially with regard to physicians. Physicians, it is said, hold the lives of their patients in their hands and, thus, to strike would be equivalent to abdicating their professional responsibility. And, in a time when physicians used to "set their own fees" and their responsibility was directly and solely to their patients, such a claim may have had considerably more standing. However, increasingly physicians are either employed or involved in contractual arrangements with HMOs, MCOs, etc. As a result, physicians are now commonly treated like workers in other industries. Increasingly, the new obligations these contractual arrangements entail may come into direct conflict with—and even threaten to re-define—that traditional fiduciary patient–physician relationship.

Thus, there is a rather severe ethical problem when it comes to strikes by physicians: on the one hand, physicians are still supposed to be committed to their patients' good; on the other hand, they are expected to behave like good employees. Unfortunately, in the latter, both the good of patients and physicians (not only ethical standards of practice, but actual working conditions) may be threatened by the demands of the organization. However, the ability to participate in setting one's own conditions of work is one of the things that distinguish free human beings from slaves.[58] Wage slavery may be as real as actual slavery insofar as the worker has little choice but to accept what is offered or starve. To give up the ability to strike—even if only as a last resort—makes one no more than the passive tool of another's interest.[57] Unfortunately, this conflict is readily and, we would argue, knowingly, exploited today by many managed care organizations.

Besides, there are workers as important to the public's health as are health care professionals. The result of a strike of garbage collectors or fire and police personnel would have far more devastating effects. Teachers striking—while the effect may not be as immediate—may, in the long turn, be equally disastrous. True, physicians are bound by an oath that workers in these other enterprises are not (this was pointed out to me by Dr. Faith Fitzgerald). But such an oath cannot be such that it sells persons into slavery!

Physicians, beyond this, are not predominantly engaged in life-saving activities. Much of what they do today is caring for chronic illness, immunizing patients and a variety of other things that can usually be safely delayed some days or even weeks. Other activities—and unfortunately increasingly more—are concerned with administrative and clerical duties: filling out forms, arguing with representatives of HMO's or MCO's, etc. A collective choice to simply care for patients but refusing to fill out forms, etc. may go a long was to bringing MCO's around—one cannot, after all, fire all physicians.

Striking may also be done for a large variety and, more often than not, intertwined reasons—some to benefit patients who suffer more and more under the rules and regulations MCO's and HMO's may see fit to institute. Some may be done because workloads have become so excessive as to preclude safe and good patient care—certainly something, which ultimately benefits patients. And some may indeed be done because the health-professionals pay and benefits have been reduced to a level felt to be unjust and not commensurate with their training or workload. There is no question that many physicians in the past were grossly overpaid and that, especially in some specialties, they still may be today. But this is becoming a lesser concern, especially in primary care fields, and, indeed, often the obverse is the case. Strikes are effective only if they bring inconvenience or discomfort to someone in a position to change matters. That may be the government, a health care institution, an HMO or MCO or, regrettably, the public and, ultimately, the patient who, in turn, will bring pressure on the employing institution.

Strikes can be incremental. The obvious rule not to abandon patients whose lives would be seriously jeopardized need not be argued for—it is too obvious. But strikes may start by stopping to do administrative duties thus causing considerable chaos; they can be continued by stopping to do non-essential other duties thus causing some discomfort but no immediate danger; and they can finally leave only a skeleton crew to take care of threatened lives. Ruling out strikes altogether seriously risks making of physicians and other health care workers mere slaves—and, even more, plays into the hands of worsening the conditions that caused the problem in the first place.

————— HEALTH PROFESSIONALS ————— AND THEIR LIVES

All health professionals besides being professionals have personal lives and individual interests. They are, no less than bakers or candlestick makers, human beings. Especially when it comes to dealing with such deeply human issues as birth, death, and illness, those involved must not allow their humanity to atrophy. Few things can be more destructive to humanity and humane function than are overtiredness, lack of sleep, or the continuous and virtually obsessive occupation with only one topic. To be a good doctor (or nurse, social worker, psychologist, or technician) requires far more than merely technical expertise in one's field. Technical expertise is the necessary but far from sufficient condition of ethical medical function. When physicians allow themselves to work 12 or 14 hours a day virtually 7 days a week and in their "spare time" peruse only medical journals, their humanity atrophies and they become automatons of health care—and bad ones at that! When physicians are overtired, harassed, and sleep-deprived, the patient becomes the "enemy:" the person who keeps them from rest, sleep, or relaxation and who, like any other slave driver, controls their every moment. If health care professionals are to be truly human (and if they are to be the good health care professionals they must be), they must arrange their schedule to allow time for their personal lives and interests. They have obligations (and can have great joy) from their family, their friends, their cats, and their dogs; they have personal needs. Reading good books, listening to good music, and walking in the woods ultimately make a person a better health care professional and subtly but surely enhance ethical function. None of these activities—since they are essential to the proper function of health care professionals—should be looked upon as "non-medical." In the beginning of this century one of the great social reformers (an anatomist at the University of Vienna) made the comment that only a good person could be a good doctor. This is a tall order. If physicians are, above all, expected to keep on the cutting edge of their field and are expected to have enough leisure to have a fulfilling family and an intellectual life of their own they above all need time. We end where we start: any medical system that we may construct needs to consider such factors and allow for them.

—————————— **SUMMARY** ——————————

The practice of medicine is a social task in which patient and healer must respect each other's personal morality and moral agency. The vastly greater power (real or perceived) of the health-care provider and specifically of the physician puts the burden of this fiduciary relationship largely (but not solely) on the shoulders of the health-care provider. While health-care providers cannot—and act ethically—impose their own personal morality on the patient neither can the patient ask physicians to violate their own personal morality. Physicians and other health-care providers cannot simply follow the dictates of their particular HMO, MCO or the rules promulgated by the government (and may, in fact, be faced with quite unpleasant choices) and blame "the system." They carry a heavy responsibility in trying to resist dictates deemed harmful to their patient. Above all they carry not simply the responsibility of accommodating themselves or resisting a system someone else builds for them but of playing their proper part in building a system which is equitable to all members of the community and flexible enough to change as do circumstances.

—————————— **REFERENCES** ——————————

1. Szasz TS, Hollender MH. The basic models of the doctor–patient relationship. *Arch Intern Med.* 1956;97:585–592.

2. Pellegrino ED. Toward a reconstruction of medical morality: the primacy of the act of profession and the fact of illness. *J Med Phil.* 1979;4(1):32–55.

3. Pellegrino ED, Thomasma DC. *A Philosophical Basis of Medical Practice.* New York, NY: Oxford University Press, 1981.

4. Amundsen DW. The physician's obligation to prolong life: a medical duty without classical roots. *Hastings Center Report.* 1978;8(4):23–31.

5. Lifton R. *The Nazi Doctors.* New York, NY: Basic Books, New York, 1986; and Jones JJ. *Bad Blood.* New York, NY: Free Press, 1981.

6. Loewy EH. Friendship and medicine. *Cambr Q Health Care Ethics.* 1994;3(1): 52–59.

7. Szasz TS, Hollender MH. The basic models of the doctor–patient relationship. *Arch Intern Med.* 1956;97:585–592.

8. Cassell E. *The Healer's Art: A New Approach to the Patient–Physician Relationship.* Philadelphia, Pa: J.B. Lippincott, 1976.

9. Cassell E. The function of medicine. *Hastings Center Report.* 1977;7(6):16–19.

10. Loewy EH. Clergy and physicians encounter medical ethics. *Humane Med.* 1987;3(1):48–51.

11. Engelhardt HT. The disease of masturbation: values and the concept of disease. *Bull Hist Med.* 1974;48:234–248.

12. World Health Organization. Constitution of the World Health Organization, 22 July 1946. *Public Health Rep.* 1946;61:1268–1271.

13. Dewey J. *Logic, the Theory of Inquiry.* New York, NY: Henry Holt, 1938.

14. Parsons T. Definitions of health and illness in light of American values and social structure. In: Caplan A, Engelhardt HT, McCartney JJ, eds. *Concepts of Health and Disease.* Reading, Mass: Addison-Wesley, 1981;57–82.

15. Glazer G. The "good" patient. *Nurs Health Care.* 1981;11(3):144–164.

16. Bhanumathi PP. Nurses' conception of the "sick-role" and "good-patient behaviour": a cross-cultural comparison. *Int Nurs Rev.* 1977;24(1):20–24.

17. Payer L. *Medicine and Culture.* New York, NY: Henry Holt, 1988.

18. Lain-Etralgo P, Partridge P, trans. *Doctor and Patient.* New York, NY: McGraw-Hill, 1969.

19. Morison RS. The biological limits of autonomy. *Hastings Center Report.* 1984; 14(5):43–49.

20. Ozar D. Patient's autonomy: three models of professional–lay relationships in medicine. *Theor Med.* 1984;5:61–68.

21. Pellegrino ED. Thomasma DC. *For the Patient's Good: The Restoration of Beneficence in Healthcare.* New York, NY: Oxford University Press, 1988.

22. Loewy RS. A critique of traditional relationship models. *Cambr Q Health Care Ethics.* 1994;3:27–37 and in *Integrity and Personhood: Looking at Patients from a Bio/Psycho/Social Perspective.* Kluwer/Plenum Publishers, 2000.

23. Kottow MH. Medical confidentiality: an intransigent and absolute obligation. *J Med Ethics.* 1986;12:117–122.

24. Rawls J. *A Theory of Justice.* Cambridge, Mass: Belknap Press, 1971.

25. Siegler M. Confidentiality in medicine: a decrepit concept. *N Engl J Med.* 1982;307(24):1518–1521.

26. Thompson IE. The nature of confidentiality. *J Med Ethics*. 1982;8:12–18.
27. Jones JJ. *Bad Blood*. New York, NY: Free Press, 1981.
28. Jonsen AR, Sagan L. Torture and the ethics of medicine, *Man Med*. 1978;3:33–49.
29. World Medical Association. Declaration of Tokyo. *Bull Am Coll Phys*. 1976; 17(6):15.
30. United Nations General Assembly. Resolution Adopted by the General Assembly. *Principles of Medical Ethics*. Geneva: CIOMS, 1983.
31. Loewy EH. Communities, self-causation and the natural lottery. *Soc Sci Med*. 1988;26:1133–1139.
32. Loewy EH. Duties, fears and physicians. *Soc Sci Med*. 1986;22(12):1363–1366.
33. Shelp EE. Courage: a neglected virtue in the patient–physician relationship. *Soc Sci Med*. 1984;18(4):351–360.
34. Ramsey P. *The Patient as Person*. New Haven, Conn: Yale University Press, 1979.
35. Drane JF. The many faces of competency. *Hastings Center Report*. 1985;15(2):7–21
36. Drane JF. Competency to give informed consent. *JAMA*. 1984;252:925–927.
37. Lidz CW, Appelbaum PS, Meisel A. Two models of implementing the idea of informed consent. *Arch Intern Med*. 1988;148:1385–1389.
38. Cassell E. The function of medicine. *Hastings Center Report*. 1977;7(6):16–19.
39. Winau R. *Das Krankenhaus Moabit*. Berlin: Hentrich Verlag, 1984.
40. Alexander L. Medical science under a dictatorship. *N Engl J Med*. 1949;241:39–47.
41. Mitscherlisch A. *Medizin ohne Menschlickeit: Dokumente des Nürnberger Ärzteprozesses*. Frankfurt: Fischer Taschenbuch, 1978.
42. Thomasma DC. Applying general medical knowledge to individuals: a philosophical analysis. *Theor Med*. 1989;9:187–2000.
43. National Commission for the Protection of Human Subjects of Biomedical and Behavioral Research. *Report and Recommendations: Research Involving Children*. Washington, DC: U.S. Department of Health, Education and Welfare, 1977.
44. National Commission for the Protection of Human Subjects of Biomedical and Behavioral Research. *Report and Recommendations: Research Involving Prisoners*. Washington, DC: U.S. Department of Health, Education and Welfare, 1976.
45. National Commission for the Protection of Human Subjects of Biomedical and Behavioral Research. *Report and Recommendations: Research Involving Those Institutionalized as Mentally Infirm*. Washington, DC: U.S. Department of Health, Education and Welfare, 1978.
46. Murray TH. Divided loyalties for physicians: social context and moral problems. *Soc Sci Med*. 1986;23(8):827–832.
47. Engelhardt HT. *The Foundations of Bioethics*. New York, NY: Oxford University Press, 1986.
48. Loewy EH. *Freedom and Community: The Ethics of Interdependence*. Albany, NY: State University of New York Press, 1993.
49. World Health Organization. Constitution of the World Health Organization, 22 July 1946. *JAMA*. 1946;131(17):1431–1434.

50. Edelstein L. Antike Diätetik. *Antike*. 1931;7:255–270.
51. Garrison FH. Life as an occupational disease. *Bull NY Acad Med*. 1933;10:679–693.
52. Terris M. Concepts of social medicine. *Soc Serv Rev*. 1957;31:164–178.
53. Lowinger P. The doctor as political activist. *Am J Psychother*. 1968;22:616–625.
54. Jonsen AR, Jameton AL. Social and political responsibilities of physicians. *J Med Phil*. 1977;2(4):376–400.
55. Childress JF. Citizen and physician: harmonious or conflicting responsibilities. *J Med Phil*. 1977;2(4):401–409.
56. Jonas H. Technology and responsibility. In: Jonas H, ed. *Philosophical Essays*. Englewood Cliffs, NJ: Prentice Hall; 1974;3–20.
57. Braithwaite SS. Collective Action by physicians that do not endanger patiemts. *Cambridge Q. Health Care Ethics 2000* 9(4):460–469
58. Loewy H. Of Health Care Professionals, Ethics and Strikes. *Cambridge Q. Health Care Ethics 2000* 9(4):513–520.

7

Genetics and Ethics

Good ethics starts with good facts—so let us first briefly review the following "facts" which are accepted today but may very well prove to be no longer quite so true tomorrow.[1] Many of our fears of dealing with genetics are fears grounded in a substantial lack of facts—our understanding of genetics and the role played by such facts in shaping individuals is, at best, in its infancy. Our fears as well as our hopes are, therefore, largely speculative and can assume an almost science fiction type of complexion. This does not mean that the problems of genetics and genetic changes should be ignored—on the contrary: it means that with our profound lack of knowledge, understanding and, therefore, capacity to make meaningful predictions, the danger of precipitous, perhaps irreversible and possibly disastrous action escalates and our caution should, likewise, greatly increase. Above all it means that we must, we think, lengthen the time between "knowing something" (whether right or wrong) and using that knowledge to develop a technology that is then often immediately and widely applied. The space between each of these steps has to be widened sufficiently so that enough time for serious deliberation (and perhaps further experimentation) is available. Above all we need time to think—and unfortunately we are a society that has little value for thinking and an inordinate value for doing. We as a human species, and especially we as a western society, are not a patient people—things must be done at once, reaction must be immediate. This—socially, politically and scientifically—has not rarely led to most undesirable results. In genetics these results could not only be unfortunate but irreversible and disastrous.

All organisms transmit "themselves" to their offspring, in part by means of genetic materials, in part by the way the social setting into which they are born or in which they are raised shapes them. We must be quite clear at the outset: We are not saying that this genetic material in and of itself determines what we will be. Nor do we separate the "intellectual" from the "physical." What we are saying is that our

141

genetic heritage determines the limits of our possibilities, which social factors then serve to flesh out. We can nurture our capabilities and maximize them, but we cannot exceed them (at least not yet) In understanding and learning to deal with genetic transmission, we may also learn how we can "push the envelope" and, given a specific genotype, alter it.

The way we transmit our biological selves to our offspring is (at least in part and at least as far as our knowledge permits us to say today) by "chromosomes," the number of which are specific to a given species. Homo sapiens happens to have 46—two that determine sex and 44 that determine other characteristics (some, but few, characteristics are "sex-linked"—that is, carried on one of the sex chromosomes). As far as we know the "chromosomes" are the carriers of heredity, albeit that extra-chromosomal factors are increasingly thought to play a large and as yet poorly understood part in fashioning who and what we actually are. Each "somatic" (or body) cell carries a full complement of two chromosomes ("diploid") and each egg or sperm carries half that number ("haploid"). At fertilization the chromosomes from each of the haploid cells meet, interact and, if circumstances are right, the development of a new organism ensues. In the process parts of one gene may move to another gene and that in turn to the donor gene. This process of crossover is one of the factors that serves to develop new characteristics. Needless to say, the change may be negative as well as positive and often will, when strongly negative, eliminate itself through natural selection

Chromosomes, in turn, are composed of genes that are made up by DNA and are located in the nucleus of the cell. Changes in the DNA structure (by mutation or otherwise) may range from a profound effect to little or no effect on the ultimate pheno- or genotype. The nature of this DNA and the way that it is transcribed (that is, translated into functional units) with the help of substances outside the cellular nucleus determine our genetic make-up. Besides the identified genes which are known to play a role, there is a vast amount of what has been called "junk DNA" which at this stage of our knowledge serves no known function but which may well prove to have a far more important role than we assign to it today. The language here is interesting: we do not know what it is and, therefore, label it as "junk" when, indeed, this "junk" may be playing a crucial role. Despite having unraveled the human "genome" we have just begun to scratch the surface of our knowledge in this field—and it is a very small scratch indeed. The involvement of the mitochondria (so crucial in converting food stuff to energy) in genetic transmission is but incompletely and poorly understood and may play a vital role in the process of forming a new organism. *We know too little—and perhaps we know what we know too soon.* One of the facts—and good ethics must begin with good facts—which we do know is the fact that we know very little indeed, that what we think we "know" may easily prove to be wrong and that we have barely begun to scratch the surface. We do not like to admit this, but as a species we are young, immature, profoundly ignorant and, unbelievably willful…a dangerous combination.

Each chromosome has a central portion (the "centromere") that carries the genes and terminal portions called telomeres. As cells divide, telomeres shorten until at a certain length cells no longer divide and death ensues. There are some cells that are "immortal"—so-called because in them such shortening does not occur. This is true of germ cells, cancer cells and the cells of some lower organisms like amoebae. An enzyme called telomerase prevents shortening. It has been possible to use this enzyme to lengthen previously shortened telomeres.[2] The possibilities and the biological as well as social and ethical problems associated with this are obvious.

Our genetic make-up determines what is called our "genotype"—the way this is expressed (the way we as individuals actually are) is referred to as "phenotype." Genotype determines the range of the possible in which environmental and other factors then determine what we are actually like. For example, we may have a gene for tallness but whether we actually reach our potential height will depend upon environmental—i.e., nutritional, environmental and social—factors. The same can be said for intelligence or any other trait. Genetics does not determine what or how we will be, but it does set the range of the possible that our environmental influences within these limits can nurture, neglect or suppress. The argument of "nurture" vs. "nature" then is probably vapid since both are needed to produce an individual and his or her talents and character. Social circumstances can maximize or minimize the phenotypic expression or our genetic predisposition. Changing the genetic make-up within a somatic cell so as to change an organism's phenotype is a far different thing from changing the genetic make-up of a reproductive cell and thus changing genotype: changing phenotype affects a particular individual; changing genotype affects subsequent generations. Furthermore changing genotype may change the way we reason about and therefore understand and are willing to deal with change.

Once the fertilized egg develops it undergoes profound changes. In the very beginning cell division produces a clump of undifferentiated cells (a blastocoel) called "omnipotent stem cells"—these, depending upon external forces acting upon them, can develop into any of the three basic tissues: ectoderm, endoderm and mesoderm. Omnipotent stem cells can develop into any tissue (including malignant tissue), organ or whole organism. Which occurs depends upon substances (most of which are ill defined and poorly understood) and circumstances (equally ill defined) that it encounters or may encounter. At our current stage of knowledge (or vastly better said ignorance) we are able to program some of these cells so as to become specific tissues. Such tissues have the great advantage of not being subject to rejection mechanisms as they would be in organ transplants from whole organisms. Likewise they may have disadvantages that we had never expected or considered—tumour formation, for example.

Omnipotent stem cells in turn produce pluripotent stem cells that have a narrower set of options—they cannot develop into whole organisms and can develop only into a narrower range of tissues. In turn such tissues contain multipotent and even more differentiated stem cells. Their range of being developed into tissues is even narrower.

It seems possible in the laboratory to "reprogram" pluripotent cells "back" to being omnipotent. But this solves few of the ethical questions and possibly raises some serious biological problems. Once such cells are again omnipotent they again have the potential for being whole human (or other) beings. We have, in other words, gone back to square one.

We have had the ability to determine sex (by ultrasound as well as amniocentesis) and, therefore, the potential (and in some cultures the actualized) ability to abort embryos of the "wrong" sex. The ethical questions this raises are profound and the cultural and social implications enormous. We in the West would condemn the practice of aborting the sex we do not want—such as happens in some developing countries in whom females are not desired. Ethically there is no question that this practice constitutes a *prima facie* example of gross sexism—but before condemning it "out of hand," one should be careful in examining the reason for such practice.

We have had the ability for some time to predict certain disabilities by sampling either amniotic fluid or some of the chorionic tissue itself. Likewise (and this is not directly related to genetic prediction) we have been able by ultrasound examination to foretell certain anomalies. This ability of course has brought with it the ethical question of what should, if anything, be done about such states. One could proceed with the pregnancy, terminate it or in some instances perform intrauterine operative procedures to try and correct it surgically. All of these options raise troubling ethical questions and have done so for some time. (See later in this chapter as well as the chapter on beginning of life issues.)

Our ability to predict an increased likelihood of developing certain diseases imposes some profound ethical questions and raises some very practical problems for patients, genetic counselors, physicians and ultimately the community. We are now able to define certain individuals with a greater propensity towards developing certain diseases—some amenable to prevention by proper measures beforehand, some amenable to cure and some, at least today, beyond prophylaxis, cure or even amelioration. All of this raises profound and different ethical questions that concern not only genetic counselors but also physicians in their respective fields.

The last decade has seen an explosion not only in what we know about genetics but also in our ability to use that knowledge to predict the presence or absence of certain "diseases" (or our potential towards developing them) in adults as well as embryos and to manipulate genetic structure. We have been able to clone beings—essentially to produce an animal (or plant) genetically "exactly" as the one from which a cell nucleus was taken. This is by no means absolutely true: the environment in which the being develops (be it uterine or artificial) likewise plays an important role. Not all animals cloned from a single cloning will even look the same: even their position in the uterus seems to affect things like the presence or absence of certain changes in their skin or fur. And although the animal may be "just like the one from which it was cloned" it will not be the same animal: its milieu, its life experiences and its social setting may be subtly or radically different. All of these factors are jointly responsible for creating a new individual many of whose characteristics may be similar but which is in no way "the same" as the cell from which it was cloned. The idea of producing another Beethoven or Hitler (or of extending "one's own life" eternally) is a gross misunderstanding of how individual beings are produced, develop and eventually turn out to be self-aware beings who not only are alive but who also develop and have a life.

Furthermore we have for some time been able to take a given gene and introduce it into an organism which does not have such a gene naturally. The introduction of a gene that causes bacteria (*E. coli* in this instance) to produce insulin is an example. But while few would argue against producing insulin this way, the potential ethical problem of producing new plant or animal species is formidable.

The use of stem cells (omnipotent, multipotent and pluripotent) has opened great opportunities and, not unexpectedly, raised profound ethical questions. Omnipotent stem cells, using proper environmental conditions, can be directed towards becoming a specific tissue or organ or if in utero may develop into a whole new organism. The use of placental blood to extract haematological stem cells has been proven to be of great medical value and has been commercially severely and (in our view) shamefully exploited. The use of omnipotent stem cells derived from blastocoels or from aborted early fetal cells has been a subject of intense ethical scrutiny and often of acrimonious debate. And it has likewise raised commercial interest and has begun to be exploited.

The term "cloning" has been used loosely and to indicate many things. To most in the scientific community the term denotes asexually reproducing an identical copy of an original. This does not indicate an identical copy of an individual being but may just as well mean producing identical tissue of any kind. To discuss the ethical issues we must be sure to be very precise as to what it is we are talking about.

When all is said and done what we know today is rudimentary. Our knowledge and understanding of the tenuous "facts" we have discovered has been prodigious but above all it has pointed at two things: first, at the tremendous amount of information that we do not know and which is crucial to even begin to speculate about many of these issues; secondly that what we are dealing with has destructive potential beyond our most dismal understanding. We are like three-year olds let lose to play in a chemists shop or in a gun depository. If good ethics begins with good facts—and we would doubt that many would dispute this—then we simply lack sufficient facts to make judgments concerning many of the issues which genetics presents.

—— THE IMPORTANCE OF LANGUAGE ——

The language we use in many respects determines how we think about a question and what we will ultimately do. Unless we have clear definitions, decisions or discussions about anything become not only difficult but, indeed, impossible. A definition must not necessarily be "correct"—but for the sake of a particular discussion all must agree upon the meaning of a given word or phrase. We need not agree that the definition of a particular word is correct—only that we all understand what, for the sake of that discussion, we mean by a given word. If we do not share a clear definition, rational discussion becomes impossible.

In genetics, as elsewhere, the question of what is and what is not considered to be "normal," and what is and what is not to be considered "disease" are critical. Normalcy or abnormalcy demands a standard of reference that is socially determined. Virtually all characteristics (be they laboratory findings or physical characteristics) have a "range of normal" which, in general, follows a Gaussian (bell-shaped) curve. Average is not necessarily normal and something that is not average is decidedly not necessarily abnormal. Determining what is a normal blood value, for example, is done by taking a requisite number of supposedly healthy persons (which in itself requires a prior determination as to what shall count as "healthy" and how many subjects must be examined)), measuring the particular thing to be determined and then discarding the upper and lower 5% of values. Thus, if the values in the supposedly healthy persons selected randomly ranged from 55 to 105, "normal" would be considered anything falling within the range of 60 to 100. That, however, does not mean that persons falling outside that range are "diseased"—and it most assuredly does not mean that all persons falling within that range are "healthy."

What is considered disease, what is considered to be sin, what is held to be a crime and what is considered to be a matter of indifference are socially determined.[3] Such definitions are subject to being used for social and political purposes and may have far reaching consequences. The language of "disease" must therefore be applied with extreme caution (see also Chapter 6). When we too prematurely or without due consideration label something a "disease" dealing with it falls within the province of medicine—it tends to give to medicine power that is not becoming to it, and a responsibility which is not really medicine's to discharge. Physicians are not social engineers. However, because health-care professionals are also citizens, they are not only entitled but in our view are obligated to participate in the dialogue. They must inform the public and their decision makers, as citizens who also happen to be experts in health-care. They are obligated not only to supply "facts" about health and disease in general but likewise are obligated to advise. Health-care professionals thus legitimately play a critical role in the public dialogue.

The same sort of language barrier comes up with terms such as "disabled," "defective," etc. Such terms carry a baggage of meaning which conditions our attitudes and responses. To be labeled as "disabled" is, in truth, meaningless: persons who wear glasses or use a crutch are "disabled" as are people with Trisomy 13 or 18. But their disabilities are radically different and the ethical questions are (one would hope) likewise radically different. We are not willing to consider—as unfortunately some of the more radical "disabled" organizations would have us do—throwing all such disabilities into one pot. Nor are we willing, from the outset, to consider "disabilities" as anything that is less than ideal function. The peculiar argument advanced by some of the more radical organizations of the disabled, (that a "disability" is really an advantage) clashes with the very word ("dis"ability) itself; the argument that members of a given disability group merely constitute a different "culture" we find completely unconvincing. That does not by any means mean that persons with disabilities should enjoy lesser rights (commensurate with their abilities) nor that the community is not obligated to do all it can to protect the disabled and to try to compensate for their deficiencies so that they can lead lives as full and complete as others. They did not ask but were born into our community, are, therefore, members of it and deserve the help necessary to give them, within the limits provided by their specific disability, a decent and fruitful life. How we treat the weak, the disabled and the powerless among us is a good measure of our civilization and proper attention to and representation of such persons is a powerful contributor to our solidarity as a civilized community.

In speaking of genetic defects or of disabilities the question of the "quality of life" comes up—as it does in so many other instances in ethics. Here we again want to re-iterate: the quality a life has can be assessed only by the entity whose life it is. We can, in more general terms, say that something (pain, paralysis, blindness) hampers optimal quality—but we cannot know what this means for a specific individual. We can, perhaps and with good reason, claim that persons who are permanently vegetative and who lack even an inkling of self-awareness are, by definition, unable to have "quality" at all and that, indeed, quality is a term misapplied to people unable to experience. But when a flicker of self-awareness is present and a patient may be said to have feelings and interests the matter cannot be that easily settled.

——————— LENGTHENING TELOMERES ———————

Using telomerase in some way so as either to lengthen or retard shrinkage of existing telomeres offers obvious social, political and ethical problems. It seems likely that if one were to lengthen (or to retard shrinkage) of telomeres, our life span would be enormously increased, as, in all probability, would our functional capacity. The distinction is critical: to lengthen being alive without at the same time maintaining the capacity to have a useful life is a quite different matter than prolonging a life which is or can be useful and enjoyable. Being able to manipulate telomeres does not seem to be a pipe dream but something that seems likely to happen and something that we should anticipate and prepare for. If left to private enterprise it is likely to be a modality available to those who are well off and out of reach for others. We shall argue that the equitable distribution of medicine is ethically mandatory—others may argue otherwise but it is a discussion that must be held prior to abandoning medical care to private profit.

In the problem of lengthening (or retarding the shrinkage) of telomeres there are various positions one can take. First of all, one can take the position that this ought not be done and base that judgment on some form of argument which claims that we, as humans, do not have the right to interfere with God's plans (as though, if there were a God, a mere human could do that!). Secondly, one can argue that such an increase in elderly people would have a severe adverse effect on society for a number of social as well as economic reasons. Thirdly, one can embrace the idea and worry about its effects when they arise—a way of going about things that we certainly could not endorse. And fourthly, we could accept such a prolonged life span as something desirable, examine the anticipated problems carefully and try to deal with them beforehand.

A longer life span in which physical and especially intellectual function can be maintained has certain personal and possibly certain social and societal advantages as well. As our knowledge increases, as there is more to be learned and more to be known, training individuals takes longer and at the present time their time to actively contribute to the community is shortened. Retirement age 65—and in parts of the world this is mandatory—may not under such conditions be appropriate for all occupations. It may not be appropriate to throw coal miners—who spend a much longer time working before they reach 65 and who work physically much harder than do intellectual workers—into the same kettle as we do University professors or other intellectual or desk workers. What is an appropriate retirement age for one may be quite inappropriate to another. Furthermore, people in intellectual activities often reach their peak at the very age at which we (even today and without lengthening telomeres) force them to retire. This is wasteful to the community and in many cases not welcomed by the individual involved. Should we ever be in a position to lengthen useful life span it would be well to plan ahead. Such planning might very well and very profitably entail training in some new field and having another career that is, furthermore, enriched by the experience gained in the first.

These considerations likewise are not unimportant today. It may very well be that retirement age should be differentially set and depend less on age than on each individual person's function and capacity. Elderly people—especially in the US— tend to be thrown on the slagheap, feel no longer socially useful or responsible for anything and in consequence go downhill rapidly. (See Chapter 11.) They are, consequently, lost to themselves as well as to society.

——— EXPERIMENTING WITH AND ———
TRANSPLANTING FETAL TISSUE

Some conditions can be treated (at least ameliorated) by the use of fetal tissues. Fetal tissues are far less apt to be rejected and are, therefore, often particularly suitable for transplantation. There seems little doubt that more and more conditions are, at least potentially, treatable by the use of such tissues. There are, of course, two questions here: the ethical propriety of using fetal tissue for experimentation, and, ultimately, the propriety of using such tissues for the treatment of patients. If we decide that using such tissues therapeutically is ethically impermissible, then, it logically follows that experimenting is likewise impermissible.

Since fetal tissues are obtained by abortion—and although spontaneously aborted fetuses are generally not suitable and not present in adequate numbers—much of the debate hinges on one's attitude toward abortion (see Chapter 10). Those who oppose abortion are generally apt to oppose experimentation with and transplantation of fetal tissue. This is not necessarily the case, however. Even those who feel that abortion is ethically impermissible may argue that it is better to at least obtain some benefit from an actualized evil than to discard any possible good that might come from it. Such an argument is a largely utilitarian one. The question here is analogous to the question of using data from the Nazi, from the Japanese Unit 731 or from the Tuskegee experiments. Persons who have no moral qualms about abortion are unlikely to oppose experimenting with or utilizing aborted tissues.

As long as abortion is a legally permissible and ethically generally accepted procedure, arguments against using the tissue of such fetuses are somewhat feeble. The question is firstly one of informed consent (can a mother who has purposely discarded her fetus consent in its behalf—which implies acting in its best interest?)

and, secondly, one of how such tissues are to be made available. Again, we are left with our own metaphysical attitude toward the fetus: can a young fetus (and a dead fetus at that) be said to have "an interest"? When all is said and done, allowing the mother to make such a decision is not altogether good (for in an induced abortion the mother does not generally—albeit she may have the fetus' best interest at heart—but it may well be the best of a number of other, more objectionable alternatives.

The fear that women may purposely become pregnant so as to utilize their tissues has two versions. In the first, pregnancy and abortion occur so as to sell fetal tissue; in the second, pregnancy and abortion occur so as to benefit oneself or a loved one (consider, for example, a young lady with diabetes and early renal damage who grows her own pancreatic tissue, or one who grows fetal tissue so as to have a father, disabled by Parkinson's, treated). What is needed is not either to forbid the procedure outright or to allow it to go on uncontrolled; what is needed is communal deliberation and action so as to formulate reasonable laws that are respectful of cultural differences and that would guard against tissue and organ sales. On a practical level it would be hard to convince a severe diabetic whose progressive renal deterioration might well be salvaged by the use of fetal tissue not to grow her own tissue to cure her disease or to dissuade someone from growing fetal cells to save a loved one from living and dying with a horrible disease.

————— GENETIC MANIPULATION —————

There has been great hesitancy to allow the development of procedures that could allow genetic manipulation. Many have felt that when man enters the area in which he "creates himself," he fulfills the biblical prediction of the devil. "Creating oneself" has an interesting history: in original Hebraic law the prohibition against the making of graven images of living things, for example, appears to denote that man should not assume the position of creating life. Others have felt that even the first step (understanding our genome, for example) inevitably will lead down a road that must end up with the attempt to "improve" the race in the service of some theoretical image of what the perfect human being should be.

We must differentiate clearly among several issues:

1. Experimenting so as to develop an understanding of the human genome.
2. Genetic manipulation of lower forms that will enable us to use them for general benefit: for example, producing colonies of *E. coli* that produce insulin or colonies of mice (not quite the same thing since mice do have feelings and can suffer) that can be infected with the HIV virus.
3. Genetic manipulation of nonhuman higher animals so as to create, for example, particularly productive cows or hens.

4. Genetic manipulation in humans. Genetic manipulation in humans must be subdivided into manipulation that affects merely the somatoplasm (which changes the way a particular person is—say, by correcting the genetic defect of a disease in that individual only) and manipulation that affects the germ plasm thus producing changes down the generations. In the latter instance, one must further distinguish between a manipulation meant to eliminate a disease (realizing that what is and what is not disease is inevitably a social construct) and a manipulation intended to change a trait like skin or hair color, height, or some other "non-pathological" (again a social definition) attribute.

Unless one were to appeal to some metaphysical "sacredness" of genes (which on a secular basis is hard to do), throwing all the ethical problems genetics presents into one pot seems indefensible. Although there have been ethical objections raised against the genome project (objections that essentially make a slippery slope or "one thing will lead to another" type of argument), understanding the human genome is merely another step along the road of understanding our environment. And (*pace* Professor Jonas!) it is difficult to argue how a species of knowledge can, by itself and without its use necessarily following, be bad purely in itself: that is, by existing.[4] There may be some species of knowledge so dangerous that one ought not to pursue them: the search for an extremely virulent, totally anti-microbial-resistant microorganism or the production of an explosive device that could destroy the earth can be argued to be examples of "knowledge bad in itself." But even here it is not the knowledge but the fact that the use or potential use of that knowledge can only have destructive consequences. The fact that control is never certain is what makes it so very potentially destructive and dangerous. A better understanding of our genetic heritage hardly falls into the same category.

It is difficult to argue against genetically manipulating lower organisms such as bacteria so that they can produce abundant materials critically needed by higher human or nonhuman animals. Moreover, as long as animal experimentation (rigorously controlled) is deemed ethically permissible, it is difficult to argue against genetically manipulating an organism like a mouse so that it can be used in truly vital experiments. Likewise genetic manipulation to produce more productive cows or hens can be argued against only by invoking a species of argument that rests on the claim that one must not manipulate natural process or selection. An argument of this sort is difficult to maintain when we artificially inseminate or otherwise manipulate the production of the type of farm animal we desire, or when we treat diabetes or when we make it possible for previously sterile persons to have children.

Unless, for reasons similar to those of Professor Jonas[4] (see Chapter 2), one were to argue that such knowledge and its use makes us, in a sense, too God-like and that it gives us powers we ought not to have (for on the one hand reasons akin to the slippery slope and on the other hand reasons that are far more mystical), it is difficult to argue against any and all sorts of genetic manipulation or the gathering of precise genetic information. Being too God-like as an argument is, in and of itself, problematic. Every time we treat pneumonia, choose the temperature of our room by setting the thermostat, build a building, or do coronary bypass surgery we "play God:" we manipulate nature so as to make it serve us and our purpose. All higher animals survive only by manipulating nature. In that sense, they "play God" and since doing so must be done within the framework of biological and material possibility, it is difficult to argue that, in a sense, doing so is not, in fact, "natural." The question is not whether we may "play God" or manipulate nature (for we have no choice but to do so). Rather, the question is what the legitimate limits of doing so are—and why. The probity of genetic manipulation in humans depends, rather, on its extent and purpose.

Genetic manipulation may merely affect the somatoplasm (those cells which are not concerned with the next generation but with the individual in whom these changes are made) of an individual: such a change affects only the individual treated and does not affect future generations. Changing an inherited gene so that the patient no longer has a severe disease would be one example. Of course, as we have said previously, what is considered a disease is a socially determined fact and, therefore, one could conceivably define all sorts of things as "disease" that we would have a hard time accepting. However, that goes equally much for non-genetic treatment. Changes that affect merely the somatoplasm and that are used to treat "disease" should be viewed largely as just another form of therapy. Arguments against using this type of therapy in some respects are reminiscent of the early objections to organ transplantation or, at times, the creation of an artificial heart. There is inevitably a supra-rational and even mystical component in such an argument.

Manipulations that affect the germ plasm produce changes in subsequent offspring. Such manipulations change not only the treated individuals themselves but also all individuals who subsequently share their genes. Such changes may be far more ominous: the possibility of changing a whole race and eventually changing humanity itself is no longer a mere fantasy of science fiction. And yet it is difficult to argue that if one could eliminate sickle cell anemia, Tay–Sachs disease, or Huntington's chorea, humanity would not be better off for it. Here, however, the slope has become considerably more slippery, and "more sand" needs to be put down: that is, the community must carefully define what constitutes misuse, and sufficient legal hedges against such misuse must be constructed. It is likely that at some point a community will find that the slope is too slippery and that an arbitrary barrier needs to be erected.

Changing human traits and engineering people so that someone's or some particular society's vision of what the perfect human should be is quite another matter. It is so because it favors and in a sense forces a racially or socially idiosyncratic vision of humans that many within such a society as well as most outside such a society would reject. Therefore, this type of manipulation is ultimately divisive. In that it shatters solidarity (national, and at least equally importantly, the solidarity so necessary for world survival) it ultimately threatens the survival of all. Such an argument does not speak against exploring the human genome or trying to better understand genetic process. In our view when fundamental changes are actually contemplated is the time that barriers and strictest regulation to the point of downright prohibition are in order.

There are many other and perhaps far more profound ethical problems than the use of stem cells or genetic counseling associated with the use of our expanding knowledge of genetics—our capacity to affect changes not only in existing individuals but in affecting individuals far into the future as well as in modifying existing or even creating new species, thereby affecting the course of natural evolution. The ability to do this has imposed a heavy ethical burden. Once made, such decisions may not be reversible. And as matters stand now we truly do not have the foggiest notion of what we are doing or what consequences our actions might have.

The fundamental question of what we consider to be "health," "disease," "normal" or "abnormal" has been discussed elsewhere (see Chapters 2 and 6)—our tendency throughout history has been to call those things we do not like "illness." Once we start down this road there is little to stop a bloodless, aesthetically inoffensive and virtually invisible holocaust. We can define being white, black, yellow or green as a disease and eliminate persons falling into that category by eliminating their genes—there are genes which control whether we happen to be black or white (albeit there are certainly none for being Jewish as Hitler believed!). In so doing we give medicine power it is neither trained to handle nor should. Furthermore, because intelligence depends, in part, on one's genetic make-up we can go about creating an elite ruling class and a class of people whose intelligence suffices to be labourers and little else. Thus we can create a stratified society in which social mobility is unlikely if not impossible and in which innovation and progress is slowed if not indeed halted. Democracy under such conditions is a farce. Diversity—which, at least in our view, does much to further human progress—slowly grinds to a halt. More frightening still—we may create a class of people docile enough to allow such a state and thus prevent what we today would consider social progress altogether.

The problems of genetics are not confined to the animal species or to humans. Manipulating the genetic make-up of plants (or of dairy cows) may have profound effects on the ecological system. It may produce a plant that produces more of a given crop and is more resistant to certain diseases. On the other hand such seemingly desirable consequences may have hidden and quite dangerous consequences as well—with such a change, for example, other changes that are destructive to the ecological system may occur. Are we at our state of knowledge ready to assume the responsibility which interfering with the natural course of evolution entails? Often it is argued that we could feed more of the world's hungry by such engineering—except for the fact that for commercial and business interests we today have often destroyed excess crops or food rather than send them abroad, such an argument might be a rather compelling one. It seems likely that the interest in genetic engineering of crops is motivated far more by business than by humanitarian interests.

GENETIC COUNSELING

Genetic counseling has become a well-recognized part of medical practice. Indeed, it is not entirely new.[5] The heredity of diseases like sickle cell anaemia, Tay–Sachs, or

Huntington's chorea have been well known for quite some time, and predictions about couples contemplating having children could, although less accurately, be made. What has changed is our capacity by various manipulations (amniocentesis, intrauterine biopsy, etc.) to accurately predict the presence or absence of certain diseases as well as the sex of the developing fetus. Parents can then be counseled appropriately and may choose to abort the fetus. Logically, such procedures with their attendant risks can only be justified if the parents are willing, at least under some conditions, to take appropriate steps. Such appropriate steps are generally but not always an elective abortion—not always, because some parents who know that their child will be afflicted with a disability may want to prepare for such an event or consider intra-uterine surgery. However such reasons are relatively unusual.

Determining sex and then aborting a child because it is not of the sex one wishes seems a particularly unsavory thing to do. If, however, one subscribes to a stark autonomy model and also believes that abortion up to a certain age should be a matter entirely up to the mother, it would be hard to argue against it. In some Third World countries, daughters, if they are to have acceptable lives, must eventually be given an appropriate (and often, considering the family's circumstances, an enormous) dowry and sons are needed to provide for parents in their old age. Thus having more than one or two daughters may wreck such persons' lives. Here, as is so often the case, changing the unduly coercive social conditions that cause parents to feel compelled to make such choices must be dealt with first.

We want to be clearly understood: we are not making an argument for the appropriateness of abortion because of sex bias. We think that doing so is at the least unsavory and probably, since it creates an inevitable imbalance in a society in which parental desires are biased culturally in one or the other direction, biologically highly undesirable. Ethically it is an expression of a form of prejudice that any reasonable system of justice would reject. Abortions for sex selection are ultimately racist (or sexist— simply a species of this type of world view) and, furthermore, threaten the species' health. Since the species' health is one of the necessary conditions for having a future, such procedures as abortion to eliminate a fetus of an undesired sex are, at least in our society, arguably immoral.

What we wish to point out is that (1) a stark autonomy model in which we only have an obligation not to directly harm other persons cannot simultaneously argue against a person's right to determine the sex of the offspring simply does not work and (2) like all else, sex selection must be seen in the context of a given culture and a given set of circumstances.[6] Sex selection through amniocentesis and subsequent abortion in developing countries—which is what is often mentioned in the literature— is, in fact, an activity few in such countries could afford. What we should (besides disallowing abortions for reasons of sex selection in the Western world) strive to do is to eliminate the conditions and thus ultimately change the attitudes that underwrite such a course of action in "underdeveloped" nations. Decreasing poverty, striving to eliminate gross inequality and, perhaps most importantly, promoting sound education and not resorting to law or ethical imperialism, are what can ultimately eliminate this evil.

The ability to test for chromosomal defects is not an ultra-recent development.

By means of amniocentesis or chorionic sampling many of the characteristics of the developing infant can be accurately foretold—thus Down's Syndrome, Trisomy 18, Tay–Sachs and many other such syndromes can be known at a stage at which abortion can be a viable option—and this then becomes a matter of the personal morality of the parents. Furthermore, the gender of the fetus can be accurately known—a fact that raises profound ethical problems. Also—and neither decisive but also hardly trivial—giving birth to a severely handicapped child whose life span may be only a few pain-filled weeks or months, which may completely disrupt a family with other children and, furthermore, cost the community hundreds of thousand of dollars does enter into the equation.

The ethical problems here are manifold. First of all, what is "normal" or, perhaps better said, what is within the range of the acceptable? And yet, this still does not help us since the "range of the acceptable," in turn, raises the issue of to whom it is acceptable: the afflicted individual, the parents or caregivers or the community. Secondly, what should be done about the particular developing being in question: should the pregnancy be continued undisturbed, should it be terminated or should it be subject to intra-uterine operative procedures which are difficult, often fall into the gray zone of experimentation and are immensely costly? Thirdly, of course, is the problem not so much if as how to tell the prospective parent. Are genetic counselors or obstetricians to serve merely as informants who also present patients with options or are they entitled to advise? On the one hand, advising inevitably involves presenting such options colored by the presenters values; on the other merely dispassionately laying out the options as neutrally as possible leaves the patient with information, but with no counseling in what can be a more meaningful and pragmatically useful sense of that word (see also chapter on autonomy). Fourthly, are healthcare providers entitled to perform a procedure carrying inherent risks merely to inform a patient who has—should the results be adverse—already ruled out all interventions beforehand? Fifthly, and not as simple a matter as it sounds, gender selection is a problem which, because of its profound cultural and economic aspects cannot simply be dismissed as one "merely" of prejudice.

The questions as to what is "normal" and what is disease, sin, crime or of no great matter at all have been discussed elsewhere (see Chapter 6). What is an acceptable state of disability is a somewhat different matter. To consider—as some radical societies of disabled persons claim—that a disability is not a negative deviation from an ideal physiological state (or to consider it merely another culture) is something we *a priori* reject. It is linguistically odd, to say the least (being "dis"-abled cannot be an advantage and having such a disability does not produce a different culture as that term is commonly understood), and any further discussion is beyond the scope of this book. That is, however, quite different from considering it not to be obligatory—for both communities and individuals—to ameliorate the effect such disabilities have on the individuals concerned to every reasonable extent. Giving the learning impaired every possibility to learn is one thing; having people who are obviously barely sentient hauled daily to schools at great public expense and with no evident profit to them quite another. Some of these issues will be discussed in the chapter on problems at the beginning of life. In genetic counseling it is the mother carrying a defective embryo and her family who, ultimately, will have to live with the situation—even with communal support such children may or may not pose a severe complication to the family and one which some parents depending upon the disability the child has, may be willing (or able) to take and others may not. It seems that such a decision at an early stage of gestation is a decision that the person carrying the child and her family should, with adequate social support, be entitled to make themselves. Absolute opponents of abortion are, of course, free to carry such a pregnancy to term but they are not ethically entitled to enforce this same choice—the result, after all, of their own personal morality—upon others.

Intra-uterine surgical procedures—when they have, as is often the case and depending upon the condition, not become an established procedure—are experimental and consequently fall under the same ethical guidelines as do any other experimental procedures. The difference here is that the result may be devastating for the life of an unborn and the family and community who (one hopes) would be prepared to accept and care for the consequences of failure. Whether experimental or not, the informed consent of the mother (and perhaps father) and some assurances that the as yet unborn would be adequately cared for if the attempt fails and a severely impaired individual results must be most carefully obtained.

The problem of informing and advising patients has dimensions far beyond pregnant women. When a person age 25 is found to carry a gene putting them at high risk of an incurable disease or making the onset of a devastating condition virtually inevitable (say a gene for Huntington's chorea) at age 40 or 50, should they be informed of their condition? If they are or are about to be married or to enter some other permanent relationship are their spouses entitled to this information? If one argues on the basis of beneficence an argument (in our view slender) could be made for withholding such information—slender because not informing is inevitably based on the particular worldview of the person who could, but chooses not to, tell. If one argues on the basis of autonomy then no argument for withholding information can be made. In many ways the results of such a test are like the results of any other examination: When we have a secret only we know about, only we can tell another. When it comes to medical information (genetic or not) the question of "secrecy" comes up. In general a secret is something I know about and am or am not willing to share with selected others. Here—while it is a secret—it is a secret about myself known by another but not known by myself. When it comes to medical information I need an intermediary (the physician) to discover it—but when all is said and done it remains my secret and legitimately my right (which I may refuse) to know.

Sometimes patients wish to have invasive tests but state that they would under no circumstances ever agree to an abortion. Amniocentesis or chorionic sampling is not without risk. Here we again encounter the problem of autonomy and the relationship of healthcare provider to patient in yet a different guise. Is a physician or other health care provider compelled to do a possibly risky procedure in order to satisfy a patient's curiosity when nothing substantive will be done about the findings? The fiduciary responsibility of physician to patient would seem to argue against this—physicians must try to balance the harm they do with the (hopefully) greater good. When the only good appears to be satisfying someone's curiosity, the balance would counsel against doing such a procedure.

STEM CELLS

Modern technology has raised the hope that omnipotent stem cells can be programmed to become virtually whatever tissue we desire. Such programmed cells could then be introduced into the body of severely ill patients to build new functional tissue. Preliminary experiments at the very least look hopeful. At this point we are not at a stage were such procedures are clinically useful—but we are at the point where research and experimentation may confirm or deny this hope to us. Such procedures offer great hope for a diversity of diseases—Parkinson's, Alzheimer's, diabetes, myocardial failure and many others. People have raised a great deal of objection to the use of stem cells for research—objections that are quite similar to those made by the absolute opponents of abortion.

Most stem cells these days are obtained from embryos created during in vitro fertilization (IVF). In the course of IVF invariably more embryos than are ultimately used are produced; those left over are either frozen, discarded or frozen and eventually discarded. They are not created for the purpose of stem cell research but, rather than being discarded, are utilized for such research. The argument against the use of such cells rests on the fact that they have (among other things) the potential to be human beings and, therefore, deserve the full protection of actualized human beings. They are—according to some beliefs—"ensouled" at the time of conception and from that time on are of equal value with any other actual person. (See the chapter on the beginning of life.) Outside of religious grounds, this is a hard argument to sustain.[7]

An argument to the effect that because the superfluous embryos would perforce have to be discarded IVF is morally illicit, is one that (while we do not agree with it) can reasonably be made and is logical within the context of some belief systems. But the argument that IVF is a morally acceptable procedure but it is morally preferable to discard rather than to use the superfluous embryos for the good of another is logically, at the very least, peculiar. Equating the use of stem cells with the Nazi children's euthanasia program or with the holocaust is not only ridiculous (the persons who were murdered certainly were self-aware and were very much interested in being alive) but obscene—especially when made by the very same Church who (except for some outstanding individuals to the contrary) remained silent during both the so-called euthanasia program and the holocaust.

There is little doubt that stem cell research will be done. The possibilities for curing or ameliorating devastating illness are too vast to permit wasting this opportunity—and, unfortunately, the danger of commercialization and the lure of profit is too strong to permit the use of such possible treatment modalities without strict supervision. Stem cell—as other medical procedures—are for the public good and, in our opinion, ought as such to be kept out of the hands of those who seek to make profit out of the misery of others. If health care is an important and perhaps even vital, social good—and we believe so, wishing to avail ourselves of the reasonable opportunity range that a society has to offer[8]—then those things needed to maintain good health and function are not properly viewed as commercial but as public goods.

Public goods may be viewed as those things humans need to meet biological and social essentials. While social goods are variable depending upon time, place and circumstance, biological needs (such as the need for water, food, shelter, etc) are universal. Thus, some social goods are necessary if one is to enjoy what Norm Daniels calls a "reasonable opportunity range."[8] In our industrialized, modern western world such things as education and health care meet this requirement; in other civilizations other goods may be considered to be necessary social goods (see section entitled "What are needs?" in Chapter 8). Health-care to maintain life and restore function and education in our society fall well within the range of being necessary social goods.

Stem cells other than omnipotent stem cells have been routinely used in some diseases. Pluripotent stem cells garnered from placental blood are used successfully in treating certain blood dyscrasias.[9] This technology is currently being used commercially and has raised certain, and in our view severe, ethical issues. Until recently—at least in most cultures—the placenta, after being examined by pathology is discarded. Suddenly it has become valuable: valuable to those who stand in need of such cells and valuable to those who stand to make financial profit out of the procedure. Mothers before the birth of their child are often asked to pay for having such blood collected and frozen so that should the child later come down with leukemia a bone marrow transplant using its own stem cells might be done.[10] In truth this is an empty promise: first of all the chance of any one child coming down with leukemia is a bit like its chance of being struck by lightning. Second, there is no factual basis to this promise (it has never been done). Third, there is a serious question as to the advisability of transfusing the very same cells that we know, in the future will, once again, became leukaemic. What is needed is careful and thorough experimentation, not commercialization. Banking pooled cells may well be a viable option and might profit many—it, as any other medical procedure, is a public good and not a private whim for someone's profit.[11–13] Our experience with blood transfusion—where commercialization has been largely forbidden—may well serve as a model.[14] Some cultures value the placenta and within their belief system need to perform certain rites or bury the placenta in certain ways. Informed consent is, therefore, necessary so as not to violate an individual's deeply held beliefs—but respecting another culture's belief system is quite a different matter than tolerating commercialization.

It can, of course, be argued that the pharmaceutical industry as well as the manufacturers of other medical devices are for profit-based enterprises and are not nationalized in any health-care system. That is undoubtedly true and it would be most difficult to change this state of affairs. It is, nevertheless, peculiar, that the cost of the same medication manufactured by the same international concerns differs vastly from country to country. The novelty of advertising prescription drugs in lay magazines and on television has added further pressure to prescribe costly (and not necessarily more effective) drugs than would have been prescribed otherwise. While it is probably impossible, practically speaking, to nationalize the drug industry, some greater modicum of public control could well help stabilize costs.

DESIGNER BABIES

It may well become possible to choose to have babies with certain but not other physical or mental characteristics. This "opportunity" when actualized would present a severe ethical problem. Aside from the fact that it would be extremely costly for a health care system to provide and would, therefore, in most probability be available only to those who are well off, it seems unlikely that parents would choose very differently—this could have serious ramifications for the future health of our gene pool. Moreover, it carries considerable potential for being socially destructive. It seems likely that most potential parents would want highly intelligent, sturdy, well built and athletically capable children—and one can go on naming characteristics we all admire and would want our children to have. If "designer babies" became a real possibility and actually were to be "made" we would end up with a society in which the rich and well born would be able to dominate not only—as they do now—by virtue of their greater political power (they can buy the candidates) but by virtue of engineering their progeny so that they could continue to maintain and expand their power. Social mobility would—at the very least—be severely slowed down and democracy would suffer severely. But, more importantly, the natural course of evolution, promoted as it is by diversity, would be severely constrained.

GENETIC INFORMATION

Once we have genetic information about individuals, rather undesirable consequences may follow.[15] There are many ways in which such information might be used unfairly to advantage or disadvantage individuals—such as screening for prospective employment, insurance and in many other ways. It seems that stringent laws restricting access to such information are essential and, perhaps, laws forbidding the release of such information even if the prospective employee or insured were to sign permission. It would be easy to coerce people into signing such wavers—"if you do not sign we cannot consider you for…."—quite easily overcomes the intent of such a law.

Patients may or may not want to know such information about themselves—living with the knowledge at age 20 that one will have Huntington's chorea at age 40 may not be something an individual desires. That does not suggest by any means that such information should be withheld, but it does suggest that patients need to be asked beforehand (and before the test is done) whether they do or do not wish themselves or their family members to be privy to such information. Furthermore, although the individual may not wish to know, his or her prospective spouse may very much want to know. Whether the community has a right to prevent such births is a very legitimate question.

Laypersons read more and more in the popular press about medicine's ability to predict disease by genetic examinations and consequently often besiege physicians to be tested. While this desire to be tested for every conceivable disease is, perhaps, understandable, patients have to be fully aware that some diseases are, while others are not, worth testing for in various population groups. There is, for example, a very high percentage of breast cancer in certain Ashkenazi Jews in whom testing may, in fact, make a lot of sense. There are other population groups in which it makes little or no sense. Like all other medical procedures genetic information must be gathered and used with discernment. In the final analysis, it is not the physician's duty to do merely as the patient wants but to practice reasonable, compassionate medicine.

THE DANGER OF NOT PURSUING KNOWLEDGE

We have spoken largely of the dangers which pursuing knowledge may bring—especially knowledge which, when transferred into action, can have irreversible and devastating results. There is a flip side to this coin. Deliberately not pursuing knowledge freezes us inevitably into the *status quo* and leaves us without a method for determining whether what we know is, in fact, correct. As a species, humans are curious about their environment. Finding out more about our make-up and ourselves becomes dangerous only when we blindly translate our knowledge (which is and always will be incomplete) into action. The temptation to convert what we know into action is admittedly great and once knowledge is generalized we are in constant danger—the best laws not withstanding—that someone will do so and start an unstoppable cascade. Our only hope lies in a thoroughgoing education of all—not indoctrination, not recitation of what are supposedly facts, but real education so that people think for themselves and are willing to discuss such issues with each other seriously. Such a dialogue will happen only when all who are capable of participating are assured the basic necessities of life, are given an adequate basic education and are shown that learning can be fun and that the discussion of serious topics is an enjoyable and fruitful experience. Only when something can be shown to be relevant to the daily lives and futures of peoples and only when engaging in such discussions is seen to be at least as enjoyable as engaging in what so many consider "fun" will this happen. But when all is said and done—educating people, eliminating economic, class and racial barriers and learning to have mutual respect for one another are the only things which can hope to address the ethical problems of today and of tomorrow.

REFERENCES

1. Hartwell LH. et al. *Genetics: From Gene to Genome.* Boston, MA: McGraw Hill, 2000.
2. Prescott J, Blackburn EH. Telomeres and telemerase. In: *Wiley Encyclopedia of Molecular Biology.* New York, NY: John Wiley, 2001;3110–3114.

3. Engelhardt HT. The disease of masturbation: values and the concept of disease. *Bull Hist Med.* 1974;48:234–248.

4. Jonas H. *Das Prinzip Verantwortung: Versuch einer Ethik für die technologische Zivilisation.* Frankfurt a/M: Suhrkamp Taschenbuch, 1984.

5. Harper PS. *Practical Genetic Counseling.* New York, NY: Oxford University Press, 1998.

6. Loewy EH, Loewy RS. Of cultural practices, ethics and education: thoughts about affecting changes in cultural practices. Health Care Analysis.

8. National Ethics Advisory Committee. *Ethical Issues in Human Stem Cell Research (Vols. 1–3)* Rockville, MD: Bioethics Advisory Commission, 2000.

9. Daniels N. *Just Health Care.* New York, NY: Cambridge University Press, 1985.

10. Gluckman C. Ethical and Legald aspects of placental/cord blood banking and transplant *Haematol. J.* 2000;1(1):67–69.

11. Saba N, Flaig T: BM. Transplantation for non-malignant disease. *J Hematotherapy Stem Cell Res.* 2002;11(2):377–387.

12. Nicole D, Otlowski M, Chalmers D. Consent, commercialization and benefit sharing. *J. Law Med.* 2001;9(1):80–96.

13. Cahill LS. Genetics, commodification and social justice in the global era. *Kennedy Institute Ethics J.* 2001;11(3):221–228.

14. Nelk D, Andrews L. Homo economicus:commercialization of body tissue in the age of biotechnology. *Hastings Center Report* 1998;28(5):30–39.

15. Dame L, Sugarman J. Blood money; ethics and long term implications of treating blood as property. *J. Ped. Haemat. Oncology* 2001;25(7):409–410.

16. Burgess MM, D'agincourt, Cannind L. Genetic testing for hereditary disease: attending to relational responsibility *J. Clin Eth* 2001;12(4):361–372.

Problems of
Macro-allocation

A "system" a political, economic, or cultural system insinuates itself between my-self and the other. If the other is excluded, it is the system that is doing the exclud-ing, a system in which I participate because I must survive and against which I do not rebel because it cannot be changed... **I start to view horror and my implica-tion in it as normalcy.**

Quoted in Barnett V: Bystanders: Conscience and Complicity during the Holocaust. Westport, CT: Praeger Publishers, 1999

─── INTRODUCTION ───

Problems of macro-allocation (see also Chapter 3 and Chapter 6) have become in-creasingly important today. It is not just that resources are shrinking, as has been so often said; rather, as technology develops, the resources needed for the care of pa-tients have escalated, and this escalation promises to continue. Further, the popula-tion here and abroad—especially in technologically underdeveloped countries with a population justly clamoring for a share of the good life—is increasing markedly, in part as a result of changes brought about in and by the medical and social sciences. Not only is the population increasing, but the number of elderly living on retirement has also increased at the same time as a much greater length of time is needed to pre-pare many of the young for their life's work. This has altered the traditional relation-ship between those in and those not in the actual work force and it has raised questions about retirement, retirement age and the utilization of the experience and talent which many of the elderly could easily bring to the community. There are more persons justly expecting to share in the available resources, a change in the resources needed and relatively fewer involved in actually producing needed resources and actively par-ticipating in the working of the community.

The human community shapes individual ethics. It is the fundamental context, the necessary stage on and in which our actions unfold. Community, just like being and nature itself, therefore, is considered to be of "prior worth"—the necessary condition for all else (see Chapter 3). Communities, like individuals, have their needs. In a smaller and a larger sense we inhabit a commons that we all share. Preserving this commons necessitates placing limits on its members so that they cannot pursue unbridled personal gain mindless and often to the detriment of communal good. "Freedom in a commons brings ruin to all."[1]

The medical commons, no less than the greater commons we inhabit, shares in the inescapable fact that absolute freedom for many courts the destruction of all. Medical resources are not unlimited, and limiting their use for patients, as well as equitably making these resources available to all members of the community who may still benefit, is one of the problems of contemporary society.[2]

In order better to understand the problems associated with the macro-allocation issue, one can use Rawls' veil of ignorance (see Chapter 3) as a heuristic device to choose among three goods. Such a veil of ignorance allows us to know what goods and services a society may offer but does not allow us to know our own condition. We must choose not knowing how old or how young, how intelligent or how stupid, how sick or how healthy, how poor or how wealthy we are.[3] Let us, under these conditions, choose among three goods, only two of which we could have guaranteed to us: (1) that all your biological needs would be met; (2) that your ability to develop your interests and talents (i.e., your education to the fullest) would be assured; and (3) that your healthcare needs would be met. The prudent chooser would in general and for the most part be inclined (even if regretfully) to leave healthcare to luck or chance: Surely one needs to have one's biological needs met and surely without the capacity to develop one's talents one's life most probably would not be worth living. One might, if one is lucky, not need much healthcare, but one cannot go without food, shelter, warmth, and the ability to educate oneself. Our thesis is not that healthcare is unimportant; rather, it is that a civilized community that has the capacity to do so is obligated to supply all three in adequate but not necessarily opulent amounts. So as to make such a goal possible, a balance between and among social goods must be struck. Not everyone (neither patient not health professional) can, it turns out, have everything.[4]

Macro-allocation (as we have said in Chapter 3) deals with the ways in which groups of people allocate resources rather than concerning itself, as micro-allocation does, with problems on a one-to-one basis. Problems of the latter kind (problems that deal, for example, with discontinuing or starting dialysis for a specific patient) necessarily follow a different set of moral rules and have a different history from the problems of the former kind (problems that, for example, concern the funding of dialysis programs). Necessarily, problems of macro-allocation (in which lives are, by and large, unidentified lives) must follow a utilitarian calculus: Decisions here must attempt to promote the greatest good for the greatest number. We must, however, be aware that these "unidentified" lives (unidentified only because we ourselves happen not to know them personally) are very much identified and personal lives to themselves and to others. Micro-allocation issues cannot be tackled in quite the same way as those of macro-allocation or distribution. Problems in which people deal with each other on a one-to-one basis (in which lives are identified lives) cannot aim for the greatest good of unspecified others but must be attentive to mutual need and historical context. Physicians, for example, in dealing with their patients must, at least in the context of our current and historical vision of the patient–physician relationship, be mindful of their patients' good above all else. Patients for millennia have expected and still expect today that *their* physician will pursue *their* best interests—should physicians primarily feel themselves to be agents of their particular health-care institution, managed care organization, HMO or state plan, then the public, at the very least, has to be aware of this. We, for our part, think that this change in traditional focus is not what the public wants or would agree to.

Ethical reasoning, when it comes to the care of a particular patient by a particular physician, follows a much more deontological line. Problems of macro-allocation or distribution where one deals with groups of largely (by oneself) unidentified lives must perforce try to bring about the greatest good for the greatest number of these lives. Using rational compassion when it comes to dealing with identified lives (not allowing one's compassion to degenerate into sentimentality, a form of behavior ultimately aimed at indulging one's emotional self rather than at improving the situation) and compassionate rationality when it comes to dealing with unidentified lives (so as to preserve and promote our common humanity) may help us forge more equitable (but undoubtedly still difficult to attain) solutions (see discussion later in this chapter).[5,6]

Macro- and micro-allocation issues are inevitably linked. Ultimately, macro-allocation allocates resources so that micro-allocation can take place, and micro-allocation, of necessity, takes place in the context provided by macro-allocation. Since this is undeniably so, the interface between the two has to be carefully scrutinized. To claim that these two concerns can each follow its own unique set of rules without inevitable conflict is to wear blinders. Analogous to a unified field concept in physics, some unity of law must exist if two systems are to operate smoothly in the same time and space.

In this chapter, problems of macro-allocation can merely be introduced. They are complex, and the literature dealing with such problems (literature that, of necessity, encompasses many fields: economics, law, sociology, medicine, ethics, to name but a few) is necessarily vast. Here, we will examine (1) problems of justice and of rights; (2) a definition of "need;" (3) types of macro-allocation decisions and the community's role in macro-allocation; (4) the physician's role in macro-allocation and physicians as gatekeepers; (5) distributing scarce resources to individual patients, looking at various approaches such as market, social value judgments, lottery, and "first-come, first-served;" (6) age as a consideration in rationing; and (7) the question of making healthcare available to all members of a just community, the role of the market, and various ways of building healthcare systems. We shall rely heavily on the section dealing with viewpoints of community in Chapter 3.

JUSTICE AND RIGHTS

Justice (see also Chapter 3), central to issues of macro-allocation, is often spoken about as one of the mainstays of ethical behavior. And yet, justice, if it is indeed to "give to each what is his or her due," is ephemeral and subject to specific definition. A definition of what is "someone's due" is not something that will be easily agreed upon. It is here that ethics and politics (as Aristotle remarked again and again) are no longer separable. Habermas in his groundbreaking works suggests that rules of ethics (and I think he means specific rules—things we refer to as personal morality) can never be agreed upon by consensus but that basic rules of justice (among which we would include rules of general ethics) can.[7] Justice would involve participation in such a dialogue by all members of the community—that is, by all those humans living within a community and by those who are weak (disabled, children, the elderly, etc) receiving special protection. This, of course, pre-supposes a functional democratic process—more than merely a political one (see below).

We conceive of doing justice in necessarily different ways when we deal with groups (in which lives are statistical or, at least by us, unidentified lives) or when we deal with the individuals within such groups (which are now identified lives or, at least, lives we identify). In dealing with individuals in a one-to-one setting (physicians, for example, dealing with their patients), justice, while necessary, is often not sufficient. At the bedside (if by justice is meant conserving resources for others by not giving them to another who is in need of them) it is, in fact, a frequently inappropriate or, at least only a minimally helpful, concept.[8] Justice, for example, stands, in a sense, opposed to generosity: A generous act is not a just one, and a just act is not a generous one.[9] To be generous is to be more than just; to be just is to be less than generous. And yet beneficence (together with technical competence)—an essential, if not, indeed, the most essential, historical component of medical practice on a one-to-one basis (and of crucial importance when individuals deal with each other in whatever setting)—implies more than mere cold justice. The use of compassionate rationality (in which compassion tempers reason) and of rational compassion (in which reason tempers compassion) can help one with many of these problems, as can the realization that, whether we deal with identified or with unidentified lives, individual lives eventually are affected.

Rational compassion is necessary if good solutions to problems involving identified lives have to be found. Our compassion is easily aroused (at least if we have not suppressed what Rousseau calls our innate sense of pity or compassion with the suffering of other creatures[10]) when we are confronted with another's problems, especially but hardly only when we know that person well. We can, as it were, feel their pain. Such compassion, however, if it is to truly help, must be tempered by reason: Healthcare professionals, for example, cannot afford to let their compassion for the suffering of their patient stand in the way of doing those things (even when they are painful) necessary to save a meaningful life a patient desires to be saved; those charged with allocating scarce resources cannot simply abandon their well-worked-out scheme to accommodate the idiosyncratic demands of particular patients they now happen to know.[5,6]

Problems of Macro-allocation

Compassionate rationality, on the other hand, is most useful when we need deal with unidentified lives in which our compassion is aroused only by our imagination. When we deal with problems of justice (or in the medical setting with problems of just distribution), especially with problems of justice in the healthcare setting, our purely rational approach to theoretical problems must be tempered by our compassion. Curiosity urges us to look about us and discover a problem that may not be our own immediate one, imagination allows us to see the theoretical problem fleshed out in its human dimension, and compassion forces us to understand that another's problem or suffering could very well be our own and prompts us to translate our feelings into consequent action. Thus we begin to see the theoretical problem as one that indeed, down the road, has flesh and blood, and to understand that eventually our solution will affect lives that, while not identified to us, are identified to others.[5,6]

Whereas compassion alone (whether used for dealing with identified or unidentified lives) easily leads to sentimentality, reason untempered by compassion can lead to quite cruel and inhumane solutions. The interplay of both is needed if equitable and humane decisions are to be made. An ethic without reason is an ethic of either reflex or pure feeling and one that denies rationality to moral agency. In that making ethical choices is, in part at least, an activity of our capacity to reason, an ethic without rationality misses the point of the entire enterprise and is not one that could be easily defended. An ethic without compassion, on the other hand, likewise misses the point: If it is to strive for humane solutions and actually help people, ethics cannot be conceived as a sterile exercise of pure reason. Ethics should be viewed neither as a game of chess nor as an exercise in hand-wringing.[4–6]

Justice, then, plays an important, but of necessity different, role in both macro- and micro-allocation. In macro-allocation it is the fundamental concept underwriting proper distribution of resources: Here the groups dealt with are dealt with as groups, and the individuals within the group are unidentified strangers. In micro-allocation issues, in issues in which individuals deal with other individuals, justice acts as perhaps a fundamental consideration but not as a satisfying condition of that interaction. The individuals, far from being strangers, are no longer faceless but are identified and known. Dealing merely justly with our patients leaves that interaction cold, austere, and devoid of its necessary human content. *Prima facie* duties, compounded in part of obligations arising out of individual relationships, deal with notions other than merely those of justice.[11]

Our conception of the standards of justice is rooted in the social context in which men find themselves. Our view of social contract and the resulting viewpoint of mutual obligation we develop decides how we choose to define what is and what is not just in a given society. Justice as a formal standard—externally applied and neither internalized nor adjusted to its social context—makes justice an immutable and unchanging concept. It therefore cannot evolve or adjust human needs or to human experience. Such justice is empty and therefore no longer justice in the sense that humans usually think of it.[12]

Our view of justice conditions our response to what we consider unjust and what unfair. If by justice we mean, by Aristotle's ancient formula, a virtue that gives each his or her due, we are left with the question of what that due is. Justice can, for example, be seen as a Kantian "perfect" duty; i.e., that not to be just cannot be universalized and that it essentially violates logic because willing injustice cannot be logically sustained as a "law of nature." To remedy unjust situations thus becomes a perfect duty. Indeed, according to Kant, justice and law are concerned with "perfect" duties to others, leaving the "imperfect" duties more optional. Unfortunate situations, on the other hand, appeal to a duty of beneficence, a morally "imperfect" duty, i.e., while it is logically possible to conceive a principle of non-beneficence, willing that such a principle everywhere should be a law would represent a contradiction of the will (see also Chapter 3). [13,14]

Duties of beneficence concern the welfare of others. If community is a free association of individuals united by more than duties of refraining, then these "others" are members of the community whose welfare is at stake. While justice and, say, beneficence stand in opposition to each other in one sense,[9] they are both due members of a community conceived of as such an association. If justice is conceived as a dynamic and evolving concept in communities that hold both freedom and beneficence to be incumbent upon themselves and that view the ethos of such communities as resulting from an interplay between these two principles, then the laws deriving from such a vision of justice cannot be seen to emerge from a regard for freedom alone.[15]

Duties of justice can be seen in many ways. If viewed consistently in a minimalist way, one model emerges; when, on the other hand, community is seen in a non-minimalist fashion, another model suggests itself: (1) The minimalist view sees in giving what is due purely a duty of noninterference ("autonomy-based justice");[16, 17] (2) the broader view sees in giving what is due more than merely noninterference with personal freedom. What is due encompasses issues of beneficence ("beneficence-based justice"), and, therefore, such communities see ensuring minimal standards of basic needs to be, at least, an ideal for which they must strive.

The relationship between the requirement of respect for personal freedom and that of serving communal needs has often been painted as a dialectic: Two opposing forces, each seeking their own goal, reach tentative compromises the result of which are expressed as the ethos of a given community. The libertarian insistence upon absolute individual freedom stands in opposition to the needs and goals of the community. It is a constant struggle in which each strives to get as close to their own goals as they can. Capitalist communities tend to value personal freedom much more highly than communal needs; "communist" communities tend to value the needs of the community far above respect for individual freedom. The two extremes end up with quite similar problems: communities in which large numbers of unhappy and dissatisfied individuals find it difficult to lead fulfilling and satisfying lives.

Seeing these two apparently opposite drives as in conflict instead of viewing them as being in a homeostatic balance may be a mistake.[5] A homeostatic balance (a concept current not only in biology or physiology but equally used in psychology, sociology, and ecology) is quite different: In a homeostatic balance diverse forces balance and modulate (not compete with) each other so as ultimately to serve the common goal of survival, learning, and development. Communities cannot develop solidarity and continue to prosper without respecting and developing the needs and talents of their individual members; individuals, on the other hand, will find it most difficult to achieve full lives outside a communal context. The myth of the asocial being (as Jonathan Moreno so aptly calls the belief that individuals are essentially free-standing and not necessarily enmeshed in a social context) is a most destructive myth.[18] These various points of view can only be briefly sketched here (see also Chapter 3). In such communities justice and mutual obligation cannot be conceived as merely attuned to the perfect duties of mutual non-harm but must pay more than lip service to the more optional visions and obligations of beneficence. It is only when justice is conceived broadly enough that great disparities among the destitute and the opulently wealthy are prevented, that solidarity and ultimately survival of such communities and of the individuals within them are possible.

Justice, as John Dewey pointed out, is not an end in itself.[19] It is a means that facilitates communal life as well as personal opportunity. As such the content of justice will vary as history and societies evolve and change. Justice, like all human activities, must be adaptive and must support survival. If justice does not do this, it is inapplicable to the human condition. It will, therefore, wither, die, and in its dying exact a heavy toll. Justice, like all other human activities, is biologically grounded in a common framework—man's perception of the good.[5,19]

Our understanding of the standards of justice arises out of the social context in which persons find themselves. Justice as a formal standard—externally applied and neither internalized nor adjusted to its social context—makes justice an immutable and unchanging concept. It therefore cannot evolve or adjust to human needs and human experience. Such justice is empty and therefore no longer justice in the way humans usually think about it.[20] Justice, as Dewey sees it, is an instrumental good: one that serves to promote social good, not a cold absolute whose requirements (whatever they may at a given time be held to be) must be blindly followed. Dewey, indeed, puts this very well:

> Justice as an end in itself is a case of making an idol out of a means at the expense of the end that the means serves. The means is organically integrated with the end it serves. There are means that are constituent parts of the consequences they bring, as tones are integral constituents of the music they serve.[19]

Our notion of "rights," likewise, is inextricably linked with our vision of the nature of community and justice (see also Chapter 3). "Rights" may be conceived as "natural" or "God-given."[21-24] Such "rights," derived from nature or "from nature's God," are immutable, fixed, eternal, and, of course, self-evident. Being self-evident, they are not subject to proof or disproof, and the concept therefore has an absolutist ring. If one wishes to dispute such "rights" (say, the right of property), one lacks a logical appeal to reason and simply stands in violation of God or nature. Such "rights" are secured to man by God or nature and therefore are not man's responsibility. In securing these rights, men are simply the agents of an unquestioned and unquestionable higher power.

The language of "rights" in and of itself is problematic and laden with a baggage of assumed meaning that at times makes it inflexible and unwieldy. In many respects, "rights" seen in an absolutist and context-less fashion are meant to preclude all further discussion.[25] When such rights clash, no method of arbitration between two conflicting absolutes, short of force, is possible. One may scoff at the notion of such rights and prefer to take a point of view that makes of all rights a social construct, promulgated and secured by communities.[15] Such rights may then be looked at as "interests" to be adjudicated between the individual and the community. Basic and fundamental interests (say, freedom) become a societal good of greater or lesser value in a hierarchy of social considerations that are the result of the interactions of unique individuals. If one adopts such a viewpoint, it is the community's and the individual's duty not only to enunciate but also to safeguard such fundamental values. Specific decisions, the product of growth, learning, and experience, are not immutably fixed but evolve over time and differ with circumstance. Analogous to freedom as a side constraint or freedom as a value, the view we take conditions our further choices.

"Rights," accepted as God-given, absolute, and inflexible, on the other hand, necessitate a static viewpoint. Eternal concepts adapt poorly to new and unforeseen conditions. If rights are looked upon as interests enunciated and secured by community, fundamental values are not, therefore, taken lightly or easily negotiated away. Rather, such a point of view affirms that what is a fundamental value is not writ large in the stars but is writ small and with much human effort and pain.[18] Persons who not only treasure their fundamental interests but also are held responsible for enunciating and safeguarding such interests will maintain a higher level of vigilance and care in the discharge of their social responsibilities.

————————— WHAT ARE NEEDS? —————————

Inevitably when physicians and other healthcare professionals decide to use or not to use a given intervention or when communities choose whether or not to allocate resources, the language of "needs" is invoked: We do such and such or allocate so and so because it is "needed." Often "need" is the key word, and deciding what to do hinges on its definition.

The concept of "needs," as Daniels has so aptly pointed out, is a slippery one.[26] In popular language a need can be almost anything: a passing fancy (I need to take a look in this store window), a desire (I need to go to concerts), or a condition of my existence (I need air!). In any case, and derived from its root of necessity, a need implies the necessary condition to a predetermined end. My need to look into the shop window can reasonably be expected to satisfy my curiosity as to what it contains; my need to satisfy my love of music is necessarily served by going to (the relevant kind of) concerts, and my desire to live requires air as a necessary condition.

Using the term "need" does not indicate the importance of that "need" in a hierarchy of values. It merely indicates that having or doing a certain thing is a necessary condition if a given goal is to be attained. In order to attain a goal, no matter how lofty or how trivial, certain things (or actions) are necessary. Their being a necessity depends not on the importance or value of the goal but on the importance of the means (the needed things) to reach the goal. In that sense the term "need" in and of itself does not indicate anything about the nature of the goal. It is somewhat like the "ought" in a hypothetical: It is an "ought" that must be fulfilled if the indicated goal is to be reached. As such, the term "need" (as the term "ought" when hypothetically used) is, in and of itself, essentially value neutral. It can be applied equally to the despicable (if you want to kill Jones you need—ought—to use poison), as it can to the commendable (if you want to save that child you need—ought—to give it food).

If needs are the necessary condition to desired ends, they may still not, by themselves, be sufficient to attain those ends. Food, for example, is only one of the necessary conditions for sustaining life: Without it life does not long continue. But food alone does not suffice; other conditions to sustain life are needed and together constitute the sufficient conditions to sustain it. Biological human needs exist in a social setting, and goals are social goals. If modern man is to live in an acceptable manner, rather than merely exist in a biological sense, conditions other than those of strict biological need must be met. Such needs are socially defined.

Saying that healthcare (or basic nutrition) is or is not a "need" demands further definition. In a sense, going to the opera is a "need" for many, and having at least a little pleasure in life is a "need" for all. But these are different kinds of needs. They are different because sub-serving them satisfies a basic desire to make life worthwhile rather than sub-serving life itself.

When we speak of "basic needs," we essentially will mean one of two things: (1) a first-order necessity, something required to sustain primitive biological existence and its goals—air, food, warmth, and shelter are examples; or (2) a second-order necessity, something required to sustain acceptable existence within a given social context so that its reasonable individual goals can be met—healthcare and education are examples. In the state of nature (H.T. Engelhardt's by now famous Ba Mbuti are an example[16]) first-order necessities are presumably the crux of the matter, and the second-order necessities, taken for granted in the modern industrialized world are either unknown and unimaginable or of little use in realizing the reasonable individual goals peculiar to the Ba Mbuti. Other socially structured second-order necessities take their place. In modern industrialized societies (for better or worse hardly in a state of nature) first-order necessities, or even the second-order necessities of primitive tribes, cannot suffice to permit a realization of reasonable individual goals. Second-order necessities that are far different than those in primitive societies, become essential.

In delineating "needs" beyond first-order needs, then, the social context becomes all-important. Even for the Ba Mbuti living their traditional life, there are "needs" beyond those of merely sustaining life; but their needs are obviously different from those of highly organized and industrialized societies. To realize access to a normal opportunity range consistent with the pursuit of an array of life plans that reasonable persons are likely to construct for themselves[26] among the Ba Mbuti (or among the ancient Greeks, the medieval peasants, or the 25th-century inhabitants of Greenland) is a different matter than from doing so in Moscow, New York, or Tien-tsin today. Although "first-order necessities" remain essentially stable throughout those societies, it is the social context that fashions the things we legitimately may want to call "second-order necessities" and those to which we may deny that standing. Except for the biological needs of first-order necessities, other needs and their prioritization are a social construct and not one that can be settled for all times or all places.

This leaves unsettled what to include and what to exclude among this category of basic "second-order needs," a category meant to include those things required to sustain at least minimally acceptable existence within a given social context so that reasonable individual goals can be met. The definition hinges on what is acceptable or reasonable as a goal within a given context. And what is or is not acceptable within a given context is ultimately, and in a changing and ongoing fashion, determined by the community. Education and healthcare certainly are legitimate second-order necessities in our industrialized world: but how much healthcare and how much education (and how they are to be come by) are not answered by labeling them as "needs." Further, there are socially accepted second-order necessities depending on individual aspirations and talents. Aspiring musicians may justifiably claim the use of a piano among their second-order needs, medical students a dissecting kit or stethoscope, and carpenters their particular tools. Each occupation or profession will have second-order needs peculiar to itself, second-order needs that are socially legitimized because they are critical to the attainment of a fair opportunity range.

First-order needs, as we have pointed out, are purely biological. They are needs because they underpin bare biological existence. Unless they are met, biological existence cannot continue, and such basic first-order needs are determined by our particular biology. They change from species to species: Essential amino acids for one species are not, for example, essential amino acids in another. Basic second-order needs, on the other hand, are basic needs because without them our lives are not acceptable: Without them we are unable to avail ourselves of the legitimate opportunity range prevalent in our particular communities.[26] They are, therefore, socially determined. Like amino acids, which vary from species to species in being or not being essential, the "basic" nature of second-order necessities changes from social structure to social structure. Without meeting second-order necessities, first-order necessities are empty; second-order necessities, on the other hand, are meaningless without initially satisfying those of the first order. One has to be alive to enjoy a social order, and one has to have a fair opportunity within one's social order if life is to be meaningful.

MACRO-ALLOCATION AND THE ROLE OF THE COMMUNITY

Macro-allocation issues are, as we have said (see Chapter 2), divisible into three parts:

1. The larger community (the state, for example) allocates its funds to segments within it; thus, communities, by whatever means, choose to allocate resources to education, defense, health care, social

services, etc.

2. At the next level, these different enterprises take the funds allocated to them and distribute them to their various subdivisions. At this level, for example, the funds allocated for health care are divided among hospitals, public health facilities, nursing homes, etc.

3. In the last of these levels, specific institutions—hospitals, for example—decide how much to spend for birthing units, operating rooms, ICUs, or outpatient departments.

Each of these levels is interconnected with the others so that the higher, in some ways, maintains at least some control over the disbursement at the lower level and so that the lower level, in turn, may bring its arguments for more funds or for a different allocation of funds to the attention of the higher level. Communities may, for example, allocate resources to medical care with an understanding that these funds will be spent in certain ways but not in others. Still, the lower level invariably maintains a certain, even if not complete, autonomy over its own budget.

Basically, a utilitarian calculus is followed at all of these levels: Communities will allocate funds according to their vision (rightly or wrongly) of what they perceive to be best for the greatest number of their constituents. If they fail to do this, accusations of pandering to special interest groups and of betraying communal interests are sure to be heard.

Basic to such considerations are fundamental communal definitions and decisions:

1. What are the societal goods (and, therefore, institutions) that merit public support, and how are they defined? What, for example, is healthcare and how does it differ from or intersect with social support? Are nursing homes social or healthcare institutions?

2. What are properly seen as subdivisions of a given public good or institution?

3. What particular department within a particular institution merits support?

Such decisions are, we have and shall argue, decisions that must be made by a democratic process within the particular corporate unit making them. Deciding what are social goods is thus a societal task; deciding what the proper subdivisions of a given social good or its executing institution is a task that must be made within the context of such an institution; and deciding what the proper departments are must be established within the institution itself. Nevertheless, such definitions and decisions are not isolated enterprises but are all made within the larger society that makes their existence possible and that will therefore have the ultimate say. Society may, for example, decide that all hospitals must have certain facilities to merit their approval.

Communities of various sorts and in various ways make the decisions that ultimately result in macro-distribution at all levels. Decisions here, of necessity, are political in that they are prone to the same decision-making process as are other communal decisions. They therefore accommodate themselves to prevalent political usage. Decisions made in the Greek *polis*, the Roman Empire, a New England village at the time of the revolution, or the United States today do not follow the same mechanisms, even though they remain communal decisions arrived at by political means. That is not to say that all political process is equally valid or that all decisions are justly made: It is to claim that decisions, however arrived at, ultimately must be, at the very least, not entirely unacceptable to the community, and that they are, in that sense, communal decisions. When communities strongly disagree, decisions within any political construct cannot long endure. Communities, it is true, may make wrong decisions (or decisions perceived to be wrong); that, however, speaks merely to the particular choice and does not invalidate the necessity and the right of communities to make choices. Decisions, made by communities today, furthermore, may not be reasonable as future contingencies change. It is, therefore, essential that such decisions are reviewed and adapted as the need arises. There are few fields in which this is truer than in health-care.

As communities have developed and as individuals have become better educated, more persons have become aware of the fact that they can shape their lives. If communities are to make just decisions, they ought to be acceptable to a broad consensus of individuals within them. The least that justice can demand is fair process. In today's society, fair process implies that all ultimately affected will have a say in decisions that affect them and that those who are effectively voiceless are maximally represented. Such fairness of process necessitates some form of democratic interaction.

To have meaningful democratic process, however, implies far more than merely political democracy. Without the necessary conditions for political democracy, political democracy itself becomes a sham. If all concerned are to participate effectively in the political process—that is, if a real and effective instead of sham and often ineffective democracy is to come about—certain preconditions must be met. At the very least, political democracy requires (1) personal democracy in which individuals will listen to and respect each other's right to an informed opinion, (2) economic (Dewey calls it "industrial") democracy in which all are assured fair access to the basic necessities of life, and (3) educational democracy so that all have a chance to be fully educated and informed. Only with these preconditions can a true political democracy exist and thrive; without these preconditions it is easy for some to gather sufficient power to deprive others of a meaningful choice between significant alternatives.[27–30]

Political democracy—so that some of the issues brought forward can be equitably decided—specifically requires that various coalitions and points of view receive an adequate hearing, a fair chance to present and discuss their concerns, and a decent opportunity, should they be able to persuade a sufficient number of their constituency, to participate effectively in the political process. It cannot function when the ability to do this hinges on the private wealth of candidates or their ability to "sell themselves" (often literally!) to various wealthy organizations or interest groups. Personal, economic, and educational democracy, together with a fair chance for divergent points of view to receive a hearing and a method of funding candidates divorced from personal private or corporate private sources, are at least some of the necessary prerequisites for a viable political democracy in which communities can be truly said to choose. When we speak of communities choosing, we have in mind communities that have such a foundation. A reasonable approximation to such communities is not a Utopian vision.[29-31]

In granting communities the right to make macro-allocation decisions, the method of arriving at such decisions is crucial. Fair process in a democratically functioning, informed, and interactive community is essential. Whatever the political underpinnings, communities in arriving at such decisions will be well advised to employ multifaceted and expert advice. In a world that has become as complicated as ours has today, a certain reliance on "experts" and on their guidance is essential. If such guidance is to be accepted, mutual trust must underpin it. Such trust is possible only when persons do not feel themselves disenfranchised or ill-equipped even to understand the advice given. This, of course, gets us back to where we were before: the necessity for a viable political democracy firmly founded on a basis of personal, educational, and economic democracy. The specifics of communal decisions, furthermore, like all other judgments, must be adaptive to changing conditions and must vary as technology and communal worldviews change. Justice, together with the community in which a particular notion of justice finds itself, in that sense, evolves and changes. Communities must decide the type and limits of their institutions—and they are likely to get (within the limits of what they can afford) those they deserve.

RATIONING, RATIONALIZATION AND HEALTHCARE: WHO KEEPS THE GATE?

First of all we must be clear that resources must be understood as constituting more than merely money. Resources, at least as importantly, are time, skill, effort, love and many other things beyond merely (or even most importantly) finances. Financial resources are the necessary but hardly sufficient resources needed to further any project. When it comes to the allocation of resources, we must be aware of what economists call "opportunity costs"—that is, what we spend for "A" in a closed system cannot also be spent for "B." In other words: the community cannot spend what we spend for health-care for other social goods it considers necessary. Secondly we must differentiate between rationalization and rationing.

Rationing and rationalization are two concepts that are often conflated with one another. Rationalization in essence refers to the elimination of waste occurring within a system: five people used to change one light bulb might be an example. The problem is that what is "superfluous," while often evident, is more often an arbitrary decision. Many of us would argue that every patient complaining of a headache does not need to have an MRI done and that doing so is wasteful of resources. Others might not. The time a patient should spend in the hospital following a given procedure not only is not fixed but medical opinion varies and changes. Questions like this are—in part—answered by outcome studies and so-called "evidence based medicine." But such studies can merely provide guidelines within which individual practitioners must be able to follow a different justifiable course in individual patients. While the repair of a hernia may in most patients require but a brief hospital stay the matter may be quite different for an 80-year-old patient who lives on the third floor in a house that lacks an elevator.

There is, nevertheless, no doubt that there is much waste within our system. Curiously enough this waste partly occurs because the physician's time for a thoroughgoing history and physical has been so curtailed that "tests" are often substituted for thought. The result is not only expensive and wasteful but is "bad medicine." Likewise, waste occurs when administrative costs are allowed to escalate—costs that initially were instituted to save money often have led to more expensive care and care that is of lower quality. The hodge-podge of arrangements we call a medical system—which in fact is not a system but various systems competing with one another—has led to an overall state of affairs in which waste is rampant and care of decreasing quality. Only when we have reasonably rationalized—that is, eliminated as much waste as possible while leaving physicians free to make proper medical judgments—should we begin to think about rationing.

Like it or not, resources are limited, health care is only one of many social goods and eventually some rationing is inevitable. Not everyone can have everything. And in fact we have been rationing all along. Although this statement has been denied, it is, call it what you may, the case. It is not, true enough, overtly done, but it is done. Rationing by ability to pay (by private means or by insurance), by race (the Indian Health Service), by disease state (the "end-stage renal disease" funding program), by age (Medicare), or by geographical region (benefits differ from place to place) is very much part of our daily lives. We have been rationing healthcare while often calling that process something else.[20,32–35]

Physicians are often charged with two seemingly irreconcilable obligations. On the one hand, they are charged with doing all they can for their patients regardless of other considerations; on the other hand, they are expected to conserve resources. We do not have in mind here performance of unnecessary tests, giving of useless treatments, or unnecessary lengths of stay in hospitals. Such things are by definition useless or unnecessary, and therefore illogical. They are, in fact, "bad medicine" and are what would fall under the rubric of rationalization when it comes to cutting cost. Rather than being done to serve the patient's "good" (a "good" that can obviously not be served by non-efficacious means), they are done thoughtlessly or are motivated by other considerations. When physicians, however, have a fair chance of serving their patients' actual "good," they cannot, within our current vision of the physician–patient relationship, be held back by considerations of costs, societal considerations, or the needs of others.[36]

That is not to say that considerations of cost or societal needs are trivial; indeed, they may and probably must in certain situations and under certain circumstances preclude the use of life-saving resources for some if not for all. There is no doubt that, from a purely technical point of view, physicians are in the best place to make such decisions. Ought they not, for that reason, be the ones to make and enforce such decisions in the context of their special knowledge of each case? Certainly a strong argument for the physician's role as primary gatekeeper can be made.[35]

To say that physicians caring for individual patients should give preference to the good of society or to the finances of their institution rather than to their patients' "good" is to do violence to our current vision of the physician–patient relationship. Expecting physicians to continue to embrace the traditional vision of the patient–physician relationship in which physicians as a first priority must serve the possible good of a particular patient and simultaneously to ration healthcare for reasons of cost to these same persons is a contradiction in itself. [37] Physicians in this situation cannot be expected to serve both of these masters simultaneously. But resources somehow must be used wisely, and decisions must be made.

If one accepts the premise that communities are empowered to make macro-allocation decisions, some of these conflicts may be resolved. Physicians can only disburse what is made available to them; resources not made available by the community, or made available only under certain conditions, are not available for distribution by the physician enmeshed in the obligations of the physician–patient relationship. A marginally effective and horrendously expensive modality may, for example, be made unavailable (except, perhaps, under restricted experimental circumstances), or a modality may be precluded for certain groups within a community (for example, communities may decide not to make ventilators available for infants under a given gestational age, may decide no longer to sustain permanently vegetative or comatose patients, or may decide to preclude the transplantation of organs into convicted murderers).

Healthcare professionals are not only expert at dealing with health and healthcare. They are also citizens of the community (see also Chapter 6) and as such they must (if they are to fulfill their obligations as citizens) participate in communal decisions. By virtue of their expertise, when it comes to healthcare, they are better equipped than most to advise communities. In that role, in which healthcare professionals are no longer dealing with identified lives to whom they are directly obligated, they can help give expert advice about such decisions without any fear of violating their and their community's vision of the physician–patient contract. Here they are advisers only, advising on medical efficacy and advisability. Their input is crucial to the final decision, which, however, must be compounded of many other factors and to which experts from many other fields must contribute. A healthcare professional under such circumstances serves as adviser to the community, which is the ultimate gatekeeper. When it comes to individual decisions made within the context of the professional relationship, healthcare professionals are then free to treat patients within a recognized framework set by the community. They cannot be expected to make available to their patients resources that are not made available by the community for their distribution.

Other models have been employed. In England, physicians function as primary gatekeepers with individual patients. They have learned to say "no," and they usually frame their denial of further treatment under the rubric of "medical advisability." Such models may work if work is defined as saving resources. But such models encourage (if not, indeed, force) physicians to participate in a basically hypocritical charade: The procedure is denied not because it would not be efficacious (dialyzing otherwise well-functioning elderly uremic patients certainly is "efficacious," if by that is meant returning them to what the patient, his or herself considers a meaningful existence) but because it would be too expensive. And often enough it is said that it in many ways depends upon who you are: one of the authors (EHL) is certainly over sixty-five and would, we feel quite certain, be dialyzed if he were lecturing in England and became uremic. The ethical dilemma faced by physicians in such situations and the danger that such a method will lead to capricious decisions and ultimately erode our current vision of the professional relationship seems obvious. Further the danger that this sort of institutionalized hypocrisy will make hypocrisy in other matters more acceptable—will create an atmosphere in which hypocrisy is accepted—is readily at hand. Communities can and should make such decisions democratically and openly and should be ready, as facts and experience change, to adapt or change these decisions. So that such a process can function well, the advice of the health professionals as "experts" is critically important.

Physicians in their arrangements with many HMOs stand to profit from work not generated. In such HMOs profit depends on not doing too many procedures, hospitalizing too many patients, or doing too much investigational work. Under some plans the physicians who are responsible for such savings share heavily in the profits they helped generate. In others they are rewarded for doing less. The pressure to do as little as possible—a pressure that may, for that matter, not even be consciously acknowledged—is ever present (see also Chapter 6).

The role of gatekeeper is, however, not always one of limiting access. Many hospitals need patients rather than, as was the case a few years ago, being short of beds. Physicians at times own or, at least, have financial interests in laboratories, free-standing X-ray and surgical units, and other medical installations. They generate the work done and simultaneously stand to gain from the work done by such institutions. Here, as contrasted to the situation in an HMO, the physician profits by doing more. In that capacity, physicians serve as positive gatekeepers.[37]

Before one can justifiably limit medical expenditure because of limited resources, gross waste has to be eliminated. The healthcare system in the United States is the most expensive in the world: Despite the fact that about 20% of the population is uninsured, and at least another 50% or more underinsured, the costs of the system are between 14% and 14.9% of the gross national product. We must realize what we are saying when we speak of being "fully insured" here in the US—when co-payments are expected for each visit, medication, procedure, hospitalization, etc. one cannot speak of "being fully insured." One of the authors (EHL) who happened to become ill had over $ 8,000 of co-payments in the last year: something he can—but many other workers cannot—afford.

More of the so-called healthcare dollar is spent on administration in the United States than in other industrialized countries. Outcomes, furthermore, are certainly not the "best in the world." Mortality rates (especially for the poor, for infants, and for minorities) are far higher than in some other countries, and longevity is less. The American healthcare system (if the current hodgepodge in the United States can be called a system!), which was but a few years ago the "best in the world," has decidedly fallen behind.[38,39] We have—and with our own hands—transformed ourselves from being the flagship of health care in the world to being the coal tender!

The costs of the system are, of course, multi-factorial. Without a doubt one of the main contributing factors is the rapid proliferation and use (most of us would feel excessive use) of technology. In the United States there has been a notorious overuse of technology: More technology is used in making simple diagnoses, more coronary bypass surgery is performed in dealing with coronary artery disease, and so forth, than is done in the rest of the industrialized world. And yet: *The results are no better here than there!* To take an example: Most U.S. hospitals today own and heavily utilize machines for magnetic resonance imaging (MRI). This technology—which constitutes one of the greatest technical advances of the last few decades—can, when appropriately used, add much to a physician's ability to diagnose and ultimately to treat. But it never is, and never can be, a substitute for thinking. Although hardly needed for dealing with most ordinary problems, it has more and more tended to be used in such a way. As of several years ago, the city of Peoria, Illinois (population about 125,000), was said to have more MRI scanners than Norway, and the city of Chicago more of these machines than can be found in all of Australia!

The explanation for such over-utilization is to a great part the market control of the medical system. Healthcare institutions as well as freestanding units compete with one another (why competition and the market are believed inappropriate to healthcare will be addressed shortly). Hospitals and other healthcare institutions advertise in the media and will often base their advertising on various claims for having the "best," the "newest," or the most "advanced" device in town. As a result, other institutions purchase a new and fancier unit. Once bought, these expensive devices need to be amortized and, therefore, subtle and not so subtle pressure is exerted on the medical staff to order such tests. The word spreads in the population and patients come to see proper medical care as using such tests.[40] It isn't long before even physicians consider it "standard practice."

Another important factor to the inflated costs of the healthcare system is the emphasis on crisis intervention and the neglect of preventive measures. Such preventive measures may be hygienic: proper diet, not smoking, the use of moderate alcohol, exercise, etc. But such preventive measures also include attention to social conditions that increase the incidence of illness: poverty, overcrowding, poor education, safe food and water, adequate prenatal care with proper nutrition, and decent conditions during the puerperium as well as ready early access to a healthcare system both for preventive care and early in the course of an illness.

Managed care was allegedly introduced as a solution to the rapidly escalating costs of medical care. Indeed costs tended to go down initially although of late costs have again steadily increased. Furthermore managed care (see Chapter 6) not only has not addressed the problem of the uninsured but has, in fact, made it worse—their numbers are steadily rising. In addition the co-payments extracted by these insurance companies have made the concept of being "fully insured" illusory. Managed care has become the principle layer in the distribution of resources—a player who calls the shots and with whom health-care professionals end up wasting untold amounts of time which (since time is one of the most critical and limited resources), in turn, they cannot spend taking care of patients. It has attempted to supervise closely (depending on the particular system) referrals, the use of certain technologies, procedures, prescriptions and—perhaps most importantly—has directly and/or indirectly attempted to limit the time physicians can spend with patients.

It is a system whose only profiteers are the stockholders and the excessively paid administrators. Furthermore—since these systems differ from and are in competition with each other—they have substantially altered the health-care professional–patient relationship. Patients—and rightly so—have begun to lose trust in their physician and feel less sure than before that physicians are primarily committed to their good and often suspect—and understandably so—that the physicians' primary loyalty is to the institution which directly or indirectly employs him or her. The time spent with each patient is often strictly limited as is the choice of physicians. A patient frequently finds that the physician who knows them best, with whom they have become familiar and whom they have come to trust is not—or is no longer—a member of the MCO (managed care organization) their employer currently provides.

The way in which most managed care organizations operate these days is that approval for various interventions, consultations or procedures must be given before what the physician deems proper can be done. Furthermore, a clerk, generally with a high school diploma and quite certainly without medical experience or training, generally gives such "permission." Cases can, of course, be appealed. But the length of time for physician and patient to go through this process assures that many will simply give up and forego the recommendation in question. This—quite openly—has been called the "hassle factor" and saves the managed care organizations thousands of dollars. How many lives it costs or stunts is unknown.

One could easily comprehend a single tiered managed care system that is one system for an entire society and in which physicians are free to practice medicine within the limits set by the system. Over usage could be retrospectively examined much as chart review is done today. Physicians may be told that they routinely overuse a given modality—but such criticism would be retrospective and not at the time the physician is dealing with a given patient.

Under the system as it exists, persons who are uninsured will find it virtually impossible to obtain proper medical treatment for mild or moderate illness. When such persons are ill they utilize emergency rooms since, under the law, they cannot be turned away but must be seen and "stabilized." What is meant by "stabilization" is an elastic concept and the fact of the matter is that, all too often, such patients are poorly treated—or not treated at all. Only when the illness has become critical (and when it is most difficult and expensive to deal with) will the patient be admitted. Then no costs will be spared so as to return him or her to the very environment responsible for having caused the problem.

The real problem, in the view of many, is that competition and the market are not appropriate vehicles for the distribution of medical care. They neither assure better quality nor afford lower costs: That much the American experience ought to have taught us! When used in areas appropriate for it, the market can indeed assure us of better goods at a lower price. Briefly put, however, the market philosophy rests on the following assumptions:

1. Consumers must have sufficient funds to enter and participate in the market and to choose among various goods and classes of goods; that is, they must have sufficient money to buy pears, vacation trips, or, perhaps, if they wish instead of a vacation trip, a new car.
2. Consumers must be well informed about what they wish to buy: They must know what for them is a good pear, a good automobile, or a fine vacation trip—including what it would be like to forego any, or all, of their options.
3. Consumers must have time to "shop around," to compare and to reflect on their options: They must be able to go to various supermarkets, car dealers, or travel agencies and then deliberate.
4. Consumers must not be coerced: They should not be stampeded into making a choice.
5. If they choose unwisely, such a choice should perhaps be regrettable but hardly fatal.

Medical care does not satisfy any of these conditions:

1. Most patients lack enough funds to buy healthcare for themselves. Generally their "purchase" is through employer-chosen insurance, and what is delivered is frequently at least in part decided by their employer as well as by insurance carriers. And employers who must choose among various offerings by insurance companies will (understandably) emphasize the costs to themselves rather than the quality of care for their employees.
2. Laypersons (or those laypersons who ultimately decide what policies to buy for their company) are not informed about the details of technical medicine. They cannot know what "good medical care" in fact is: Persons know that they would like to be well but the means of attaining that goal are not means that they can knowingly choose.
3. Patients who are ill or worried do not have the time and leisure to "shop around": Even if they knew what means were appropriate, they could only occasionally "comparison shop."
4. Sick patients are not normal persons "with the knapsack of illness strapped on their back." They are in pain, worried, and troubled. Their autonomy is reduced, and their pain, worry, and fears are a form of coercion.
5. Persons who choose wrongly in the medical arena and who are seriously ill may well end up dead or seriously compromised.[40]

Libertarians will, of course, deny all of this. To them, not only is the market the proper way of distributing everything—including medical care—but entrepreneurialism is touted as the legitimate basis of ethics: medical and otherwise. Such a point of view is predicated on the libertarian notion of a world of "moral strangers" (briefly reviewed in Chapter 2) in which freedom is the condition and not merely a high value of the ethical life and in which beneficence has only negative moral standing. Within such a framework, only a free market for the distribution of medical care is possible.[41] (See the discussion of various ways of conceptualizing and arguing for various ways of forming a healthcare system in Chapter 11.)

—— SOME WAYS OF DISTRIBUTION ——

When we must decide to allocate resources to individual patients or choose groups of patients to whom resources should or should not be allocated, we have several options of choosing. In general, the choices of macro-allocation will be made prior to micro-allocation: The decision to provide or not to provide funds for renal dialysis is an example. There remain decisions that, although individual decisions for identified lives, nevertheless introduce a severe quandary. When, for example, medical conditions have been met, there still may not be enough organs to serve all who may benefit. Decisions made at the communal level (decisions, for example, that would exclude axe murderers from being considered as organ recipients[42]) may not suffice. A residue of eligible candidates clamoring for an individual scarce resource (the notorious last bed in the ICU example) will inevitably persist.

While no firm answers can be given, ways of proceeding with such allocations need to be examined. Briefly, five methods of allocation (or a combination of these) have been suggested: (1) "need" and chance for benefit; (2) a market approach; (3) a lottery; (4) the queue or first-come, first-served, often (and, we believe, erroneously) lumped together with the lottery; and (5) social value judgments or "judgments of merit." The market approach, in which resources would be for sale to the highest bidder, has often been suggested and has lately seemed to gain in popularity. If community is seen as constituted of individuals united merely by a duty of refraining from harm to one another and in which beneficence is not an obligation (see also Chapter 3), an argument for this can certainly be made. Freedom, in such communities, is an absolute condition, and market price alone may control availability. Though note, parenthetically, that what this entails in conditions of scarcity is that the unfettered freedoms of the stronger will necessarily and always trump the allegedly unfettered freedoms of the weaker. If, however, communities are conceived as cemented by obligations of beneficence, this may not be the case. We—as individuals in a community—owe the less fortunate among us those things reasonably necessary to maintain life and to strive on as level as playing field as possible for the good things in life.

A modified market approach (one that makes resources not necessarily available to the highest bidder but precludes them for those who cannot bid at all) is, in fact, largely the way that medical care is distributed today.[38,39] Persons who lack independent funds as well as insurance may have no way of entering the healthcare system until it is too late. It is not only the indigent (for often the indigent are at least theoretically "covered" with Medicaid) or the elderly who find themselves in this position; it is, above all, the underemployed or minimally employed (the rent-a-cop, checkout clerk, or domestic). In living the fiction that life-saving care is, in fact, available to all and in generally insisting on the truth of this fiction, the community tacitly expresses its sense of obligation even as it fails to discharge it.

Intuitively we feel that making vital resources available only as an expression of market forces violates the duty of respect and caring that beneficent communities owe their members. Introducing an auction approach for resources in which resources are either available to the highest bidder or unavailable to those who cannot bid at all reduces allocation to a "trial by combat" in which the weapons are economic.[42]

Using "need" and chance for benefit is not as simple as it seems. It is obvious that persons who "need" a given intervention to return them to useful function must constitute the pool of possible recipients. Moreover, it seems obvious that those with the most urgent need should have preference over those whose need is lesser. Likewise is it obvious that a patient who has a 90% chance for long-term improvement would be chosen over one who only had a 1% chance. To that must be added the presumed length of benefit: Should a person (generally younger) who might benefit for 25 years from a given intervention take precedence over another (usually older) who might benefit for only 10 years (or 1 year or 1 month)?

Consider two candidates for a heart transplant: One, age 20, will die within hours or days if no organ is found but—in part because of a long waiting period—he already has severe damage to other organ systems making survival far less likely than without such a complication; the other, 55 years of age, is not quite as critically ill—he or she could perhaps wait a bit longer but would, during that time, run a substantial risk of severely damaging other organ systems and lessening their chances. In clinical medicine cases are difficult to compare: The variables are too vast. Such cases, within a general framework set by the community and a specific framework set by the particular institution and its staff (both of which should be publicly known) will have to be individually (and, undoubtedly, never to the satisfaction of all concerned) adjudicated on a case-by-case basis.

The lottery approach has often been suggested as a "fair" method for giving to one what cannot be given to all. It is the method by which occupants of lifeboats traditionally choose those who must be jettisoned in order to save the others. Leaving allocation to a lottery may be fair in the sense that choice has randomly fallen and that the decision to choose in this way was made prior to the time that anyone could possibly predict the outcome. Refusing to make a choice prior to exhausting all possible avenues of seeking out entitling differences is, however and in fact, making a choice. It is a choice that favors caprice over reason and says either that all reason has been exhausted and that no reasonably pertinent or relevant entitling characteristics remain (a rare situation and one in which only a lottery or force remains) or that we have deliberately chosen not to trouble ourselves to make the agonizing choices that we must if we are to live up to morality in the human condition. To hold that all reason has been exhausted and that no relevant entitling characteristics exist in many if not most instances violates common sense; to choose not to trouble oneself denies responsibility.[42] There may or may not be relevant (or apparent) entitling differences among passengers in a lifeboat situation, but the same is rarely true in the medical setting.[43]

Queuing—a first-come, first-served approach—has often been equated with the lottery.[43] In this approach, allocation decisions for groups have been previously made, and claimants who present themselves are the only eligible ones. Those queuing are within the groups. If physicians must do all they can for the identified lives under their immediate care, they cannot reasonably be asked to defer their present patient's good for the potential good of a possible (or even probable) later one. Objections to queuing that say that the time of queuing is often a social factor (since the more sophisticated and more affluent patient usually presents earlier than the untutored or the poor) are unquestionably true but fail to provide a solution: Moving the poor *qua* poor ahead is just as morally wrong as giving preference to the rich.[42]

Social value judgments, judgments that hold different individuals to be of different social worth, are underwritten by the belief that values can be judged as better or worse by some acceptable standard.[44] Such judgments obviously empower the allocator (or the community if the framework has been socially set) to superimpose his or her standards on others. On an individual basis and when dealing with identified lives, such judgments are an obvious violation of our current vision of the professional relationship; they are bound to be arbitrary and to lead to capricious abuse. They are, however, quite different from the making of such judgments by communities for groups of their members. Communal judgments of this sort are, at the very least, judgments made for groups of people by communal (rather than by personal) standards, and they are made by communities of which the claimants are a part and in whose values the claimant more or less share. Such communal judgments, while far from being non-problematic, are less likely to be capricious and arbitrary than are individual decisions. (See Chapter 9 for a discussion of social value judgments.)

The community might, for example, decide that convicted felons or felons convicted of particularly heinous crimes will not be entitled to have organ transplants unless superfluous organs are available. As things stand now, convicted persons lose their freedom and certain civil rights but are promised food, shelter and medical care. Under such conditions denying transplants to convicted murderers, for example, violates this promise. While it would not serve as a deterrent for crime it would spell out what a convicted felon can and what he/she cannot expect and not leave such decisions to the caprice of individual institutions or physicians.

A special case of social value judgment is the judgment that would penalize persons held to have "caused their own illness." These, it is said, do not merit care as much as those not implicated in their own illness. While there can be no doubt that habits and lifestyle have a profound influence on sickness and health, translating this fact into allocation judgments is quite a different matter. If we hold persons responsible for knowingly choosing their own lifestyle, we may be left with the conclusion that such persons are largely responsible for their own fate. We may, then, end up blaming the victim and washing our hands of much previously assumed obligation.[45] Our answers here will depend on the vision of the natural lottery and of community (see Chapter 2) as well as on our perception that causality is not so simple or so easily defined as one might think.

Persons are born into communities and therefore into situations not of their own making. Values of their family and of their community, their schooling, and their life experiences condition them to do certain things in certain ways. Free choice in the context of self-causation is a complicated thing. The external forces that tacitly condition our choice of lifestyle are too complex to be subsumed under "personal choice." Social forces and advertising, to name but two evident forces, are too powerful to ignore. Further, self-causation is too indistinct a concept to be useful. Where do we draw the line: smokers, drinkers, eaters of excess salt or fat, drivers of fast cars, the sedentary, those who fail to air their houses or to get enough sleep?[46] Social value judgments of this sort, although enticing, seem too complex to be useful.

———————— AGE AND RATIONING ————————

When we speak of rationing healthcare, we can think of such rationing as rationing by types of procedures or interventions (say, limiting the use of extremely expensive, exotic, and marginally effective interventions) or rationing by some other criterion. Rationing by such other criteria is largely what we do today. Some have proposed that rationing may be by age, social utility, etc. Some of these have been proposed not only as a last resort but also as a first line. Age—a simple and easily determined criterion and therefore one that has a great deal of appeal to many and especially to bureaucrats—has been proposed as a proper way of proceeding.[47, 48]

Age has been used because as people live longer they not only consume more resources but also consume resources in whose production they no longer participate. The amount spent on healthcare has, in part, increased because people live longer. Some feel that there is a "natural life span" beyond which no healthcare other than the minimal ought to be provided. They base their argument on a perception of what is and what is not a "natural" lifespan and would limit access to medical care based on an arbitrarily fixed cutoff point.[47] Others, who also are inclined to conclude that age might well serve as a limiting factor, reach this conclusion from a Rawlsian type of argument: They argue that behind a "veil of ignorance" most of us when given the alternative, would choose to spend our resources earlier in life, thus assuring for all at least a reasonable life span, rather than providing the chance of having more life at the end.[48]

Many (including ourselves) are uncomfortable with both of these arguments. In the first instance those who oppose age as an independent variable in medical decision making feel that using age in this manner is a statistical artifact and does not speak to the individual and his or her needs and capacities. One 85-year-old is not like another 85-year-old in intellectual or physical capacities. A natural life span for a species is not necessarily a natural life span for an individual organism within it. The fact that many persons at age 85 are beyond enjoying their life, are, perhaps, senile and bed-ridden or incapacitated by other illness, does not speak to the individual. Many, indeed most, at that age live vigorous, enjoyable, and productive lives. Throwing all into one pot merely because an arbitrary length of time has passed seems capricious.

Furthermore, it is impossible (at least at this stage of the game) to say what is and what is not a "natural life span." Even if this could be determined for the species, individual variations in all species are profound enough to make one hesitate in imposing what is again a statistical fact on an individual consideration. Even if one could determine a "natural" life span, there remains the question as to why a "natural" life span should constitute an ethically relevant fact. All of human activity, in one way or another, has involved adapting and, at times, extending the apparent framework that nature has provided. Prior to effective therapy, the "natural" life span associated with a variety of disease states was short indeed: yet we do not deny patients a treatment that would, often dramatically, change it. After all and among other things, that is what medicine is all about! Persons with "early onset" diabetes died prior to the discovery of insulin—their natural life span was limited indeed! Does that mean that we should not use insulin and thus change their "natural" to an unnatural but productive and meaningful life?

The fact that many aged persons are permanently vegetative, comatose, or so severely impaired that their current or future capacity to recognize or communicate with their environment is virtually nil is beyond dispute; so is the fact that many more persons at age 90 than at age 40 find themselves in such a condition. But what matters, ethically speaking, about Ms. Jones, who is ill, is not that many like her are permanently vegetative, permanently comatose, or incapable now or in the future of meaningful interaction. What matters, no matter what her age, is Ms. Jones' diagnosis and prognosis. Persons who are self-aware, and not merely *are* but "*have* a life" are members of the human community who, by virtue of that fact, are entitled to its benefits.

Those who oppose the "veil of ignorance" argument do so for a number of reasons. First of all, the young can and do have little conception of what they would want or not want at a more advanced age. To a 20-year-old person, age 60 is an advanced age and not quite imaginable. The veil of ignorance, those who are uncomfortable with this argument feel, is too thick to allow informed choice. Secondly, it is not at all certain that prudent choosers would make this choice if given sufficient facts. If healthcare to prolong life at a more advanced age were to be juxtaposed to having a facelift at an earlier age, many would hesitate. Healthcare, it seems to many, is too broad a concept to be easily encompassed under one umbrella—especially an umbrella that would "shield" against it!

One note of caution: Age, like any other physical fact, has to enter into the equation when it comes to deciding the prognosis for critically ill persons. There is ample evidence when one looks at various empirical data, for example, that triple organ failure in persons beyond a certain age today has no measurable chance of being reversed whereas such does not appear to be the case in those who are considerably younger. When one must make patient care decisions, the disciplined use of such data is not only ethically acceptable but is, indeed, as mandatory as not ignoring some other relevant piece of data.

To limit access to healthcare for the elderly—rather than for those of whatever age who can no longer benefit from it—seems an arbitrary decision and one that, among other things, would clearly deny equal protection to an arbitrarily chosen group of people. A decision to limit care for all the elderly above a certain arbitrarily determined age should not (ethically speaking) even be considered before what constitutes waste is not communally defined and, as far as possible, eliminated. Society would be hard put to defend a decision to support young permanently vegetative or comatose patients (or to buy and use more than the medically essential technology) but to deny therapy to functioning persons merely because they had attained a particular chronological age.

——— BUILDING A HEALTHCARE SYSTEM ———

A system is something that has some sort of internal coherence and controlling elements. The cardiovascular system or the educational systems are examples. In the health-care system as it exists in the United States today the only internal coherence and the only controlling element is a theory of the free market—and not even that is entirely carried into practice. Before we can even speak of building a health care system, certain basic concepts, terms and language must be agreed upon. Many terms are loosely used and need to be defined. What follows is an attempt to define some of these terms.

The term "socialized"—since it is bandied about rather freely—must be understood. Socialism, first of all, is a term often equated with communism. This is untrue and inaccurate. Communism denies the right to private property; socialism recognizes the right to private property but insists that the fruits of labor ought by right go to those whose labor it is—*i.e.*, the worker—thus worker ownership of, for example, United Airlines or the Saturn Car Company are, in a sense, a form of socialism. Socialism (and this is where the term "socialized medicine" comes in) furthermore, and most importantly, holds that certain goods and institutions essential to the community should be owned and controlled by it. Persons are persistently taught that democracy necessarily entails capitalism and that capitalism furthers democracy. Nothing could be further from the truth. Capitalism is an economic and socialism a political system and, while economic and political systems should preferably fit together, they are not synonymous. Capitalism or socialism can exist in a monarchy, a dictatorship or a political democracy. The philosophical basis of capitalism is the freestanding, largely asocial individual whereas the basis of socialism as well as of democracy is shared community. Social democracy is a democracy that emphasizes democratic process and accepts social responsibility; democratic socialism is a system in which the means of production are predominantly in the hands of those who have a part in creating the product and in which decisions are made in a democratic fashion. In democratic socialism, private capital exists but the community controls those things basic to communal life (things like health-care, education and public utilities).

Most national health care systems are not "socialized." A socialized system is one in which the state from general taxation creates, maintains and operates a health-care system. Many, if not most, systems in the industrialized world that provide at least basic health-care to all citizens do not meet such a definition. They are operated by and through various usually government supervised insurance schemes; but they are not, in the true meaning of the word, socialized.

If one is thinking of creating a health-care system one first of all must decide whether such a system should be single- or multiple-tiered. Although the two terms are often used as though they were synonymous, a single payer system is not synonymous with a single tiered system. In a single payer system there is one agency (by it government or private) which pays out "benefits"—conceivably this could be a large insurance company selling different policies to different persons: *i.e.*, one payer but with different payment arrangements for different persons.

Various countries have adopted a variety of health-care systems. In all of these countries there is one common denominator—they all provide at least basic health-care coverage to virtually all residing within their borders. The United States, as has been said, is unique in not doing this. The Scandinavian countries differ among themselves but have two important features in common: they are exclusively publicly funded and they use primary care physicians as gatekeepers. German, Austria and to some extent Switzerland are funded *via* mandatory employer/employee contributions, have a strictly regulated (but becoming less strictly regulated) insurance system and provide insurance for those who would be otherwise uninsured. The United Kingdom has a multiple-tiered, nationalized system with the national health care sector publicly funded. In that sense, albeit multiple-tiered, it is socialized. Canada's system is single-tiered and nationalized with public funds distributed among the provinces. France has a mixed system.

Different systems spawn different ethical problems. A system in which, as is generally the rule in the United States, the same physicians care for their patients both inside and outside the hospital, has somewhat different or at least differently shaped ethical problems than does a system in which ambulatory and in-hospital care are strictly separated. A capitated system offers different incentives from one that is fee for service. Physicians who must deal with private insurance companies face different ethical problems than do physicians who are paid directly by the government.

The language we use conditions the way we think and often determines the way we feel and act. In the last few decades there has been a gradual shift in language that has both caused and been caused by these other changes in physician–patient relationships. Physicians have become providers; patients have become first clients, then consumers and now, even worse, customers. Often they are, in insurance jargon, simply spoken of as "lives," or even worse, "units of care!" This shift in language (one fairly unique to the United States) is, in our view, by no means accidental or trivial—it is a shift at the very least encouraged by those who stand to gain by the disruption of an ancient relationship. It is one health-care professionals buy into at their peril.

Although our media would equate them, "single-payer systems" are not necessarily single-tiered. Single-payer system simply refers to the fact that there is one source of payment, be it the government, an insurance cartel, or whatever. Likewise, as we alluded to before, there has been a tendency to equate "democracy" with "free-market capitalism" and to equate "socialism" with a form of government that of necessity is not democratic. It bears emphasis, once again, that democracy is a political system; the free market and socialism, on the other hand, are economic systems. Although economic and political systems must, so to speak, be compatible with each other, they are not identical. Democracy is ultimately founded on a strong belief of a need for a well-functioning community and a respect for all individuals within it; the crasser forms of capitalism, on the other hand, are predicated on a philosophy of stark individualism in which lip service is paid to the individual "free" to starve or succeed merely by his or her own devices. A form of democratic socialism in which a homeostatic process aiming at survival, learning, and growth adjusts the goals of the individual to the needs of the community and sees the needs of the community as necessitating the prosperity of the individuals within it is, despite of all that is said against it, an eminently reasonable and viable choice.[4,5,15]

When we look at healthcare and healthcare providers in America today, we are seeing a system and a profession in transition.[49] In former times, physicians and hospitals were far more ready and far more able to offer care to the indigent than they are today. Resources needed to provide care were fewer. Obligations, furthermore, could be met by charging paying patients sufficiently more so that caring for those who could not pay was not as burdensome (sometimes called the "Robin Hood principle"). That is not to say that the poor invariably received proper care or that conditions were necessarily better than they are now. Rather, it is to make the obvious statement that society has undergone critical changes and that the delivery of healthcare is merely one of these.

Poverty in America is one of the realities of life.[50] It is well known that 20% of our people—many of them fully employed—are beneath the poverty level, that one out of three children goes to bed hungry, and that at least one out of seven persons (many of them children and half of them employed) lack access to medical care.[50,51] About 25% of children go hungry a significant part of the year. Under such circumstances, the question of whether communities are obligated to provide a decent minimum of necessities to their members takes on new urgency. Healthcare is felt by many to be such a need, since without it individuals cannot "maintain normal species functioning" and thus maximize the full range of opportunities.[48] Those inclined to this point of view will feel that the provision of healthcare is a basic necessity in today's world and that the lack of access to such healthcare constitutes a serious flaw. Others may argue either that healthcare is not, in today's world, a basic necessity, or that communities are not obligated to provide at least a minimum of basic necessities to their members.

Our viewpoint toward the idea that a decent minimum of healthcare (or of other necessities) is a human "right" in a just community depends, of course, on our prior viewpoint of community and of justice (see Chapter 3). If one

1. Believes that the "natural lottery" operates in selecting who will and who will not be ill and that it is the working of blind chance
2. Holds that the definition of community entails no necessary duties of beneficence
3. Maintains that "rights" are "natural" or "divine" and are therefore to be discovered and not constructed, and
4. Sees freedom as the sole necessary condition of communal life and not as a value to be traded on the marketplace of other values, one will hold that communities have no obligations to provide healthcare.

If, on the other hand, one

1. Believes the results of the "natural lottery" confer obligations on the members of a community because (a) the undeserved misfortune of a member of the community automatically confers such obligation or/and (b) that the "natural lottery" is, to a significant degree, a social construct.
2. Holds that the definition of community entails not only duties of refraining from harm to one another but likewise powerful obligations of aid to one another.
3. Maintains that "rights" are not discovered and therefore "natural" or "divine," but, rather, the product of human choices and values; and
4. Sees freedom as a fundamental value of a just society but not as the necessary and absolute condition of their existence, then one will affirm that just communities have an obligation to provide a decent minimum of essential needs for their members. They will do this because their view of what is just—what it is to "give each his or her due"—is grounded in these assumptions.

Most of us today would want to pay at least lip service to a view of obligation and community fashioned on some version of the latter model. Even those committed to a thoroughly individualist or libertarian philosophy will hedge their bets and hold that it would be meritorious or "nice" to help the unfortunate, even though without "creating a straightforward obligation on the part of others to aid those in need."[41] In a sense, we all feel committed not to let our neighbor starve or go without medical care (even though many of our neighbors do, in fact, starve, and even though at least one in five do, in fact, go without medical care). And having said this, many would feel compelled to provide at least a decent minimum of essential needs to all members of our community. That leaves notions of "need," "decent," "minimum," and "essential" undefined. Such definitions will vary from society to society and from community to community depending on time, circumstances, and values. Definitions, therefore, can be seen not as immutably fixed but as evolving and changing. They, like many of the specifics of ethics, reflect the values of the community by which they are constructed.

There are two ways of looking at "entitlements." One can adopt what has been called a "poor law philosophy" in which entitlement is the result of belonging to some particular group within a community (say the poor, those with end-stage renal disease, etc.), or one can adopt what has been called a "welfare philosophy" in which entitlement is the result of being a member of a given community.[51] In truth all communities are a blend of both: All of us (at least theoretically) are equally entitled to police or fire protection, but only some of us are entitled to food stamps. Just as there is not a community built entirely on a libertarian or an absolute communitarian model, there is none that does not have a mixture of these ways of allocating entitlements. In general, communities that lean toward the libertarian way of thinking will tend more toward having entitlements follow a "poor law" than having entitlements provided from a "welfare" point of view.

When all is said and done there are three possibilities of fashioning a healthcare system:

1. A so-called single-tiered system in which all are provided with a certain level of healthcare and none can buy more. Such a system will not make it possible for the wealthy to "buy" different physicians, different staffing of hospital wards, different instruments, different drugs, a different waiting period, or any other services that make a direct contribution to outcome. In such a system it is quite possible that one would be able to buy "luxury" items that do not affect outcome: a private room with pretty curtains, a gourmet meal, a bottle of wine at dinner, etc.

2. A multi-tiered system in which "basic services" are provided to all members of the community and those so inclined (or affluent enough to do so) are free (through insurance or out of their own pocket) to buy more.

3. A strict market system in which consumers buy (or receive, as part of their employment benefit package) various types of insurance. Such a system, in fact, is what is currently used in the United States—with the lowest tier getting nothing at all.

Except for the United States and South Africa, there is no nation in the developed world that lacks some sort of healthcare system providing at least basic healthcare coverage to all its citizens. Many of these nations (Great Britain, Germany, and Austria among them) have some sort of multi-tiered system; others (Canada and the Scandinavian nations, among others) have single-tiered systems. The United States today has one overwhelming advantage among other nations: It is so far behind that it could and should learn from the experience of various types of healthcare systems and then attempt to adapt the best and avoid the worst of all of these systems for its own conditions and culture. Obviously, one can no more easily "transplant" a healthcare systems into a culture into which it does not "fit" than one can transplant organs into a host not ready to receive them. The social or physical immune system of the nation or the organism would reject them.

Those who argue for multi-tiered systems generally argue in the following way:

1. Although basic healthcare should be supplied to all, purchasing healthcare above this level ought to accord with the values and plans of the individual being covered.

2. Persons should be able to express their values by freely choosing among diverse goods: more elaborate healthcare in case of severe illness, expensive cars and vacation trips, or a plusher lifestyle.

3. Persons who have worked hard, lived moderately and saved properly should be able to use such savings as they see fit. Why

should such persons pay for more than basic healthcare for those who have been lazy, lived extravagantly or refused to save? At first blush, such an argument has much to recommend it. After all, free people ought to be able to express their own values in the way they choose and allocate their income as they see fit. Respect for persons would suggest that this is true.

When one examines this argument more closely, several obvious fallacies are evident. Persons whose jobs do not give them a large income but who may be working very hard, living frugally, and saving as much as possible (and often that is nothing at all) would not be in a position to buy additional insurance, let alone pay for services out of their own savings. They may have a very high value for healthcare but of necessity have a higher value for providing sufficient basic necessities and sufficient education to themselves and their families. For such people, medical care above the "basic" would go begging. Furthermore, should a person, even one who was lazy, wasteful, or who failed to save, be punished by not having a life- or function-saving procedure when their neighbor can? Should such a person's child or other dependent be punished? Would we, as a people, conceive as just a society in which life was "for sale"?

The British system gives a perfect example of a multi-tiered (and socialized) system in which a great disparity among services for the generally covered and for those with private insurance exists. And that has had a distinct influence on outcome. In one study, the time from tentative diagnosis to cardiac catheterization and then the time from positive findings to surgery was measured in two ways: What was the case at the time measured (1988), and how had this waiting time changed over the last decade? The time from diagnosis to catheterization and the time from confirmation to surgery were both about two weeks for the private sector and almost a year for the public. Moreover, waiting time had remained stable in the private but sharply increased in the public sector.[52]

This should come as no surprise. Inevitably persons who have the greatest might in a society make laws. Members of Congress or Parliament do not generally come from the poor. They are usually in the class that could well afford to buy supplemental insurance or to pay for some care out of their own pocket. Ultimately it is the better off who will be paying for the healthcare of those who cannot afford it. It is in the interest of the wealthy (and, therefore, of the powerful) to consider "basic" to be as little as possible.

A single-tiered system is one in which everyone gets the same things (affecting outcome) and no one can buy more. It is the system largely in use in Canada, the Scandinavian nations and in many respects Holland. Arguments for such a system are largely based on the idea that we as a community owe those things available and needed to sustain life as well as adequate function to take advantage of the opportunities a given society may offer to one another. While in a single-tiered system all members of the community have equal access to those things that affect outcome this does not mean that all will, in fact, avail themselves of that opportunity. But it is there.

We shall (if we end up with any organized system within a foreseeable time at all), most probably, end up with a multi-tiered system in the United States. Such a system in an extremely individualistically oriented culture can probably be safely predicted—at least, one hopes, as a first step. It is not, in truth, the system we, the authors, would advocate, but there is no doubt that it is better than nothing.[53] The problem is that healthcare systems, like all major institutions, are basically difficult to change once they are put in place. In building a healthcare system for our country it would be well to sit back, consider, and think. First of all, being clear about where we are (our resources, our possibilities, and our problems); secondly, being clear about our values, goals, and hierarchy of needs and interests; and thirdly, interconnecting this by an appropriate means, would be a first step. In a well-functioning democracy such a decision ultimately is a political one. But a process of deliberation in which experts from diverse disciplines (the health-care professions, hospital managers, sociologists, economists, ethicists and so forth) work together free from political pressure should occur prior to the inevitable and necessary political decision. From such a process a variety of possible options could emerge, only then to be decided upon by the political process. Those of us who would or could be affected should all have a voice and those who cannot participate (some disabled, children, etc.) are represented—in a democracy through a political process but one that is informed and truly representative. Thus it would seem essential that in building such a system communal dialogue plays its part. From such a process a variety of possible options could emerge, and then to be decided upon by the political process.

Such decisions should be made by a well-informed and interactive public well advised by appropriate experts and transmitted to their elected officials and then inevitably acted upon and subjected to the political process. It is a complicated task and not one that can be accomplished by relying on gut feeling or rhetoric. It is in forging such critical decisions (whether they involve the creation of a decent healthcare system or the accomplishment of some other vital communal task) that the value of a true and the failure of an inadequate democracy manifest themselves. One cannot expect a nation that lacks the underpinnings of genuine political democracy to perform these tasks well. Moreover, decisions of this sort are decisions that must be made in view of the "opportunity costs" which allocation necessarily presents us with. These are hard choices which must be made by the community and translated into actuality with compassionate rationality: compassionate so that all can and do envision themselves as affected and as possibly—even if they are not now—affected at a time of greatest economic and personal vulnerability; rational because the facts and possibilities need to be clearly seen and dispassionately examined.

Medical costs have escalated. There will be no easy solutions. But solutions can and must be found. We have a plethora of examples in the form of other nations. These cannot, of course, be simply imported but must be fitted into a particular culture and into its particular needs and values. But we can and must learn from these other experiences, from their successes and from their failures.

The health-care system in the United States is the most expensive, the most inequitable and the most bureaucratized in the world. As good as the care of critically ill patients is in the United States even that is no longer the best there is. In the United States today we have become very skilled at remedying crises we could, with a modicum of foresight, have easily prevented. Often we remedy an acute crisis only to send patients out into the very same situations that have produced the crisis in the first place. Not only is this ethically problematic—it is, in the long run, economically unwise.

The most erudite discussions of the finer details of justice or the professions of despair by the medical community at the number of uninsured are pointless without political action. This is not a new observation: Aristotle long ago saw politics and ethics as firmly entwined. Questions of ethics ask questions directed at courses of action—action which when it comes to systems can only be modified within a political context. It is our thesis that those persons associated or concerned with the ethical practice of medicine have an obligation which transcends that of the ordinary citizen to take an active role in creating a system in which ethical practice can take place. Such an obligation is one which (and with particular force) ethicists who are supposedly the most concerned about ethical practice should eagerly embrace. Doing one's job as well as one can—including teaching the finer point of ethical theory—is pointless if the constraints of the system force one to practice in a way that one readily recognizes as being ethically problematic.

——— GENERAL AND ——— ECONOMIC PROBLEMS

In speaking of the problems of health-care ethics, of new knowledge in genetics or biochemistry, of advances in diagnostic modalities or in therapy or in developing new technologies we must be aware of one of the main ethical problems we are facing today: a problem we try to sweep under the rug and one that remains largely unmentioned. These new advances benefit a small fraction of this earth's population. We in the so called "developed nations" are developed in good part because of our systematic exploitation of vast areas we today please to condescendingly label as "underdeveloped" or as "third world" countries. Even in our so-called "developed" world there are large areas of poverty—often the result of exploitation and the abuse of resultant power. As we develop new scientific and medical advances we tend to forget that many of our fellow-citizens lack access to any but capricious health care and are dependent on charity while most of the rest of us are really inadequately insured: severe or prolonged illness can extract more in co-payments than many can afford and long term care for the elderly is a national scandal. The question before us is quite simple: do we as a community have the right to expend our resources on developing new and exotic means of treatment for the few who can afford it or are we—as fellow humans and as those who exploited and gained—obligated to use these resources to level the playing fields and bring at least basic health care to all of our fellow humans? Millions die of quite eradicable diseases—malaria and schistosomiasis are examples. Infant mortality (including infant mortality and unnecessary premature births in the United States) is unnecessarily high. Overpopulation and under-education aggravate these problems progressively. The Dickensian specter of Christmas future and the two children, whose names are want and ignorance threatens us still and, population growth, if not vigorously and rather immediately addressed, threatens to engulf the whole world.

We are not suggesting that our research into genetics or into new methods for treating disease should be stopped or that high technology medicine should not be used. But we are suggesting that a part of the monies used to support this "rich man's medicine" be used to fund massive programs of development, of education and of birth control in the nations which we have had a hand in impoverishing. Clearly most of us alive today are not culpable for slavery, exploitation or imperialism yesterday; but clearly we benefit from this past. Our comfortable and, compared to the rest of the world, opulent life style is possible only because of severe social injustice in the past, just as the miserable conditions in the lands and peoples we have historically exploited are our doing. We are, however, suggesting that part of the resources used to finance these advances be set aside to begin to level the playing field and to speedily bring about at least basic social and medical changes. The basic assumption underpinning these statements is quite simple: we are all human beings whose status, place of birth, sex or race are matters of "moral luck" and whose social standing is at least in part also attributable to the same "moral luck." We, if we are to maintain solidarity within our human community and together strive for the good of all, need not only to stop exploiting but, indeed, to help those whom we as a society have historically exploited. And we need to do this, among others, for self-serving reasons for without solidarity our future will be one of perpetual war of all against all.

REFERENCES

1. Hardin G. The tragedy of the commons. *Science*. 1968;162:1243–1248.
2. Hiatt HH. Protecting the medical commons: Who is responsible? *N Engl J Med*. 1975;293(5):235–241.
3. Rawls JA. *A Theory of Justice*. Cambridge, Mass: Harvard University Press, 1971.
4. Loewy EH. Kant, healthcare and justification. *Theor Med*. 1995;16:215–222.
5. Loewy EH. *Moral Strangers, Moral Acquaintances and Moral Friends: Connectedness and Its Conditions*. Albany, NY: State University of New York Press, 1996.
6. Loewy EH. Compassion, reason and moral judgment. *Cambr Q*. 1995;4:466–475.
7. Habermas J: *Moralbewußtsein und kommunikatives Handeln*. Frankfurt a/M, Deutschland; Suhrkamp Taschenbuch, 1992.
8. Cassel E. Do justice, love mercy: the inappropriateness of the concept of justice applied to bedside decisions. In: Shelp EE, ed. *Justice and Health Care*. Dordrecht, the Netherlands: D. Reidel, 1981.
9. Hunt LR. Generosity. *Am Phil Q*. 1975;12:235–244.
10. Rousseau JJ. *Du Contrat Social (R. Grimsley, ed.)* Oxford, UK: Oxford University Press; 1972.

11. Ross WD. *The Right and the Good*. Oxford, UK: Clarendon Press, 1938.
12. Loewy EH. AIDS and the human community. *Soc Sci Med*. 1988;27(4):297–303.
13. Kant I, Beck LW, trans. *Foundations of the Metaphysics of Morals*. Indianapolis, Ind: Bobbs-Merrill, 1978.
14. Kant I, Ladd J, trans. *The Metaphysical Elements of Justice*. Indianapolis, Ind: Bobbs-Merrill, 1965.
15. Loewy EH. *Freedom and Community: The Ethics of Interdependence*. Albany, NY: State University of New York Press, 1989.
16. Engelhardt HT. *Foundations of Bioethics*. New York, NY: Oxford University Press, 1986.
17. Nozick H. *Anarchy State and Utopia*. New York, NY: Basic Books, 1974.
18. Moreno J. The social individual in clinical ethics. *J Clin Ethics*. 1992;3(1):53–55.
19. Dewey J. *Theory of the Moral Life*. New York, NY: Holt, Rhinehart & Winston, 1960.
20. Loewy EH. Communities, obligations and health care. *Soc Sci Med*. 1987;25(7):783–791.
21. Locke J. *Two Treatises of Government*. Cambridge, UK: Cambridge University Press, 1960.
22. Brown SM. Inalienable rights. *Phil Rev*. 1955;63:192–211.
23. Hart HLA. Are there any natural rights? *Phil Rev*. 1955;64:175–191.
24. Frankena WK. Natural and inalienable rights. *Phil Rev*. 1955;64:212–232.
25. Churchill LR, Siman JJ. Abortion and the rhetoric of individual rights. *Hastings Center Report*. 1982;12:9–12.
26. Daniels N. *Just Health Care*. New York, NY: Cambridge University Press, 1985.
27. Dewey J. The public and its problems. In: Boydston JA, Walsh BA, eds. *John Dewey, the Later Works 1925–1953*. Carbondale, Ill: Southern Illinois University Press, 1988.
28. Dewey J. Creative democracy: the task before us. In: Boydston JA, Sharpe A, eds. *John Dewey, the Later Works 1939–1941*. Carbondale, Ill: Southern Illinois University Press, 1991.
29. Dewey J. Human nature and conduct. In: Boydston JA, Baysinger P, eds. *John Dewey: the Middle Works*, Vol 14. Carbondale, Ill: Southern Illinois University Press, 1988.
30. Campbell J. Democracy as cooperative inquiry. In: Stuhr JJ, ed. *Philosophy and the Reconstruction of Culture*. Albany, NY: State of New York University Press, 1993.
31. Stuhr JJ. Democracy as a way of life. In: Stuhr JJ, ed. *Philosophy and the Reconstruction of Culture*. Albany, NY: State of New York University Press, 1993.
32. Fuchs V. The rationing of medical care. *N Engl J Med*. 1984;311(23):1572–1573.
33. Schwartz WB, Aaron HJ. *The Painful Prescription: Rationing Hospital Care*. Washington, DC: Brookings Institute, 1984.
34. Churchill LR. *Rationing Health Care in America: Perceptions and Principles of Justice*. Notre Dame, Ind: Notre Dame Press, 1987.

35. Thurow L. Learning to say "No." *N Engl J Med*. 1984;311(24):1569–1572.
36. Levinsky N. The doctor's master. *N Engl J Med*. 1984;311(24):1573–1575.
37. Relman AS. Dealing with conflicts of interest. *N Engl J Med*. 1985;313(12):749–751.
38. US Bureau of the Census. *1984 Current Population Survey*. Washington, DC: U.S. Government Printing Office, 1985; for latest: http://www.census.gov/hhcs/poverty1.pdf
39. Blendon RJ, Altman DE, Kilstein S. Health insurance for the unemployed and the uninsured. *Nat J*. 1983;22:1147–1151.
40. Loewy EH. Of markets, technology, patients and profits. *Health Care Analysis*. 1994;2(2):101–110.
41. Engelhardt HT. Morality for the medical–industrial complex: a code of ethics for the mass marketing of healthcare. *N Engl J Med*. 1988;319(16):1086–1089.
42. Loewy EH. Drunks, livers and values: should social value judgments enter into transplant decisions? *J Clin Gastroenterol*. 1987;9(4):436–441.
43. Atterbury CE. The alcoholic in the lifeboat: should drinkers be candidates for liver transplants? *J Clin Gastroenterol*. 1986;8:1–4.
44. Caplan AL. How should values count in the allocation of new technologies? In: Bayer R, Caplan A, Daniels N, eds. *In Search of Equity: Health Needs and the Health Care System*. New York, NY: Plenum Press, 1983.
45. Allegrante JP, Green LW. When health policy becomes victim blaming. *N Engl J Med*. 1981;305(25):1528–1529.
46. Loewy EH. Communities, self-causation and the natural lottery. *Soc Sci Med*. 1988;26(11):1133–1139.
47. Callahan D. *Setting Limits: Medical Goals in an Aging Society*. New York, NY: Simon & Schuster, 1987.
48. Daniels N. *Just Health Care*. New York, NY: Cambridge University Press, 1985.
49. Starr P. *The Social Transformation of American Medicine*. New York, NY: Basic Books, 1982.
50. Physician's Task Force on Hunger in America. *Hunger in America: The Growing Epidemic*. Middletown, Conn: Wesleyan University Press, 1985. see also http://www.familiesuse.prg/930censusdata/clease.htm
51. Barry B. The welfare state vs the relief of poverty. *Ethics*. 1990;100:503–529.
52. Marber M, MacRae C, Joy M. Delay to invasive investigation and revascularization for coronary heart disease in South West Thames region: a two-tier system? *Brit Med J*. 1991;302:1189–1191.
53. Reinhardt UE. Future trends in the economics of medical practice and care. *Am J Cardiol*. 1985;56:50C–59C.

9

Organ Donation

INTRODUCTION AND BRIEF HISTORY

In the latter half of the 20th century, the possibility of transplanting organs has become a reality. It has been made possible not only by advances in transplantation technology but especially by new insights into immunology. Transplantation has been helped along the way with the introduction of the concept of brain death as the point at which death may, officially, be declared. The propriety of arbitrarily redefining death so as to make organ retrieval more easily possible is only one of the many ethical questions that have arisen since organ transplantation has become feasible.

The issue of transplanting organs from newly dead or from living donors may, if our hope in growing organs from omnipotent stem cells materializes, be relegated to the dustbin of history. Regrettably at present this is not the case. It is, therefore, likely that some patients critically in need of certain organs will meet certain, agonizing and perhaps unnecessary death.

At this point in time, some of the earlier ethical objections and quandaries, such as whether organ or tissue donation is ethically permissible in the first place or whether selections made by groups of people (called because of their function "God squads") were a legitimate option, have largely become irrelevant, either because funding, science or public consensus has resolved them. Some very critical ones remain—but these deal more with issues of allocation than they do with the substantive question of the probity of transplantation in the first place. Among the most important to be dealt with today are questions dealing with allocation—not to individual patients *per se* but allocation to centers which, in turn, allocate to individuals.

Despite the fact that death in most of our western societies (a problem to be discussed more fully in the "end of life" chapter) has been legally equated with death of the whole brain (instead of, as heretofore, with the cessation of heart action), serious obstacles remain in the way of utilizing transplantable organs as fully as possible. The problem is examined from the following perspectives: (1) a brief history; (2)

organs as public resources; (3) objections raised to organ transplantation, and their counterarguments; (4) possible instruments of donation; (5) the donation of organs from live donors and the ethical problems of using either relatives or strangers as donors; (6) allocation to transplant centers; (7) allocating organs fairly and the use of social value judgments in transplanting organs; and (8) the practice of organ sales as a method of distribution, and the sale of organs by poverty-stricken persons in under-developed areas. The ethical question of making brain death equivalent to legal death is addressed in Chapter 11, which deals with end-of-life issues, and the related questions of transplanting fetal tissue and of using anencephalic infants as organ donors is discussed in the chapter dealing with questions at the beginning of life (Chapter 10).

———— HISTORICAL CONSIDERATIONS ————

The idea of transplanting organs is not new, even though the reality of doing so is.[1] The miraculous transplantation of organs was spoken about in medieval times, and a 16th-century picture by Fernando del Rincón, hanging in the Prado, shows a sacristan, whose leg had become gangrenous, receiving the healthy leg of a black man, presumably a slave (the instrument of consent and the outcome of this venture, are, unfortunately, not revealed!). A probably apocryphal story speaks of Pope Innocent III's being transfused in 1492 with the blood of two youngsters conveniently sacrificed for this purpose.[2]

There is no clear-cut evidence that tissue transfer took place prior to the 17th century, when Richard Lower of England first transfused blood from one animal to another.[3] Shortly thereafter, Denys in Paris transfused animal blood into man, but a failure of one such procedure and a suit brought by the patient's widow (who, it was later found, had murdered her husband and blamed the physician) soured physicians on this procedure.[4] No further attempts were made to transfuse blood until Blundell in 1818 at Guy's Hospital first transfused blood from man to man. From then on, transfusions (tissue transplants in their own right) have been carried out with variable success—and their not infrequent failure made them a measure of truly last resort or one not even attempted. It was not until Landsteiner in 1900 described blood groups (and until he and Wiener further refined our understanding by finding the Rh factor some decades later) that at least some of the disasters that not so rarely befell recipients were understood and could now be avoided.[5]

Successful skin grafting and early transplantation awaited the late 19th and early 20th centuries. Ullman in Vienna and Alexis Carrel in New York first successfully transplanted the kidney of one animal into another in 1902, and occasionally successful experimental transplantation of other organs from animal to animal soon followed.

Transplantation of kidneys from one identical twin to another was first done in the middle of the 20th century. Successful transplantation from identical to identical twin became a successful operation but one which was only rarely available. Successful transplantation in humans from non-identical-twins, other family members or from non-related donors awaited a better understanding of the immune process and, where needed, its relatively safe but still difficult, expensive and long-term suppression. Tissue transplants other than those of the cornea (not usually subject to rejection) and blood (transfusable with relative safety since the development of blood grouping) have become a viable but still problematic option only in very recent times.

In speaking of "organ transplantation" we are, in fact, speaking of a variety of tissues. Blood transfusion is a form of organ transplantation different in its being renewable and in the fact that it is given as a stop-gap until the patient's own bone marrow can again produce blood. It also, unless an error in matching occurs, is not associated with any form of the rejection phenomenon. Transfused red blood cells maintain their function for approximately 100 days: in that sense they do not "implant" the way other tissues do. In considering organ or other tissue transplantation one must differentiate among: (1) paired but not renewable organs or tissues of which one can be taken from a living person without necessarily causing them grave harm (kidneys, for example); (2) unpaired and not renewable organs or tissues that are essential for any persons survival (the heart, for example); (3) renewable organs or tissues that will regenerate in the donor after some of the organ or tissue is removed; (4) tissues or organs whose removal (a) is associated with little short-term danger or serious discomfort, (b) is associated with considerable immediate danger or discomfort, and (c) can be a threat many years after and (5) the use of placental stem cells.

Thus, transplanting corneas is one thing, kidneys another, and hearts or heart–lung transplantation a quite different matter. It had formerly been thought that removing a kidney from a person whose kidney function was excellent posed slight immediate but virtually no long-term risk. This has, unfortunately, not been true, and this fact has subtly changed the way we look at the ethics of using live donor kidneys.[6,7] Blood and bone marrow can be taken from a donor with little discomfort (more discomfort with bone marrow with which anesthesia is required than with blood, however) and with virtually no associated short-or long-term risk. Lately it has been shown that when pieces of liver are removed from a live donor the organ will regenerate and the portion of transplanted liver will, if the transplant is successful, form a new and functioning liver in the recipient. Here there is substantial immediate risk for the donor: inevitably much blood is lost, and many transfusions are needed.[8,9] Since, however, the organ fully regenerates, the long-term risk is probably slight.

Using fetal stem cells (omnipotent or pluripotent) has opened a new avenue of transplantation. In that it is or could be in large supply and in that risk of rejection is minimal, the use of pooled hematological stem cells has become a reality that—largely because of the question of "ownership" and of commercial exploitation—has become a practical, social and ethical problem in its own right. It will be alluded to here but more fully discussed in the chapter on genetics. Furthermore some of the scientific questions ("how long can such cells be maintained in a frozen state" or "are, if that is possible, a person's own stem cells useful in treating the later onset of leukemia in such a person?") remain unsettled.

In the case of harvesting corneas, much more time is available to harvest the eye than there is for the harvesting of other tissues. Storage for a reasonable time is possible, and rejection is rarely a problem. Kidney transplantation is unique in that dialysis, as an alternative to transplantation and as a stopgap when rejection occurs, is readily available. On an individual basis this makes the problem less immediate and makes rejection less devastating than rejection would be with liver, heart, or lung. Organs for transplantation are best obtained from patients whose physiological homeostasis is maintained up to the time of harvesting and when permissible storage time is brief. In other words, organs are removed during biological (albeit entirely artificially maintained) life. Rejection is very much a problem here. Heart and heart–lung transplants must likewise be harvested while the organ is functioning, and time from harvest to transplantation is even more constrained. These technical differences elaborate subtly different ethical questions.

———— ORGANS AS PUBLIC RESOURCES ————

Organs are a different kind of resource than are, for example, ventilators, drugs, or ambulances. First of all organs or tissues are, in our current state of ignorance, not resources we can supply at will or by investing more funds, a fact which, if stem cell research proves to be clinically useful, may well change. In live donation, donors are truly giving of their very self and this has practical as well as symbolic value. Cadaver organs, likewise, have overtones that we intuitively feel make them different from all other resources. There is, rationally speaking, good reason for this. Cadaver organs, among other things, differ from other resources in that they (1) like live donor organs cannot be renewed (although they could most certainly be made more available); (2) are of vital use to persons whose organs they are but, except under most unusual circumstances, are of no use to anyone else (except, perhaps, as articles of food!); and (3) are not, once a person is dead, the property of any specific person except, perhaps temporarily, for purposes of burial or organ donation.

Besides these essential differences cadaver organs very recently were an organic part of a living, breathing, thinking member of some human association, who was loved (or hated) by others, and cadaver organs are, therefore, of symbolic value to others as well as peripherally, to the community at large.[10] While the newly dead whose organs are suitable for transplantation have symbolic worth to those who knew them as well as to the larger community, they are of critical secondary (or material) worth to other members of community in need of a transplantable organ. Organs from the newly dead are thus different in being not only un-renewable but also in having been the functioning part of a person connected in fact and in symbol with the community in which such transplantation occurs. A legitimately greater communal as well as familial concern for the proper fate of organs exists than for renewable and potentially (given sufficient funds) plentiful objects that were never the actual body parts of persons.[11]

Organs voluntarily donated by the living are somewhat different. Many are un-renewable (kidney, for example) but others like bone marrow or liver—will regenerate. Organs are an integral part of and use to a person who has socially recognized ownership rights at least for his or her own person. Such "use" may be a real need (as in heart or lung) or the loss of such an organ or piece thereof may be a nuisance and, at times, a small risk, In either case; when transplanted they constitute something that morally rightfully "belongs" (if anything does) to that person and which that person (for whatever reason) can ethically and legally only voluntarily surrender. The voluntary and free donation of such organs by the living to the living is thus an example of altruism and is, ethically speaking, a supererogatory act. The donation of such organs must be hedged by considerations dealing mainly with issues of free consent.

—— OBJECTIONS TO TRANSPLANTATION —— OF ORGANS AND COUNTERARGUMENTS

The thought of transplanting the organs of the newly dead into the living makes some people uncomfortable. As the practice has proliferated and proven to be often highly effective and as religious objections have largely been put aside, objections have tended to be not as much against the practice as against associated problems. Transplanting organs from the living donor into the needy recipient often meets with other objections—objections that are concerned with risk, discomfort and the possibility of subtly—or even not so subtly—coercing the donor. Many of the objections to transplantation in the past were enmeshed in a not unreasonable fear of technocracy and today of the philosophy so prevalent that says what can be done must and, therefore, ought to be done. A sustained religious tradition, which endows the physical remains with mystic qualities beyond the symbolic, often also underpins some of these objections.[12]

Among others, three main philosophical objections have been raised: (1) the fear of "mutilation" and of disturbing wholeness; (2) the concern that making body parts between individuals interchangeable might serve to make individuals looked upon as organs of the community to be disposed of at will (in Kantian terms, using one as mere means to another's purposes); and (3) the misgiving that diluting the respect for natural symbols will weaken a necessary communal respect for symbols.[13] A practical objection has likewise been raised: the fear that critically ill but not yet brain-dead persons may have the type of care they are given influenced by their being potential donors. As a result, it is feared that attention may shift too early to preserving organs rather than to sustaining the ill person.[14]

The first objection, the argument from wholeness, essentially takes the following form:

1. Capriciously removing a part of an organism (say, my hacking off my ear merely for the sake of doing this) not only is irrational[15] but is "mutilation" and unacceptable. Persons are their body's stewards and compelled not to treat their bodies in injurious ways. (A hidden premise, beyond the rational, is often a religious one: that persons are ultimately the property of God and that treating one's body in an injurious way is damaging what one ultimately does not own and is, therefore, offensive to God.)

2. Persons, since they are merely stewards of their body, are justified in removing a part of their body only if by so doing they preserve the integrity of the whole. If, however, a part is removed so as to preserve the integrity of the whole, then, in the context of stewardship, such "self-mutilation" is not only permissible but, since it ultimately promotes wholeness, mandatory.

3. Mutilation of the body by removing a part is impermissible for any reason, even that of helping one's neighbor, other than to preserve the integrity of the whole body of which it is a part (this, again, has religious rather than rational roots).

4. The idea of totality to be preserved intact when man dies persists (this, of course, hearkens back to the religious issue of resurrection).

Allowing the invasion of one body for the sake of another, according to the second objection, creates a society less mindful of individual rights. It is connected with the argument from wholeness and takes the following form:

1. A person's wholeness may be disrupted only for the sake of preserving that individual's own personal integrity.

2. Communities and their members relate to each other in ways substantially different from the relationship of individuals to their parts.

3. Although one might hope for more, individuals and their communities are bound together merely or at least predominantly by negative duties of refraining from harm to one another and not by any other duties, such as the positive duties of beneficence, or doing good.

4. If one allows persons to be invaded for the sake of other persons or of the community, one is apt to produce a state of affairs in which the individual becomes merely another organ of the community and is disposable for the needs of the state.

The newly dead serve, among other things, as symbols for the living. A decent respect for symbols is one of the things that unite communities and make them what they are. The fear, here, is that customarily salvaging tissues might lessen communal respect for other symbols and, hence, for the reality they portray.[12,13]

Some physicians have expressed the fear that customarily salvaging organs may make the healthcare team tend to look at critically ill patients more as organ donors whose organs are to be preserved at all costs than as salvageable patients to be healed. Efforts may, therefore, be bent to do things calculated to preserve organs (giving large amounts of fluids, for example), rather than doing some very different things to save patients (restricting fluids in cerebral edema, for example). Customarily salvaging organs, so the argument goes, produces a mindset that would favor salvaging organs from a possibly, but not yet certainly, moribund patient rather than making every effort to save the patient. Presumably, by creating a conflict situation for the attending physician and tending to jeopardize the patient, making organ donations routine is inimical to the traditional vision of the professional relationship.[14]

The counterarguments to these objections often fail to meet them entirely because many of these objections are underpinned by religious or quasi-religious (and, therefore, supra-rational) feelings. They predicate themselves, for instance, on the concept of "wholeness" as a totality to be resurrected or the body as, ultimately, the property of the Deity. Justifying objections by an appeal to a supra-rational system of beliefs excludes these objections from rational argument. That does not, by any means, make these objections trivial. In a pluralist society, respect for other belief systems underpins the possibility of peaceful coexistence. But it does mean that the persuasiveness of such an appeal is limited to a specific moral enclave and to a form of personal morals that cannot, therefore, be rationally translated to the wider community.

Many of the arguments opposing organ donation (especially donations from the newly dead) are couched in terms of symbols. Symbols come into being as epiphenomena of a reality that they come to represent. Symbols, wherever found, relate to reality. They may outlast it, or they may be distorted and hard to recognize. Nevertheless, symbols must either represent our perception of reality or, occasionally, be derived from yet another symbol, itself ultimately grounded in such reality. Symbols comfort the bereaved and allow the abrupt transition from life to death to be softened, and, in a sense, they mitigate the moment of loss. They are thus important to our discussion—even if we think we cannot allow them the same standing as we give to the reality they represent. Confounding symbols with the reality they represent, or holding them to have the same, or even higher value, ultimately sells out reality. When symbols rather than the things they stand for assume primacy, honest sentiment is replaced by mere sentimentality.[11,15] Exercising compassionate rationality (that is, tempering our compassion with reason) may help us along the way.

Not surprisingly, many of the objections to organ salvage are couched in language that expresses a conclusion while posing a question. Speaking of "mutilation" introduces a repulsive metaphor that inclines one against the act; it carries unwarranted and unnecessary connotational baggage. People facing gallbladder surgery do not think of themselves as being "mutilated," and even amputation is not generally couched in such terms! "Wholeness," on the other hand, which "mutilation" is said to disrupt, carries a pleasing sense and one that one generally would not wish to disturb. Once again, as in so many other issues, the language in which we frame an issue or situation determines how we see a problem and, consequently, what we count as solutions.

The practical objection, the fear that critically ill patients who may be potential organ donors may be treated primarily as organ donors rather than as sick patients, is a fear grounded on the possibility of a very real conflict of interests. Such a conflict of interests, however, can be largely avoided. Under established protocol, the team treating the patient must be distinct from the team dealing with transplantation. There is no doubt that even with two teams operating there is bound to be some overlap: it is unrealistic to claim that the treating team is entirely unaware that they are treating a potential organ donor, and it is unlikely (indeed, it is impossible) that no contact and consultation or at least conversation between the two teams takes place. Nevertheless, it is the treating team's obligation to treat fully and completely until the patient is indeed "brain dead." This argument, moreover, confuses the newly dead (or, at least, those who have been declared "brain dead") with the not yet newly dead (or those not yet declared "brain dead"). Treating the not yet newly dead (or not yet "brain dead") as though they were newly dead (or as though they were "brain dead") is not only a category error. It is, in fact, malpractice and, therefore, subject to the same controls and sanctions applied in all such cases. Abandoning practice because, occasionally, malpractice may occur, does not seem an altogether wise move!

The fear in using what has come to be called "beating heart donors" (a practice in which an otherwise organically dead patient is sustained until time of organ removal by artificial means and purely for that purpose) has raised similar problems and given rise to a variety of procedures, protocols and instruments of consent which seek to minimize such risks. One of the conflicts—and one that seems obviously not an ethically viable option—is that sufficient analgesics be withheld so as to have "more perfect" organs. Well worked out protocols, differing teams and continued surveillance should do much to reduce (if not entirely obviate) this new problem.

Most faiths today have accepted organ transplantation as a legitimate medical practice. Some few religious groups continue to have a lingering distaste for organ transplantation, and a very few sects would forbid it entirely. In terms of secular ethics, our duty to the newly dead (who, since no longer capable of being harmed or benefited, are no longer of primary worth) is one of respect for what has been and for the value that others place in them. Largely, then, the newly dead are of secondary (symbolic as well as material) worth. Our main ethical obligation now is to others: family, loved ones, members of the healthcare team, and others in the greater community, both those who may stand in critical need of viable organs and those who don't. All of these others have legitimate interests for whom arguments can be made, and such arguments will need to be adjudicated. However, our respect for the living who may, by proper transplantation, regain health and function must weigh heavily against an appeal to sentimentality. In a sense, it is a clash between symbol and reality: the symbolic standing of the newly dead may, at times, conflict with the actual standing of persons who are very much alive and likely to be kept alive if a successful transplant can take place. Instruments of donation become an essential part of this discussion.

——— INSTRUMENTS OF DONATION ———

Communities may decide that the needs of the living properly preempt the rights of the newly dead and of their families.[16] In such communities, instruments of donation will not be thought necessary and routine salvage of all available organs will occur. Salvaging organs routinely, presupposes that (1) a dead person can no longer own anything; (2) while persons can will some of their possessions to others and while testators then own this property, organs are a different matter; and (3) a society that freely uses the organs of the dead to benefit the living is a more mature, generous, and humane society than one that does not. However, the routine salvage of organs, in our current society, is intuitively felt to violate the dignity of the newly dead and of the family with its still poignantly close connection to the deceased. Respect for persons, we often feel, demands that we acquiesce to, facilitate, or at least, not hinder autonomously made rational decisions. Treating the newly dead merely as a means to another's end and ignoring the family's expressed wishes offends that respect. Some means of attaining the goal of maximizing the salvage of organs while maintaining a decent respect for others must, we feel, be possible.

The other available option is some sort of voluntarism in which either consent for or refusal of donation is presumed. A distinction is currently drawn between voluntarism that establishes refusal as the norm ("presumed refusal") and voluntarism that makes consent a norm ("presumed consent"). Both of these subscribe to the essentially voluntary nature of donation: merely the presumption differs as to what persons would wish to have done.

In much of the United States today, refusal is the presumed norm. Individuals prior to death, or families after the death of the next of kin, must specifically give their consent so that the retrieval of organs may proceed.[17] Those inclined to find the entire enterprise of organ transplantation more or less distasteful will, in general, support this type of volunteerism since it is, indeed, hedged with the greatest number of safeguards against removing organs without iron-clad consent. Further, as Ramsey argues, this type of voluntarism affords an opportunity for expressing human generosity. The impulse for generosity, he feels, would be thwarted by other methods of donation.[13]

In theory, it is the will of the newly dead that controls even if the family is opposed. Unfortunately in our view and perhaps legally challengeable, donation does not proceed unless the family also consents. In the 50 states today, there is no legally valid reason why a legally executed donor card should not have priority over the family's opposition. As things stand, few hospitals, even in the face of a legally valid donor card, will proceed to harvest organs over the objection of any family member. And while the fear of litigation drives such a hesitation to comply with the deceased's clearly expressed wishes, one could argue that allowing families to override the clearly expressed wishes of the deceased violates an implied contract. But, ethics aside, this does not necessarily solve the question of litigation: it is not inconceivable that a family member, outraged because the clearly expressed wishes of the newly dead were not honored, could bring suit.

Stimulated or encouraged consent actively promotes the signing of donor cards and encourages the active solicitation of organs from suitable donors and from their families. Indeed, in some states physicians by law must certify in all suitable cases that an attempt to gain consent was made. Recent legislation in many states has made requesting organ donation mandatory. When patients who might be suitable donors die, health professionals are required to ask the relatives for their consent but still often fail to do so or do so in a perfunctory manner that ensures that consent will be withheld. One of the problems, of course, is that the concept, "suitable," can be very loose indeed. Objections to such laws are quite similar to the objections raised against presumed consent—especially the objection that this would convert possibly viable patients into premature organ donors.

Objections of this sort are, as Caplan has pointed out, largely wrong: "In enacting required request legislation, our society has indicated its collective desire that people routinely be given the option of organ and tissue donation." Required request laws thus are, as Caplan points out, an alternative to presumed consent statutes and "reflect a sensitivity to the key values of voluntarism, altruism, and informed choice."[18] At the present time, and with current instruments of consent, however, only 10% to 15% of suitable donors ever become actual ones, and even these rarely end up donating fully. Donation is looked at as a supererogatory act.

In some other parts of the world as well as in an increasing number of states within the United States, voluntarism takes a different form.[19] Called "presumed consent," this form of voluntarism presumes that most persons, when the chips are down, will be inclined to donate an organ useless to themselves to another who critically needs the organ.[1, 20, 21] Under such rules, persons or, should no prior wishes by the patient be extant, their surviving relatives have a right to refuse donation, and they must be explicitly given that chance. In the absence of such refusal, however, organs are taken. In presuming that when given the chance most, in fact, would be willing to donate their useless organs to a neighbor in critical need, presumed consent makes of organ donation a species of beneficence with all the ethical overtones that this implies. Far from thwarting a generous impulse, presumed consent presumes that individuals nurtured by generous communities that take the obligation to help their neighbor as an implicit norm will themselves act generously. Instead of decreasing the impulse to be generous, presumed consent makes generosity the implicit norm and fosters it. Countries in which implied consent is the norm, incidentally have a much higher salvage and transplant rate than do others.

A recent proposal would link the chance to receive an organ (or, perhaps, blood) to one's own willingness to donate after death. Persons would be free to donate or to refuse to donate any or all organs and tissues. Such a willingness or refusal would have to be expressed by a stipulated age—say, 25 or 30. It could be recorded on the driver's license, in newly emerging medically available computer banks or by some other suitable legal instrument. The request of those who refused to donate would be honored, but a person who refused to donate would, should the need arise, not be eligible to receive the type of tissue or organs they themselves refused to donate. It is a proposal that has a certain symmetry and inasmuch as it demands universalizability (I cannot expect another to do something for me which I under the same circumstances would refuse to do) has a certain amount of conceptual appeal.[22]

No matter which form of consent is used, problems with staff cooperation persist.[23] Suitable potential donors are often victims of sudden illness or accident, and only rarely succumb to a chronic process where death is expected. In such cases an advance directive stipulating that its carrier wishes to have their organs donated is, of course, most helpful—and unfortunately rarely present. Such sudden events often occur in outlying hospitals in which transplants are not performed and in which the staff lacks the stimulation of an ongoing and evidently successful transplant program. Healthcare professionals in such institutions do not see the survivors of transplantation leading happy and productive lives: To them they are unknown, strangers whom they had never met and would otherwise never have met. In such institutions healthcare professionals merely see the newly dead and the bereaved. Not having personal experience with the benefits of transplantation makes it even more difficult to approach the family engulfed in grief, a grief in which the staff shares and one that is inevitably compounded by a sense of failure. Therefore, organ donation often goes by the boards.

When all is said and done, we are still confronted with the problem of wasting as little tissue from the newly dead as possible, while meaningfully showing respect for the symbolism so dear to members of our community. Wasting tissues and thereby wasting lives or function that might otherwise easily be saved is in itself a troubling ethical question. Designing instruments of consent that strike a proper balance between the needs of the community and the peculiar interests of the individual is, therefore, a pressing concern. Thoughtful communities make such decisions aware of their priorities, values, and history. They are, like most things, not issues that lend themselves well to ad hoc decisions nor issues that can be settled once and for all.

———— LIVE ORGAN DONORS AND ———— STRANGERS AS DONORS

In the last few years (and since the first edition of this book was written) live organ donors—including non-next-of-kin donors—have become acceptable in most centers. According to transplant policy today, live donors are acceptable only if (1) their consent is freely given and the implications of their act are, as fully as possible, understood by them; (2) the donor's state of health allows donation; and (3) there is a good chance for a successful outcome.[24–27]

The first requirement, informed consent (the cornerstone of presupposing autonomous action) is necessary if we are to speak of "donation" rather than of piracy. The second requirement, demanding that the general health of the donor permit safe donation, is similarly clear. Only if freedom were considered an absolute would a community allow donors to seriously jeopardize themselves by donation. One would, furthermore, be unlikely to find physicians within the current vision of our patient–physician relationship who would consent to take such a risk. The third reason—a successful outcome—is more problematic. Surely no one would quibble with not transplanting a kidney from a healthy donor into one whose chance of success was almost nil. The problem of "where to draw the line," the problem of what constitutes a "good chance," remains. What if a parent wishes to donate to an already severely neurologically damaged child in need of a kidney when the possibility of success is only, perhaps, 50%?

There have been serious international organ donation problems. Using persons executed for some criminal act (and what that act is varies from country to country and, it is not unthinkable, may be readily adjusted to economic or other convenience) has become a serious problem in some nations—notably, it is said to occur in China. The ethical problem is not only that of the country involved or the practice of scheduling executions to fit the convenience of recipient, institution or transplant team. The problem is also that of the recipient who, fully well knowing the circumstances, engages in what amounts to a commercial transaction that sacrifices one life to save another—it is understandable but ethically problematic. Likewise the problem arises for physicians in the recipient's home country who must feel compelled to continue the treatment of donors under such circumstances.

There are other problems for healthcare professionals: the prospective donor is not ill (in point of fact, according to the criteria for allowing such a donation, he or she must be quite healthy) and any intervention the physician makes can only serve to put the patient at risk of harm. May physicians together with other healthcare professionals play a necessary role in permitting a patient (for the donor too is now a patient) to take a serious risk? This, of course, presupposes that "risks" or "benefits" are entirely physical or at least that a physical risk outweighs a psychic or emotional benefit. Donors benefit by being donors: whether they donate to a loved relative or friend whom they thus play a part in saving or whether they contribute to the salvation of a stranger they are given the opportunity to discharge an altruistic impulse. That much has been empirically shown. In fact, physicians are constantly "helping their patients take risks." Many sports events are associated with not inconsiderable risk and require a prior examination by a physician; physicians regularly examine inductees into the Army (an activity that, at least in wartime, is not without risk) as well as coal miners and other workers whose work includes the taking of calculated risks. Although there is, as always, a serious question of how much risk one may cooperate in causing, taking an organ from a healthy and willing patient to give it to another who has a high chance of meaningful benefit from such a transplant should not represent an insurmountable problem to a conscientious physician.

Finally, we need to keep in mind: transplanting a kidney differs materially from transplanting a liver, a heart or a heart and lungs. There are methods for substituting for kidney (dialysis) or pancreatic (insulin) function for a considerable length of time. That is not the case for heart, liver or for heart and lungs. Furthermore, while paired organs may be transplanted from live donors; there is consensus (another assumption we will not argue for or against) non-paired organs such as hearts or hearts and lungs may not. Although we readily accept the sacrifice of one life for another in war, we fail to condone such practices when it comes to donating organs. In terms of being obtainable from live donors, livers are more akin to kidneys than to hearts. Parts of a liver may be transplanted from donor to recipient because regeneration of an adequate liver in both donor and recipient is most likely to result. Again there is a difference here: the short-term risk to the donor of harvesting a kidney is minimal; the long-term risk may be a bit greater. In liver transplantation the opposite pertains: the immediate risk to the donor is by no means trivial but the long-term risk is extremely negligible. Such background information is essential.

Of late there has been some interest in what is called "the beating heart" donour. These are persons artificially sustained, on a ventilator and with no hope of any recovery. When the decision to use them as a donour is made, they are kept alive with the main interest being the preservation of organs for a recipient. This has raised a rather understandable debate. Medication is given or withheld so as to preserve the organ about to be donated, not because the still "living" patient would benefit but because it optimizes the chances of a good outcome in the recipient. One of the problems and serious concerns—rarely addressed—is that whatever else is started or stopped and even though the presumption is that the patient no longer feels anything the donour must, in case of the slightest doubt, be given sufficient analgesics and narcotics up to the time of their legal death.

ALLOCATION TO TRANSPLANT CENTERS

Organs to be transplanted are scarce resources. But they differ from other scarce resources in that they are (no matter how much one might be willing to spend) not renewable: in many cases the decision to give to one person precludes giving to another, another who may by that decision very well be doomed to death. They are also different in that they were once a part of another being and, therefore, have not only a material but also a very important symbolic value.[11]

In addressing any ethical problem (whether it is at the bed-side or whether it is about policy issues) we tend to ask what shall we do? We would suggest that this, as a first question, is the wrong question. It presumes that we who are concerned about the proper answer are entitled to make such decisions. In truth, the proper first question is procedural: not what should we decide but who, within a given set of circumstances, is entitled to decide and how shall such a decision be translated into praxis. We have generally evaded this question and have tended to fall back on expertise or, quite bluntly, on strength. We have made ethical decisions at best *for* and not *with* the weak, the sick, the disenfranchised. It is little wonder that most of our concerns in ethics have been directed towards the problems of what one of us (EHL) has elsewhere called rich man's ethics, the ethical questions we, who are lucky enough to have good access to medical care, trouble ourselves about.

Before we can decide how organs should be allocated (whether allocation involves micro- or macro-allocation) we need to be clear about who legitimately should decide. It is not obvious that experts make such decisions best. Certainly transplant physicians must be involved. But transplant physicians by themselves (or even the transplant team) are hardly legitimate decision-makers. After all and quite legitimately, transplant physicians and their program have their own financial as well as personal and professional vested interests. Transplantation is a complex enterprise involving many areas of expertise. Among others, transplantation involves transplant physicians, economists, psychologists, social scientists, social workers, ethicists and many others each of which have some corner on a specific and essential piece of knowledge. One could conceive of constituting committees of such persons and allowing them to formulate criteria that in turn would serve as a basis for discussion. Such criteria would then be publicly discussed and through elected representation become law.

Allocation has been done locally, regionally and nationally. There is no ethically compelling reason why our considerations of community should stop at a given (and often arbitrary) border be it local, regional or national. The first consideration (as always good ethics starts with good facts) is the viability of explanted specific organs and the ease with which such organs can be transported from one to another region. Viability differs among the various organs and likewise changes over time. As better methods of preservation are developed, viability is lengthened. In the European Community a joint program for sharing organs among the various EU (and some non-EU) countries is in place. In the United States there has been and is a brisk discussion as to the priority to be given to local over regional over national needs.

The decision as to how organs are to be distributed (the question of macro-distribution) would logically seem to be the first order of business. Before we can allocate fairly to individuals in need of organs, we must first create a way of supplying organs to various transplant centers and regions. All of us would agree that such distribution must be fair and equitable. Our disagreement stems from the definition of these terms. In order to establish a base of fairness it would seem an elemental requirement of justice that all those affected have a fair say in what, after all, may critically affect them. All of us are potential organ recipients and all of us, therefore, have a vested interested in seeing that the procedures established and the way they are implemented supports our conception of what is fair.

Currently in the United States the vested interest of diverse transplant programs has tried to keep distribution on as much of a local level as possible. Various proposals have been made, enacted, modified and re-modified generally with a great deal of pressure exerted by local programs eager to assert their own parochial interests. Great inequities have resulted.[23] Waiting times differ enormously from location to location. In some areas patients who are less critically ill or who have only a slim chance of a successful outcome are transplanted so as not to lose a transplant for a particular program. This has resulted in a number of social inequities. While we pay lip-service to the idea that lives must not be measured in terms of money, our health-care system and with it our transplant programs have shown the opposite to be the case. In the United States one gets what one can (out of pocket or by insurance) pay for—those without insurance lack fair access to basic health care, not to speak of access to transplantation. But even those fully insured do not fare equally: if one can afford to register with various transplant programs, one's chance of expeditiously receiving a transplant is greatly increased. Registering with several programs, traveling there and incurring all of the other out of pocket expenses associated with such a venture is a very expensive thing to do. It is not something an ordinary worker on an ordinary income can readily do. Until a public which has good cause to be interested in the way organs are distributed engages in a brisk dialogue and translates its will into legislative action by its representatives, inequities stimulated by vested interests will continue to occur.

It remains to decide how to define a "community" in practical terms. Are we speaking of cities, towns, counties, states, nations or a world community? In part the answer to this is technically determined—it depends on organ viability and ready access to rapid and reliable means of communication and transportation. As the length of viability increases and as transportation becomes more reliable the affected community grows. We have—as a species—moved from polis to metropolis, from nation-state to economic community and will, if we endure and prosper, move to a truly multi-cultural community whose possibilities embrace the world. For now we are still within a national borders frame of reference—albeit that transnational arrangements are more and more the order of the day. In practical terms today we see decisions in the United States as being on a federal level and eventually perhaps extending to neighboring states. For now realizing that we are indeed a community within this nation and not a loose association divided by petty vested interests will be a step forward.

———— ALLOCATION TO RECIPIENTS ———— AND SOCIAL VALUE

The problem of allocating organs fairly, while somewhat different from allocating renewable resources, nevertheless falls under the same rubric and has been discussed in Chapter 8. Here, however, a discussion of social value judgments is in order. The question of social value judgments entering into the distribution of scarce resources has been debated for many years. When recipients are selected on medical or "technical" criteria a large a number of potential recipients remain. Many will not receive a transplant and many will, in consequence, either die or be forced to live considerably stunted lives. Technical judgments, furthermore, are problematic in themselves: not only do they change with further data and leave yesterday's decisions "wrong," but they are also themselves informed by an often tacit web of assumptions and values (for example, the decision not to transplant hearts into patients beyond a certain age is not made on technical grounds alone). Tissue typing is, of course, important. But it is by no means certain that the elaborate tissue typing done today in search of the near-perfect match contributes much to organ (or patient) survival. Indeed, some transplant surgeons feel that in a sense it represents an elaborate (and rather expensive) lottery.

The attempt to hide criteria of social worth and call them prognostic, psychological, or environmental is an ever-present danger. To decide that some people are less socially desirable than others and to allocate life-saving resources on the basis of "social worth" has justifiably been in disrepute. Not only do such judgments disrespect all notions of primary worth, but they also easily open the door to arbitrary value judgments in which national origin, race, religion, and social class become determining factors.[28,29] In Kantian terms, all persons by virtue of being persons deserve absolute respect. If we allow social worth criteria to protrude into our medical judgments when it comes to the allocation of scarce resources, we are indeed violating the respect for sentient beings that forms one of the cornerstones of contemporary ethics. But does that totally and completely eliminate all social value judgments?

Arbitrary value judgments are just that—arbitrary. They allow the next step into the evidently impermissible (race, religion, sex, etc.) that much easier.[30] Slippery slope argument have, of course, much to recommend them. But slippery slope arguments, as important as they sometimes are, are in themselves questionable. Most everything we do or refrain from doing may lead to our doing or not doing something else. Having a superb dinner may be the first step to gluttony; having a good bottle of wine, the first step to drunkenness. In real-life decisions, we are constantly confronted with what could be rather slippery slopes. When we encounter such a slippery slope in our daily lives we resort to two things:

1. Realizing that the slope is slippery, we put down sand—that is, we exercise particular care and take careful precautions before we utilize that piece of road.
2. If the road is steep or slippery enough, we erect a barrier or choose not to use the road at all. Slippery slopes counsel great care, but the fact that doing something is a slippery slope without knowing the particulars cannot stand as a sufficient reason not to proceed at all.[31]

If an escaped and multiply convicted rapist or axe murderer arrives in the hospital at the same time as a well-respected member of the community and both need a given intervention (say, a ventilator, blood, or an organ) when only enough for one patient is available, we have to make a choice: we can utilize a lottery or flip a coin and let "fate" decide, or we can make a social value judgment. While at first glance, flipping a coin might seem to get us off the hook it is, in fact, evading our ethical responsibility to choose compassionately and wisely, to make clear distinctions, and to take responsibility for them. The decision here would not seem to be too difficult: one of the choices is clearly socially destructive (which, granted, is a judgment made within the context of communal values—but aren't all judgments?), while the other is, at least, not destructive.

There is already a problem with what "destructive" is: it is, unfortunately, a social decision made in the context of a given society and it may prove to be highly problematic (Are prostitutes "destructive"? Are Socialists or those who have not accepted Christ as their personal savior "destructive"?). Such a choice is bad enough, but the problem becomes far more critical when it is a problem of choosing among degrees of merit. The slope is slippery enough when it comes to judging "destructiveness" that we may have to either put down sufficient sand and proceed with the greatest of caution or to stop making further such judgments altogether. Most of us would feel that judging degrees of merit is socially so very dangerous that it should not be done. While those who have been destructive are persons and are deserving of respect, they nevertheless have forfeited the additional respect we give to members of the community who have not been destructive.

Such considerations come up with the very grave problem of transplantation in our prisons. Prisoners have been deprived of their freedom of action. In return we as a community have promised them a number of things: food, shelter, safety (unfortunately a promise poorly kept), and medical care. It must be a decision made by the community how far such care should go, and the limits of such care should be clearly known and shared by all. For example, a community may decide not to offer transplantation to certain types of incarcerated criminals, a decision that, from then on, would be known by potential criminals and would constitute a part of their expectations. It is not that this would act as a deterrent to crime—but it would be known beforehand and not constitute and evasion of an implicit promise.[31]

In allocating organs (as distinct, for example, from allocating kidney machines), the uniqueness of organs must be borne in mind. Not only are they irreplaceable (which, given sufficient funding, kidney machines or any other manmade devices are not), but they were also the organs of a member of the same community enunciating their values. Criteria in organ donation must, first of all, be criteria of medical acceptability. Such criteria, inevitably laden with value judgments, must be kept as free as possible of at least the coarser elements of such judgments. Once these changeable and difficult criteria with all their problems of arbitrary cut-off lines are satisfied, groups using other criteria must be enunciated. Ethically speaking, only when all other attempts at solving problems are exhausted may considerations of social worth finally and still hesitatingly be factored in. Such criteria may be just or unjust, but their justice or injustice is at least determinable by and appeal-able to the recipients' community, whose values they know and, at least to some significant extent, share.

In making allocation decisions, physicians and others rightfully involved in making such decisions, will, within the current vision of the health care professional–patient relationship, do the best they can for each individual patient regardless of a particular patient's social qualifications. This "best," however, is the best possible within an array of possible options determined by the social community in which health care professionals operate: they can only offer that which is made available. The choice—what should be made available—is a choice a democratic community guided by their health professionals and informed by communal values must make prior to allocation by physicians to individual patients. When, however, communities make allocation decisions, social value judgments may have to intrude into macro-allocation decisions that offer to practicing physicians the array of possible options they can use (see a discussion of this in Chapter 8).

—————— SELLING ORGANS ——————

Most nations do not permit organ sales. In some areas blood can still be sold, but that practice (which used to be routine a few decades ago) has for ethical and practical reasons been largely abandoned.[31] Yet allocating organs in the marketplace still finds some who recommend it, and organ sales are known to occur in other parts of the world and have become a grave problem.[32–33] The issue deserves to be examined closely. In condemning organ sales as they appear to exist in the poverty areas of underdeveloped nations are we condemning (1) the practice itself; (2) the person selling the organ; (3) the person buying the organ; or (4) the economic conditions that cause such sales to be made?

At first blush allowing autonomous persons to sell a kidney (some libertarians would go as far as arguing even their heart) or their bone marrow would simply be another freely entered commercial transaction: their consent has been "freely" given. If one assumes a libertarian perspective, envisioning a world in which our obligations begin and end with not limiting another's freedom and scrupulously adhering to freely entered contracts, such an argument has moral force. If, on the other hand, we assume other obligations, if we acknowledge that free action among other things necessitates a lack of coercive material conditions and takes place in the embrace of a particular community and of its social values, a quite different answer will emerge. Whether one finds the practice of selling organs morally acceptable or not depends on one's fundamental worldview of our obligations to one another.

There is no doubt that a person who does not have sufficient means to support his or her family, let alone send even one child beyond mandatory education, will be not only sorely tempted but, indeed, coerced (and some, we would argue, morally justified if not indeed obligated) to sell a kidney in order to have the means to sustain their family and perhaps enable one to receive sufficient education to have a life worth living. In such a situation, speaking of "free consent" is a sham. One could even make an argument that a person unwilling to sell an organ to save his family is failing to discharge an obligation.

On the other hand, there is evidence that organ sales have brought little benefit to those selling them. In a recent paper the authors state that: "Average family income declined by one third after nephrectomy (P < .001), and the number of participants living below the poverty line increased. Three fourth of participants were still in debt at the time of the survey. About 86% of participants reported a deterioration in their health status after nephrectomy. Seventy-nine percent would not recommend that others sell a kidney." ... "Among paid donors in India, selling a kidney does not lead to a long-term economic benefit and may be associated with a decline in health. ..."[34] The issue is unclear but what is not unclear is that a system that may coerce people into selling an organ or body part is ethically indefensible because it advantages the wealthy and disadvantages the poor in a way destructive to communal solidarity. Blaming the seller is blaming the victim of a pernicious social system. It is, we believe, the system that perpetuates such deplorable practice and not the practice that ethically stands in need of change.

Persons who vitally need an organ so as to live a full life are likewise under considerable pressure. While it is easy enough to blame them for buying an organ on the marketplace, it is an "armchair" type of condemnation made by those not in critical need. One could argue that buying an organ from a poverty-stricken person enables that person and their family to work their way out of an intolerable situation. From this perspective, condemning the person buying the kidney is again a form of victim blaming.

What is left is a condemnation of the economic conditions that have caused this entire problem. Although some individuals or nations are immensely wealthy while other individuals or nations are abominably poor, this problem is the result of historical and ongoing exploitation of the weak by the strong. The problem of organ sales from the poverty-stricken to the wealthy is not one addressable by condemning the victims. It can be addressed only by recognizing that, in cases such as these, the victims are the very poor themselves and the aggressors are those nations and individuals who by their exploitation have brought this problem about. At least one ethically sound answer is to work toward redistributing the goods of this Earth in a more equitable and fairer manner.

————————— **REFERENCES** —————————

1. Loewy EH. Waste not, want not: communities and presumed consent. In: Thomasma DC, Monagle JF, eds. *Medical Ethics: A Guide for Health-Professionals*. Rockville, Md: Aspen Publishers, 1987.
2. Joughin JL. Blood transfusion in 1492. *JAMA*. 1914;62:553–554.
3. Garrison FH. *An Introduction to the History of Medicine*. Philadelphia, Pa: W.B. Saunders, 1929.
4. Singer C, Underwood EA. *A Short History of Medicine*. New York, NY: Oxford University Press, 1962.
5. Garrison FH. *An Introduction to the History of Medicine*. Philadelphia, Pa: W.B. Saunders, 1929.
6. Duraj F, Tyden G, Blom B. Living-donor nephrectomy: how safe is it? *Transplant Proc*. 1995;P(1):1164–1165.
7. Rosner F. Is living kidney donation still justifiable? *Chest*. 1994;106(2):334–336.
8. Jurim D, Schakleton CR, McDiarmid SV, et al. Living donor-liver transplant at UCLA. *Am J Surg*. 1995;169(5):529–532.
9. Segre M, Blumstein JF, Caplan A, et al. Partial liver transplantation from living donors (article and four commentaries). *Cambr Q Health Care Ethics*. 1992;1(4): 305–325.
10. Loewy EH. Drunks, livers and values: should social value judgements enter into transplant decisions? *J Clin Gastroenterol*. 1987;9(4):436–441.
11. Loewy EH. Presumed consent in organ donation: values and means in the distribution of a scarce resource. In: Loewy EH, ed. *Ethical Dilemmas in Modern Medicine: A Physician's Viewpoint*. Lewiston, NY: Edwin Mellen Press, 1986; 133–154.
12. May W. Attitudes towards the newly dead. *Hastings Center Report*. 1973;3(1):3–13.
13. Ramsey P. *The Patient as Person*. New Haven, Conn: Yale University Press, 1970.
14. Martyn S, Wright R, Clark L. Required request for organ donation: moral, clinical and legal problems. *Hastings Center Report*. 1988;18(2):27–34.
15. Feinberg J. The mistreatment of dead bodies. *Hastings Center Report*. 1985; 15(1):31–37.
16. Perry C. The right of public access to cadaver organs. *Soc Sci Med*. 1981;15(F): 163–166.
17. Sadler AM, Sadler BL, Stason EB. The uniform anatomical gift act. *JAMA*. 1968;206:2501–2506.
18. Caplan AL. Professional arrogance and public misunderstanding. *Hastings Center Report*. 1988;18(2):34–37.

19. Stuart FP, Veith FJ, Cranford RE. Brain death laws and patterns of consent to remove organs for transplantation from cadavers in the United States and 28 other countries. *Transplant.* 1981;31(4):238–244.

20. Dukeminier J, Sanders D. Organ transplantation: a proposal for routine salvaging of cadaver organs. *N Engl J Med.* 1968;279(8):413–419.

21. Muyskens JL. An alternative policy for obtaining cadaver organs for transplantation. *Phil Public Affairs.* 1978;8(1):88–99.

22. Loewy EH. Of community, organs and transplantation: routine salvage with a twist. *Theor. Med.* 1996 (in press).

23. Loewy EH. Who should receive donor organs? Chapter 5. In: Shelton W, Balint J, eds. *The Ethics of Organ Transplantation, Advances in Bioethics, Volume 7.* Elsevier Science, Oxford, UK, 2000;125–147.

24. Fellner CH. Organ donation: for whose sake? *Ann Intern Med.* 1973;79(4):589–592.

25. Fellner CH, Schwartz SH. Altruism in disrepute. *N Engl J Med.* 1971;284(11):582–585.

26. Sadler HH, Davison L, Carroll C, et al. The living genetically unrelated donor. *Semin Psychol.* 1971;3:86–101.

27. Steinbrook R. Unrelated volunteers as bone marrow donors. *Hastings Center Report.* 1980;10(1):11–20.

28. Annas GJ. The prostitute, the playboy and the poet: rationing schemes for organ transplantation. *Am J Public Health.* 1985;75(2):187–189.

29. Atterbury CE. The alcoholic in the lifeboat: should drinkers be candidates for liver transplants? *J Clin Gastroenterol.* 1986;8(1):1–4.

30. Loewy EH. Drunks, livers and values: should social value judgments enter into organ transplant decisions? *J Clin Gastroenterol.* 1987;9(4):436–441.

31. Titmus RM. *The Gift Relationship: From Human Blood to Social Policy.* New York, NY: Vintage Books, 1972.

32. Bermel J. Organ sales: from market place to jungle. *Hastings Center Report.* 1986;16(1):3–4.

33. Manga P. A communal market-place for organs: why not? *Bioethics.* 1987;1(4):321–338.

34. Chandra P. Kidneys for sale. *World Press Rev.* 1991;38:53.

10

Problems at the Beginning of Life

INTRODUCTION

For a number of reasons, questions dealing with ethical issues at the beginning of life are often viewed differently from questions at life's end, even when such issues deal with analogous problems. Unconsciously we tend to associate youth with hope and with unfulfilled but fulfillable opportunities, whereas we associate old age (or often just older age) with decline and hopelessness. We invest the young with a range of symbols (youth as a symbol for innocence and future) and metaphors (youth as spring-time and hope) quite different from the symbols and metaphor we use for later years. Such symbolism and metaphor easily get in the way of rational thought. When we encounter the anencephalic infant, the young child who has become permanently vegetative after an accident, or the child with Werdnig–Hoffman disease, youth is not the question: the absence of cortical function or the fact that life span is severely limited is. We find the absence of cortical function or the hopelessness of illness intolerable in one so young. And often, and against all reason, we tailor our ethical judgments accordingly. Even when we deal with patients later in their lives (say, with a 30-year-old lady dying of breast cancer), youthfulness, against all reason but consistent with symbol and metaphor, obtrudes into our ethical judgments. Having a devastating and necessarily fatal illness and being young (even very young) does not alter the medical facts as the necessary facts on which ethical judgments must be based; it merely makes the situation far more tragic and far more difficult to cope with psychologically.

PROBLEMS OF ABORTION

This issue is made troublesome above all because of its religious overtones and because of the emotive response it brings forth in many people. No attempt is made here to deal with the issues of religious morality: i.e., no effort is made (or, in fact, reasonably can be made) to apply reasoning to religious strictures (see also the Preface and Chapters 1 and 3). In what follows, we will examine the history of the problem, outline some of the points made about abortion by secular ethics, and briefly address the troubling and separate issue of legislating moral codes that are peculiar to specific enclaves to entire communities.

The opponents of abortion under any and all circumstances (even if the presence of the fetus threatens the mother's life) are persons who hold the life of the mother and that of the developing fetus to be of equal worth. Therefore, and in order not to do anything that might be immoral, they knowingly fail to intervene at a time they could have. There are, of course, at least two apparent fallacies of such thinking: first of all when one fails to act when one could have acted one has, in fact, acted—that is one has become a link in a causal chain which leads to a particular outcome: in the case above a (probably but not certainly) living fetus and a probable dead mother. Secondly, especially early in pregnancy, they fail to distinguish between *being* alive and *having* a life: for surely the mother is and has a life, whereas (at least early on) the fetus *is* alive but (at least as yet) does not *have* a life.

Historically, our attitudes toward abortion have varied. Law and usage have, at times, prohibited abortion for three reasons: (1) to protect the mother's life, (2) to protect an unborn life for its own sake, and (3) to protect the community or state. As technology has progressed, protecting the mother's life by prohibiting abortion has made no sense. Under modern circumstances, protecting the unborn is the usual reason given. Nations plotting war (Nazi Germany, as an example) often have an interest in creating as many able-bodied men as possible and, therefore, have an interest in protecting and in fact in creating the unborn. Not surprisingly abortion was illegal in Nazi Germany and even, for a time and for the same reason, in the Soviet Union. At other times, the interest of the state in protecting the lives of its members is appealed to. Such an appeal, however, begs the question, since it fails to stipulate criteria for the kind of "life" entitling its bearer to such protection. The interest of the state in protecting all life is obviously far from absolute. States arm their police, prepare for or wage war, and execute criminals.

Among the Greeks, the father's right to decide his children's fate was upheld throughout the "golden ages."[1] In the Hippocratic tradition, physicians were enjoined against performing abortion just as they were prohibited from "cutting for stone." This prohibition probably resulted from the profession's fears of discrediting the profession itself (the main interest of the Hippocratic code seems to have been that; see Chapter 1) by causing the death of the patient by a hazardous procedure (or, in the case of cutting for stone, of causing sterility), or it may be part of an essentially Pythagorean "right to life" viewpoint that informs so much of the Hippocratic oath. Abortion done by midwives was not prohibited. Most physicians in ancient Greece, furthermore, were not Hippocratic physicians nor were they bound by an oath that apparently only a few took.[2,3] The practice of abortion, done by midwives or by physicians, appears to have been widespread[3] despite the fact that prior to an understanding of sterility or infection and prior to a solid understanding of anatomy, abortion was a very hazardous procedure indeed.

The influence of Aristotle on subsequent attitudes toward abortion in as well as outside the Christian church has been profound.[4] Briefly speaking, Aristotle felt that the embryo was initially "vegetative" (none-animated) and only later became "animated" as it was entered by the soul. Killing the non-animated fetus was an act of destroying life but not murder; killing the animated was an act of murder. Aristotle equated animation with "quickening": a criterion that has since been appealed to persistently but one that, with today's understanding, is no longer viable.

Aristotle's concept informed the early Christian church. The critical question—"When does the soul enter the body?"—was usually answered along Aristotelian lines. Today's question—"When does personhood begin?"—asks the same question in secular terms. A very few early theologians (in the third century Tertullian, for example) felt that the soul entered at conception, but they were in the minority. Most theologians agreed with St. Augustine (fifth century): Killing the ensouled was murder; killing the un-ensouled was not. In the 13th century Pope Innocent III decreed that a priest who had been "party to a miscarriage" after quickening must refrain from serving mass but one involved beforehand could continue to do so.

In the 13th century, St. Thomas Aquinas reaffirmed what is still the Catholic view today: Killing an ensouled fetus is murder. What has changed is the view of ensoulment: The Aristotelian view that the soul entered the body at "quickening" (i.e., when movement is first felt) was, until recently, a view commonly held in Catholic Church circles. Pope Clement VI in 1708 fixed the date at which Mary's soul entered her body as being 9 months before her birth date, but abortion prior to quickening was not officially condemned. In 1854, in fixing the date for the feast of the Immaculate Conception, Pius IX reaffirmed that the soul of Mary entered her body at conception and in an 1869 pronouncement the distinction between the ensouled and the un-ensouled fetus was removed. All souls entered at conception and, therefore, abortion was murder. The fact that papal infallibility was promulgated the same year complicated the issue.

The technical advances of the 19th and 20th centuries, the greater understanding of infection, anatomy, and the control of hemorrhage as well as the development of safe anesthesia, enabled abortion to become a practical alternative rather than a desperate last move. The risk of abortion to the mother when done in a proper medical setting could no longer serve as an excuse to outlaw abortion. Abortion was practical and safe, but was it ever morally permissible and, if so, when, where, and why? Laws, both in Europe and America, created a situation in which the well to do and well-connected had relatively easy access to safe abortion services; the poor and ignorant continued to use painful, dangerous and often fatal methods on themselves, or when financially able used back-alley abortionists or they were forced to carry to term unwanted children they often could not care for. Death or the desperate illness of the mother or the abandonment or neglect of children were the frequent consequences. "Oversleeping"—that is the mother "accidentally" asphyxiating the child lying next to her—was a not at all uncommon practice up to the end of the 19th century. Many were the sermons held against it. Gradually, laws and attitudes changed and today abortion under certain circumstances is legal in virtually all western nations.

Since "rights" are usually held to be virtually absolute, since they are usually regarded as "trump cards" of moral reasoning, and since, therefore, an appeal to such "rights" is meant to settle the question, turning to "rights" of mother or fetus in arguing the ethical stand of abortion does not help. If, however, instead of trying to adjudicate claims to clashing "rights," we look at "interests," a more flexible point of view results.[5] A flexible point of view, of course, is the very thing that those who oppose abortion under all circumstances would contest. Flexibility is not possible when absolute positions are taken. (For a more thorough discussion of "rights" and "justice," see Chapter 3.)

In the view of those absolutely and completely opposed to abortion under all circumstances, the "right" of the unborn child (no matter how unformed) is inalienable. There are, in fact, not many who would hold such an absolute view. When confronted with a situation in which a 12-year-old girl's life is threatened by a pregnancy that was the result of incestuous rape and who now carries a fetus known to have a dominant gene for a psychiatrically devastating and horribly painful condition, many of those who previously affirmed their absolute opposition would falter. But absolutism of this sort does not permit faltering: The moment an exception is granted, the edifice of the "absolute" falls. For the purposes of this discussion, we shall assume that an absolute opposition to abortion under all circumstances cannot be justified except by an appeal to the religious principle of a particular group of believers—in other words to a species of personal morality which, as discussed before, may satisfy all who are believers, but will fail to persuade non-believers.

In most pregnancies the interests of mother and fetus coincide; only in the unusual case does a conflict of interests exist. Such conflicts, then, are in need of moral adjudication. Making the interests of either binding under all circumstances settles by arbitrary fiat rather than by moral deliberation. Interests may clash in two ways: (1) the mother may wish to terminate the pregnancy "against the interests of the fetus," or (2) the mother, for whatever reason, may refuse to terminate a pregnancy when its outcome would be "against the fetus' interests." This latter situation, which has led to "wrongful life" suits, presupposes that persons have a legitimate interest in being born unhampered by serious and often tragically devastating congenital defects.[6] The view that persons have a legitimate interest in not being born as terribly damaged beings is a view based on the further belief that, at least under some circumstances, it is better not to have been born than to live with a devastating defect. In a sense, such a belief is analogous to the belief that it is better to be dead than to be forced to live life in certain ways.

Wrongful life suits and the ethical ideas that underlie them (and here we do not speak of a wrongful life suit brought against a physician for negligent management of a pregnancy but for suits which are brought by severely crippled children against their parents for not having been aborted) are difficult to grapple with. The idea of a non-existent person having any interests is an internally difficult one to envision unless one argues the whole matter on a symbolic level: An interest is only something that one who exists can have. Does the severely handicapped who brings such a suit now have "an interest in not being alive"—many others in the same position would say otherwise. How is one to predict such a thing? Beyond this, a person who really wishes not to exist has other options.

The question could be formulated otherwise: Do pregnant women have obligations toward their developing fetuses that not only include obligations of not causing them damage but also include obligations of not sustaining potential beings whose future would most likely be one of intense suffering? This is a problem we shall discuss later on when we briefly deal with maternal obligations. (For a discussion of maternal responsibility as well as a brief discussion of the wrongful life issue see the relevant section later in this chapter.)

In dealing with the ethics of abortion, two polarized views form the boundaries of the argument:

1.   The fetus is not an instance of human life but is merely a part of the mother's body (no more than "a wart on the nose," as some would have it) and may therefore be dealt with as such without any moral concerns.
2.   The fetus from the moment of conception is, under all circumstances, of equal intrinsic value as the mother and its destruction is, therefore, murder.

The argument that the fetus is not biologically human life does not hold water. The chromosome count and DNA are right and a sufficient number of other criteria for both "life" and "human" are met to leave no doubt as to this. Likewise, the argument that the fetus is merely a part of the mother's body to be disposed of without moral concerns is flawed: By virtue of its potential, the fetus is uniquely different from other body parts. Distinct from other body parts, it has a potential for being that merits concern; the colon, lacking this, does not. Wanton destruction of the colon may harm its host; wanton destruction of the fetus concerns both the host and, in an as yet inchoate way, the fetus. There is therefore a legitimate, and growing, *prima facie* reason for not destroying the fetus that does not apply to other body parts.[1] Moderate positions are often drowned out by the ensuing rhetoric and are further threatened by a stridency that threatens to enshrine personal morals as universal law.

But, as in dealing with end of life issues, we must differentiate two senses of life: (1) "being alive"—a biological condition and one that necessarily is the foundation for (2) "having a life"—a psycho-biographical state in which not only is the subject alive but he or she is also the knowing subject of that life.[7] (See also Chapter 11.) Merely "having a life" is, however, insufficient to invest an entity with primary or even secondary worth. A tissue culture of colonic mucosa growing in an incubator meets similar criteria for human life and yet could not sanely be considered inviolable. It is therefore not just life (we kill animals and eat spinach) and not just human life (we throw out tissue cultures) that some would hold "sacred." What, it seems, we hold as having *prima facie* rights against wanton destruction are entities endowed with having either primary worth (those of value now or in the future again to themselves) or secondary (symbolic) worth, or prior worth (those of value to another in themselves, those of value as representative of something held to be of value, or those things like community and being necessary for the very process of valuing itself). (See also the chapter on Ethical Theory.) Fetuses, therefore, certainly have both material and symbolic worth to their parents and to the community. But such a value can be negative as well as positive (a pregnancy carried by a rape victim often does not have positive value). And, undoubtedly, fetuses are potential (but as yet merely potential) persons (see later in this section).

The second point of view, that fetal life is of equal intrinsic value as the life of the mother and that its destruction is therefore murder (and that such a murder is the moral equivalent of murdering the mother), rests on several premises:

1. The fetus is human life and therefore "sacred."
2. The fetus is "innocent" and therefore deserving of full protection.
3. The potential of being a person (the potential for being of primary moral worth) endows the fetus with the same rights as if it were a person (or of actual primary moral worth).

The concept of something as "sacred" is frequently appealed to. However, as a religious term, how it is applied depends on the particular religion we are speaking about: Cows in some parts of the world are "sacred;" humans, in ancient times, were offered as a "sacred" sacrifice; and human life is "sacred" in some other views. Why one type of "sacredness" should trump another remains unclear. And, in fact, religions that speak of the sacredness of human life seem not necessarily to oppose other forms of countenancing unnecessary death: war, capital punishment, hunger, and starvation—to name but a few. In general, the appeal to the "sacredness" of human life is a most unconvincing argument.

The argument from innocence is a more straightforward one. The term "innocent" is a technical term first used by the Catholic Church in sorting out "justifiable" reasons for killing. Early Christianity saw all killing (at least all killing of humans) as wrong—even killing in war, as punishment, or in self-defense. The Roman Empire certainly could not countenance the right of citizens to refuse to fight in its armies. If Christianity was to be fully accepted and if it was to become, as it ultimately did, a state religion, that problem had to be solved. The language of "innocence" was invoked: "Innocent" persons could not be killed; those who were not "innocent" could be. Only those who were not engaged in fighting in an unjust war or those who had not broken a just law ("just" being, of course, in either case defined by the state) could be held to be "innocent." Killing the non-innocent was, regrettably, a permissible act.[7] Curiously enough the interdiction against killing in self-defense persisted up to the time of Augustine. When, therefore, "innocence" is invoked in the abortion debate, it inevitably carries this historical baggage.

Moreover, the term innocence as applied to a fetus is a peculiar concept. In general, innocence as opposed to guilt implies either that (1) desire and/or opportunity to do something existed but was not put into action; or that (in a far more obscure sense) (2) something was never done because there had been no opportunity. I can be innocent of murdering my wife, for example, if despite opportunity and desire I have failed to do so; or, I can be innocent of murdering Charlemagne since I never had the opportunity (or desire) to do so. The first example of innocence might hold me exempt from punishment; the second would make even the idea of guilt or the thought of punishment ludicrous. The fetus, in the first sense, lacks either opportunity or desire, and in the second would seem to have failed to earn special rights. (In a religious sense as "being innocent by virtue of being free of original sin," the fetus is no more free than is the infant, the difference, of course, being that the infant has, for the Christian, hopefully, been baptized.) In its original Church meaning, as not having broken an unjust law or not fighting in an unjust war, the fetus is "innocent," but it is innocent by virtue of lacking the opportunity (and being incapable of having a desire) and not by virtue of having made a deliberate choice. Innocence seems to add little to a fetus' moral standing.

The argument from potentiality (resting, as it does, on the zygote's probability for being) is noteworthy.[8] The human zygote most certainly is a form of human life with the potential for being of primary moral worth (in the usual case, it is of secondary and of symbolic moral worth because of the immense value to the parents, but when abortion is the issue this is not the case). It is, in that respect, similar to an acorn that has the potential of being an oak. But, as Thompson has pointed out, "It does not follow that acorns are oak trees or that we had best say they are."[9] In real life we do not treat the potential as though it were actualized: We do not give students of medicine the right to write prescriptions or perform surgery. Residents in training are not full-fledged specialists, but as training proceeds we progressively begin to treat them as though they were. The closer the potential approximates the actual, the more "respect" (and the more obligations) due the actualized do we give it.

Zygotes, among other things, have the potential for being spontaneously aborted or for being malformed, as well as for becoming villains or saints. One cannot know whether the developing fetus might be another Martin Luther King, Schweitzer, or Beethoven or whether it might become a Hitler, a Stalin, or a Pol Pot. Chances are that it will become a normal and reasonably happy and functional member of a community, but those are chances that depend upon circumstances, environment, and luck. Although a zygote has a high potential likelihood of becoming a normal child, it might with less likelihood but quite possibly develop into a chimera: two or more fertilized eggs resulting in multiple persons.[10] The potential to be, then, is the potential for being many things, and it is unknowable. Potentiality (that the fetus has the potential for being of primary moral worth is, it seems, beyond reasonable dispute) certainly carries more moral weight than does the lack of that potential. The obvious question, however, is whether such moral weight is as entitling under most circumstances as is that of an actual entity unquestionably endowed with primary worth.

Comparisons between the fetus and an anesthetized or temporarily comatose patient are sometimes made because in both sets of circumstances a potential for sentient being exists. The difference between these two situations is that in the anesthetized or temporarily comatose, new sense experience will be integrated into preexistent memory and old values and goals will be unchanged or integrated into new experience, in the fetus, prior to function of the neo-cortex at least, no preexistent memory exists. Anesthetized or comatose persons can be said to have had a preexisting vested interest (a stake) in resuming their life with its rich biography, its unrealized plans, hopes, and aspirations and therefore in their recovery, which fetuses do not. Anesthetized or comatose persons have an identity beyond the spatio-temporal continuity of the physical body: Their identity concerns continuity and connectedness of personality, memory, and other mental phenomena.[11] Anesthetized or temporarily comatose patients "have a life" in addition to "being alive." In fetuses, this is not the case.

If the right (or interest) of the fetus not to be destroyed is seen as a *prima facie* right (or interest), it must be adjudicated against other claims. The force of the fetus' claim, as well as the force of the mother's, depends on the existential status of the claimants. The claim of the fetus gains in force as the central nervous system (CNS) develops. This is the case because until sufficient development is present, the notion of present benefit and harm (the notion, that is, of being currently of primary moral worth) cannot rationally be upheld. At 8 weeks, brain waves begin to develop, and it is here, some suggest, that a life worthy of protection emerges.[12] The emergence of brainwaves, however, does not denote the possession of faculties but, rather, the physical potential for their development. Thus there seems to be a slowly growing force to a claim for life that reasonably starts at conception as a rather weak *prima facie* condition against being frivolously destroyed, and grows especially after the CNS proceeds to function. The CNS and its function form, according to that point of view, the necessary condition for the emergence of primary worth but do not constitute it. In other words, primary worth cannot be reduced to mere biology. As the fetus grows, ever-weightier reasons to set aside that *prima facie* right would seem necessary.[13]

Logically, the child a day prior to delivery is morally hardly of less concern than it will be a few minutes after birth. An infant at birth has no sense of personhood: It cannot dissociate itself from its environment, is entirely dependent on the constant nurture and support of others, and has not yet come to recognize itself—biologically or psycho-socially—as an entity in its own right. That comes a few months later. An infant at birth is what Engelhardt calls—with some justification, we think—a "person in a social sense:" that is, we have arbitrarily decided to endow such a being with full moral standing at that particular point in time.[14] We want to be clearly understood: We are hardly arguing that infanticide should not be condemned morally. We are arguing that the moral standing from fetus to birth is a continuum of ever-growing fetal interests and that we have justifiably but nevertheless arbitrarily endowed such a being with full moral standing at birth.

If one examines the issue, one is forced to come to two alternative and mutually exclusive conclusions: either (1) abortion under all circumstances is morally wrong, or (2) there are instances in which abortion (albeit it still is far from being a praiseworthy act) can be clearly seen to be the lesser of two evils. If one subscribed to the second alternative which, in fact, most persons do, what remains is to enunciate those conditions under which maternal interests are weightier than those of the fetus. One can, for example, retreat from an absolutist position and permit abortion only when maternal life is threatened. Such a stand is already a concession to the point of view that there are circumstances under which abortion is not only morally defensible but, indeed, the lesser evil of two unsavory alternatives.

Attitudes toward the ethical permissibility of abortion will vary depending on the person making this judgment. If one is to look at the matter in a non-absolutist way, denying that either the mother's right to control her own body at all times or the infant's claim to have the right to live is absolute, one will be forced to apply certain criteria. As with end of life issues, one will have to differentiate between being alive (a biological statement) and having a life (a biographical statement denoting a capacity for hopes, aspirations, and social interconnections). Furthermore, one will have to acknowledge that while fetuses certainly are alive, they, at least early on, do not have a life (although they potentially do). Early on, at the very least, fetuses are not of primary moral worth (although they have the potential to be of primary worth) but pregnant women are. The moral worth of fetuses inheres in (1) their potential to be persons, (2) their secondary worth (which may be positive or negative) to the mother and perhaps to others, and (3) their symbolic worth as future members of humanity.

Whether abortion is, at least under some circumstances, considered ethically defensible and whether a given health professional feels morally entitled to participate in performing an abortion are two different questions. Deciding to participate or not to participate in an abortion becomes a personal balancing act in which such considerations must be weighed and sifted. Patients as well as health professionals are moral agents who must have the freedom to act within their own moral framework and cannot be forced to violate their own ethical beliefs, or to superimpose them on each other. Given the power potential which inevitability exists, physicians are once again in a gray zone in which their personal moral chastity vies with the patient's unquestioned moral agency. Remember also that failure to act (by counseling, abortion or refusing to refer) is indeed a form of acting—acting in the human condition is inescapable.

An important issue remains: Can communities, or majorities within them, enunciate a moral position outlawing abortion and, accordingly, promulgate binding laws? Should abortion, in other words, be a political football subject to the rule of a (possibly slim) majority or, perhaps, should such decisions be left to private conscience? And is such private conscience to be given free reign at every stage of pregnancy or only at certain times or under certain conditions? This question goes far beyond the issue of abortion. Ultimately, generally agreed-upon moral propositions are enshrined in law: laws against murder, theft, and the breaking of contracts are examples. There are, however, many generally accepted moral propositions that have wisely not been codified into law: Laws were initially made to "keep the peace" and things like murder, theft, etc.—in that they led to private revenge, feuds, etc—disturbed that peace. Thus laws must do more than merely stipulate accepted and unaccepted ways of behaving because a consensus exists—they likewise must threaten what initially was called "the King's peace." It is, for example, generally agreed that lying to one's wife about having been out with one's mistress is not a morally defensible act. Yet it is not illegal. The ultimate decision has to rely on moral consensus: not a slim majority, not unanimity, but a broad (and that is admittedly hard to pin down quantitatively but is nonetheless quite real) moral consensus, as well as on the potential disturbance of the peace within a community.

The argument that outlawing abortion is analogous to outlawing slavery (that is, that abortion at any stage of pregnancy is an evil that is clearly enough an evil as well as ultimately a threat to communal peace) has often been made. Some argue that since many within a community felt slavery to be morally permissible, outlawing slavery has a similar standing to outlawing abortion. Abolitionists and anti-abortionists are seen to be similarly concerned with advancing moral views and with safeguarding individual "rights." Slaveholders, or those who would leave abortion to individual choice, are likewise equated in this calculus. On the face of it, the argument has logic: If we equate the black man or woman with the embryo, the conclusion follows. But the argument is flawed. The basic assumption equating the status of slaves with those of embryos begs the question and, thus, cannot withstand closer scrutiny: Slaves have actual, rather than merely potential, primary moral worth; fetuses (at least early on) do not. Slaves (or disadvantaged persons) can suffer. They actively feel the whip (or sense the discrimination against them), and they know the pain and fear of slavery or disadvantage; fetuses do not. Slaves, furthermore, are part of the community that holds slaves: They are enmeshed in a social nexus of family, friends, and community. They do, in fact, have the immediate ability to participate in judgments but are artificially and arbitrarily disenfranchised. They stand mute—but not because they cannot speak. The muteness of the fetus is of a different order.

Outlawing slavery is necessary to safeguard actual others who may come to harm. It is of the same order as laws against speeding, spitting on the sidewalk, or murdering teenagers. Outlawing abortion, on the other hand, is predicated on a unique and not demonstrable belief system that claims actual value for potential attributes. It is of the same moral order as going to church, keeping the Sabbath, or having sex with another consenting adult. Enforcement of such matters within particular moral enclaves or religious persuasions belongs legitimately within the social sanction of specific groups or belief systems. The community that enforces such belief systems in legal form arguably disrupts its own peace.

——— SEVERELY DEFECTIVE NEWBORNS ———

Some years ago a child with Down's syndrome and pyloric obstruction was born at Johns Hopkins Hospital. The parents, after considerable agony, refused surgery. Consistent with their wishes, the baby was not given nutrients or fluids and was allowed to die.[15] Similar cases in future years—some with Down's syndrome, others with far more severe mental and/or physical deformities and with various degrees of hopefulness or hopelessness—became ever more troubling issues. The recommendation that such decisions should be jointly and solely handled by physicians and parents[16] not only confronted communal sensibilities but also ran the severe danger of being capricious or arbitrary; the attempt of the government, essentially by "dumping" all such defects into a single pigeonhole, to settle the issue by *fiat* likewise proved to be even more unworkable and extreme.[17-20] It may be of interest that the pediatric resident assigned to that case was so troubled with it that he became one of our great and humane ethicists specializing in pediatrics, one who died at a young age and who is thoroughly mourned.

This issue is indeed one exacerbated, if not indeed created, by modern technology. Until fairly recently, most children with severe defects either succumbed to their primary defect or died as a result of a number of relentless and untreatable inter-current illnesses. Today's technology has made many of these conditions treatable without in any way affecting the underlying basic problem. Settling such problems by developing dogmatic formulas on the one hand (and thus to reduce moral decision making to a cookie cutter approach) or, on the other hand, purely on the whim of those immediately concerned without reference to guidelines (and thus to leave such weighty decisions to a process in which the benefit of the patient may be given short shrift) is superficially tempting but begs the question. A middle course more conducive to finding just answers needs to be developed.[21]

The problem has at least several presentations:

1. Those in which children are mentally presumably normal but suffer from severe (albeit, perhaps, partially correctable) physical handicaps (neural tube defects are the main example).
2. Children with a life threatening defect as well as a permanent irremediable handicap (Down's syndrome with duodenal atresia would be an example).
3. Infants born who appear normal, but are inevitably destined to die of a relentlessly progressive degenerative condition with or without severe discomfort long before reaching an age of more than a few years. Such infants may maintain normal mentation throughout their period of decline (as, for example, those with Werdnig–Hoffman) or may be fated to have their initially normal mentation as well as their physical state deteriorate (as, for example, those with Tay–Sachs).
4. The severely premature, hopelessly respirator-dependent newborn with, perhaps, severe intra-cerebral bleeding (and there are several sub-varieties of this condition).
5. Newborns with little or no higher cortical activity (the anencephalic or the near-anencephalic child).

At first glance, there seems little in common among these considerations. In all of these cases the prognosis is bad, but it is bad in different ways and for different reasons.

In the first instance, unless complicated by other cerebral conditions, the problem is almost purely physical. But at times the physical part cannot be ameliorated without prolonged, often severe, and persistent suffering—severe, prolonged and persistent and often without success. Chances for leading a normal life range from slim to fair, depending on the particulars of the condition encountered. It is, however, a severe and persistent suffering that those of us who are lucky enough not to be afflicted with may often tend to exaggerate. There are many such persons who, surviving into adulthood, would far rather be alive as they are than dead. Quality of life judgments made by one person for another are notoriously dangerous judgments to make. To ask ourselves what such a child would want is useless; to ask what it would surely not want may bring us a bit further.

In the second case (Down's syndrome with, say, duodenal atresia), the prognosis of the underlying disease is quite acceptable provided the rather easy-to-repair defect is, in fact, repaired. In such cases, one may be left with a near normal child, with a child able to feel pleasure and pain even though not at a level we are wont to consider "normal" or we may be left with a severely mentally impaired child. Unfortunately the condition is quite variable and virtually impossible to predict at the time of birth. There can be no question that many such children, since they are capable of being benefited and harmed, are of primary moral worth and that considerations of their secondary worth (their value to others) must take a backseat. Furthermore, a decision not to repair an eminently repairable lesion is what ultimately will cause death: The life or death of such an individual would then depend on the accidental coexistence of another lesion. In such a case the parents' responsibility to refuse a procedure needed to sustain the life of a defective child is, at the least, questionable: In our society, however, the parents are the ones inevitably left with the ultimate problem. When such a repair is done against the family's wishes, it seems ethically question-able to then saddle them with the outcome of a procedure they did not want. Once again, it is difficult to deal with this problem on an individual basis out of context with the society in which the problem exists: If, when families were unable or un-willing to assume the long-term care of such a child, communities were able to of-fer such children adequate support, the decision to allow parents to decide might very well be different.

In the third example, infants are initially mentally normal but either mentation, their physical state or perhaps both will rapidly and relentlessly decline until they die. While both Werdnig–Hoffman and Tay–Sachs children will die an early death, chil-dren with conditions like Werdnig–Hoffman will be able to feel pleasure or, at least, pain, and will maintain primary worth up to the very end, while children with condi-tions like Tay–Sachs will not. In effect, children who maintain their mentation up to the very end must, for that reason, be considered quite differently from those who do not. When the ability to feel pleasure and pain is maintained, parents stand in place of the child in helping to determine what is in the child's best interests (in other words, they act as surrogates and the child's best interests remain central); when, however, the ability to be benefited or harmed is no longer present, the "best interests" of the child assume symbolic significance.[22] Arguably, parents and the best interests of the parents and of the rest of the family now move much more into the center stage of ethical decision making.

In the fourth instance, encountered all too often in the neonatal nursery, an in-evitably hopeless condition often with little or no mentation is present. There are gra-dations of this condition. When the ability to feel pleasure or pain in adults is severely and permanently limited in the face of little chance for physical recovery, attempts to maintain their existence are seriously questioned and are often abandoned (see Chap-ter 11). In children, emotive and symbolic considerations in analogous cases often sway us to make starkly different (and, at times, difficult to defend) judgments.

A variation on this theme is the vexing problem of low-birth-weight infants. It is handled somewhat differently in different Western countries. In the United States there has been generally a cutoff point (often 500 g) above which all and below which no treatment has to be provided. In other cultures there may be other cutoffs and, above all, there may be a lower, an upper, and a gray zone area: Above a certain weight, all infants are treated; below a certain weight they are not and between these two cutoff points treatment is optional and decided on an individual basis. Low-birth-weight infants are, furthermore, prone to have severe inter-current problems: generally either intra-cerebral bleeding or an immaturity of the lungs that will make them into permanent and sometimes permanently ventilator-dependent pulmonary cripples. No strict rules can be argued for: Empirical data have to help guide the ultimate decision. There is, however, a hardly negligible problem. Such infants will spend months and sometimes years in a neonatal ICU at a cost of about $1,500 per day. When seen on a national level, this is a vast expenditure. Further, many of such disasters could have been prevented by good prenatal conditions and care. Mothers living in poverty or unable to afford decent medical care are especially at risk. Ms. Jones' low-birth-weight infant who was born to a poor mother lacking adequate prenatal care cannot be entirely separated from the social conditions that caused the birth.

In the fifth instance (that of the anencephalic infant), there is no hope for physical survival and no mentation. Such children, in our terminology, lack primary moral worth albeit that secondary and symbolic worth endures. A situation analogous to that of the vegetative state exists (see Chapter 11). Dealing with such infants differently merely because they are infants is rooted in emotive and symbolic considerations. (The problem of using such infants as organ donors is briefly addressed in Chapter 9.)

In dealing with problems concerning impaired infants, then, a number of problems exist. First of all, such infants cannot give or withhold consent: Decisions are made by surrogate judgment (in which the parents, acting as surrogates, speak for the children) or by substitute judgment (in which health professionals or the courts substitute their own judgment). In adults, there is a sharper differential between surrogate and substitute judgments: Surrogates, by virtue of acting as surrogates, are felt to have closer ties with the patient's prior world view, and such surrogates therefore would be more likely to judge through the eyes of the patient's values. Substitute judgments, perforce, are based on the judgment of strangers who inevitably will choose through their own values. The infant's values and worldviews are inchoate, unknown, and unknowable; their probable judgments are opaque. The difference between surrogate and substitute judgments, therefore, is more blurred and tenuous. Nevertheless, such infants will, at least in their early years, have much the same values as their parents. It seems likeliest that a parent can therefore speak more accurately to what "the child would wish" than could an outside agent. This does not preclude that such parents would do well to counsel with healthcare professionals, clergypersons, ethicists, or an ethics committee, but in the final analysis choices that are far from clearcut and that are, as it were, in the "gray zone" should be up to the parents to make.

Secondly, when judgments are made for infants they are embodied in a set of preconceived emotional and symbolic considerations quite different from those that operate in adults, especially in elderly adults. It is often difficult to keep the disease and the patient's actual condition and prognosis, rather than the age of the patient, in mind and to judge accordingly. Again, emotion cannot be allowed to substitute for reason; reason alone cannot be allowed to make judgments entirely bereft of compassion. A process of compassionate rationality in which one supplements, corrects, and is in conversation with the other must come into play.

Thirdly, the calculus between risk and benefit is even more obscure: We tend to judge benefit or pain from our own vantage point. Pain, on the one hand, may be easier to endure by virtue of being understood; for the infant or young child, lack of ability to understand and therefore terror may also enter the equation (see also the section in Chapter 11, on limiting treatment in demented patients). On the other hand, children are far more "elastic," and once pain and suffering are over are far more apt to put pain and suffering aside.

Fourthly, defining and ultimately even predicting success or failure is extremely difficult in the neonatal period. Viability varies as technology changes, and the definition of what is and what is not "viable" is, in itself, a difficult one. If by viability is meant biological survival, it is one thing; if by viability is meant biological survival permitting an acceptable life, it may well be another. Here more than in most other places, the "ethics of uncertainty" must be taken into account.[22]

There is one other factor, speculative though it is since we have no direct empirical data. As the work of Damasio, LeDoux and Roth have shown very early emotional memory is laid down in infants and remains virtually permanent. The psychological effect of what is done or not done to such an infant in that early period may (we do not know this) have far-reaching consequences that need to be considered.

The so-called and now largely defunct (at least on the federal level) Baby Doe regulations were historically first made to prevent arbitrary non-treatment of defective newborns. Initially, such regulations forced the treatment of all defective infants no matter what the defect and for as long as treatment was technically possible. The wishes of the parents or the opinions of health professionals were quite irrelevant. Hospitals were forced to post the telephone number of a "hotline" and anyone was encouraged to report the "mistreatment" of such an infant. Flying squads were set up and hospitals could, at any time, be inspected and their routine disrupted by federal officials. The results were disastrous and the courts eventually declared this venture to be unconstitutional.

The regulations today have been somewhat modified.[22] Instead of being tied to civil rights considerations, they are now tied to federal funding and controlled by the states: States that wish to participate in funding must comply. At the present time there is considerable option, although there is still the danger of unwarranted intrusion: In making decisions, neonatologists in many states ostensibly may not consider the child's probable future impairment and quality of life.[22] Neonatologists, in general, have been very sensitive to this issue. In a recent large-scale survey neonatologists felt that the regulations even today not infrequently result in unwarranted, useless, and, at times, even cruel treatment.[23] Quality of life issues and considerations dealing with the future impairment and function of such children, far from being prejudicial, are largely the essense to the cases being judged.

—————————— **ANENCEPHALIC CHILDREN** ——————————

The issue of using anencephalic infants as organ donors has recently been raised. In part this interest is due to the greater success that transplantation in more mature humans has had; in part it is due to the perceived necessity of extending our frontiers of knowledge further; in part it is due to parents of such prospective children wishing to see at least some good resulting from their tragedy; and in part it is because of the publicity that the media have given to this issue. Using such infants has been hailed by some as providing more badly needed organs and thus saving lives but decried by others because of the alleged affront to human dignity of extending the biological existence of a newborn merely for the sake of harvesting organs. The battle has been joined when infants of this sort actually began to be used for this purpose.

Using infants (or other humans) who are actually brain dead as organ donors offers few problems to most people today. Such (previous) persons are legally and ethically acknowledged to be dead, and only those who would object to organ transplantation, cadaver usage, or the temporary preservation of biological process until organs can be harvested would object. And those objections are a different story.[24] Anencephalic infants offer a problem because criteria for brain death under a week of age are not firm[25] and because some persons are affronted by keeping such infants on a respirator merely to harvest their organs. (See also Chapter 9.)

The anencephalic child is not, in the sense the term is used legally today, brain-dead: To be brain dead requires death of the entire brain. The anencephalic infant, however, is bereft of the necessary substrate that is acknowledged to underwrite any possibility of perception or self-awareness. Anencephalic infants, now and in the future, lack the capacity for experience, and preserving such a life as a condition for experience is therefore futile. In many respects their condition is analogous to the permanent vegetative state or perhaps to permanent coma. They have a functioning brainstem, albeit the function of that portion of the brain for reasons that are poorly understood will function for only a brief time. Such children, in other words, lack the capacity to suffer and, therefore, lack "primary worth" or the potentiality to develop "primary worth." (See Chapter 3.)

As noted before, a large body of experimental evidence indicates that the capacity to suffer is intimately connected with the presence of a neocortex. It is the neocortex that in those biological organisms known to us underwrites thought and memory. It is memory (the ability to recognize and to have sustained perception) that underwrites thought, here defined as the integration of external or internal sense perception into memory. And it is, among other things, the capacity to suffer, underwritten by the neocortex that endows entities with primary moral worth, that makes them fitting centerpieces for the physician's moral consideration. Anencephalics lack the most primitive condition necessary for suffering. They do, however, have great secondary worth: material value as potential organ donors and symbolic value to parents, families, caregivers, and ultimately the community.

Communities, which value anencephalic and brain-dead persons in a symbolic way, also include those members of the community who are in need of organs to sustain their existence. Since anencephalic infants cannot be harmed or benefited in themselves, since by our definition they are not of "primary worth," the ethical question is not as much concerned with such infants as it is with the benefit or harm that our doing or not doing things to such infants has for others. In the first instance, this would be a question of parental values. When the parents, however, freely agree to the use of their child's organs (and in some instances at least perceive that some good for another may come out of that tragedy to themselves—a perception that can be of great solace to them), communal values come into play. Communities, which ultimately must judge the morality and legality of using such infants, now must decide whether they will value symbol or reality more: the symbol of humanity represented by the anencephalic child, or the reality presented by members of the community who may live because of that symbol. Valuing symbols more than the reality they portray can lead to a dangerous undervaluing of reality and, ultimately, to its distortion. (See Chapter 9 for a further discussion.)

When we deal with anencephalic infants, however, we are not dealing with the brain dead as we do when we harvest organs. These infants have a rudimentary brain-stem and have some reflex activity. Inevitably and rather rapidly, however, they deteriorate until they can be supported only by artifice. In general such children often and indeed usually must be maintained on ventilator support so that their organs can develop sufficiently. They are, in other words, being treated at that point merely as means and not as ends in themselves. Since they are not actually brain dead, it is not quite the same as maintaining the brain dead. Or is it? In both cases our decision not to support by artifice is grounded on the conviction that such artifice serves only to prolong a state of being that, by no reasonable measure, now or in the future, can be considered to be self-knowing, self-realizing, and, therefore, to have the capacity for suffering.

In such cases, what is being maintained is a highly complex and intricate tissue culture bereft of those things that give it, itself, meaning. What meaning there is, is meaning for others who themselves are self-knowing and self-realizing. Anencephalics cannot, now or in the future, experience suffering or joy; they lack "primary worth." To fear that maintaining such infants by artifice until donation can be done is using a person deserving of respect purely as a means and not also as an end to conflate symbol and reality. Such infants do have value, but it is a value expressed by others (the parents as well as the community) who value. Such infants, in themselves, cannot be benefited or harmed: Only the valuers can be. The parents, the community, and the person desperately in need of an organ to maintain life or restore function form the legitimate centerpiece of our concern when it comes to the disposition of such infants.

Members of the community, those with and those without a need for organs, form communities because they share an underlying set of values. Among these values are the symbols they cherish, the conventions they adopt, and the regard they have for each other's weal and woe. The value of living persons whose life may be extended or made more tolerable by transplantation of needed organs must be contrasted with the value of the symbol that the anencephalic or the newly dead represent. Neither anencephalics nor the brain dead are entities of primary worth, but neither are they without any value at all. It is valuing done by others that gives anencephalic infants both symbolic and secondary worth: The valuing done by the parents and the community gives symbolic worth; the valuing done by those in desperate need of transplantable organs gives them material worth. If those for whom the anencephalic has symbolic worth are willing to use this symbolism to help another member of the community, the community can have no reasonable objection.

Some have argued that anencephalic children as children have a particular moral standing since they are of such immense symbolic significance. Our heart goes out (or at least it is argued that it ought to go out) to such a child, and using that child as an organ donor is an inhumane and ultimately callous thing to do.[26] It is claimed that infants have (or ought to have) special moral standing by virtue of being infants. But this, among other things, conflates healthy with anencephalic infants—almost a category error. This argument is specious and points out the difficulties with what is called the "care ethic:" Caring decides the issue, and there is no way of arbitrating among various ways of "caring." Parents who are willing to see the organs of their anencephalic children used are said not to be caring appropriately—a curious statement since it is the parents who bear the actual brunt of the tragedy! In this issue, where doing harm to the infant is at most doing it symbolic harm, parents ought to be free to make such a tragic choice.[27]

There is one other troubling, and perhaps more important, issue that has, perhaps, not been sufficiently raised: the question of using newborns, healthy but for the adequate function of a single vital organ, as experimental objects under these circumstances. Where a reliable human track record exists or, at least, where a comparable animal model has been shown as workable, doing such procedures may be justified. Even here, the problem of surrogate informed consent may be a troubling issue. Where, however, a track record is lacking and where animal models are not sufficient to light the way, such experiments are highly problematic. It is clear that no absolute certainty exists and that progress depends on a careful forging ahead into the unknown; but reasonable, careful progression of experimentation must and should precede the extension of a procedure to patients. Innovation without this seems ill conceived. That, to us, is one of the central problems in transplantation from the anencephalic. Experimentation must not be conflated with therapeusis, no matter how devoutly effective treatment is hoped for. (See the following section on experimentation.) The psychological benefit which in some parents may accrue by the use of such organs—their realization that at least some good may have come out of their terrible experience—may help them along the road of recovery. Far from being callous, they are compassionate, generous and receive some gratification from knowing that at least some other or others have been or could be helped and, perhaps, even given a life.

———— EXPERIMENTING WITH AND ———— TRANSPLANTING FETAL TISSUE

It is probable that some conditions can be treated (at least ameliorated) by the use of fetal tissues. Fetal tissues are far less apt to be rejected and are, therefore, often particularly suitable for transplantation. There seems little doubt that more and more conditions are, at least potentially, treatable by the use of such tissues. There are, of course, two questions here: the ethical propriety of using fetal tissue for experimentation, and, ultimately, the propriety of using such tissues for the treatment of patients. In some—but not in all respects—the ethical issues are similar to those for the use of omnipotent stem cells (see chapter on genetics)

Since fetal tissues are obtained by abortion—and since spontaneously aborted fetuses are generally not suitable and not present in adequate numbers—much of the debate hinges on one's attitude toward abortion. Those who oppose abortion are generally apt to oppose experimentation with and transplantation of fetal tissue. This is not necessarily the case, however. Even those who feel that abortion is ethically impermissible may argue that it is better to at least obtain some benefit from an actualized evil than to discard any possible good that might come from it. Such an argument is a largely utilitarian one. The question here is analogous to the question of using data from the Nazi or from the Tuskegee experiments. Persons who have no moral qualms about abortion are unlikely to oppose experimenting with or utilizing aborted tissues.

Shall we use—or would it be more proper to use—data obtained by clearly ethically unacceptable experiments or should such data be ignored or discarded? This is a question which antedates the Nazi experience—much of what we know and knowledge we routinely use in the practice of medicine was obtained by means we would today find impermissible: no informed consent, coercion, bribery and the use of unanesthetized animals was the order of the day. If we foreswore the use of data obtained by such clearly unethical means our entire structure of medicine would collapse. The fact that we would have to forego all knowledge gained in this way and that this would almost completely hamstring medicine in no way excuses previous practice. It does make it all the more necessary to have clearly defined criteria for and supervision of experiments with living organisms—but it cannot, in a very pragmatic sense, willfully put us back hundreds of years and, thereby, allow many existent lives to be wasted.

As long as abortion is a legally permissible procedure, arguments against using the tissue of such fetuses are somewhat feeble. The question is first one of informed consent (can a mother who has purposely discarded her fetus consent in its behalf—which implies acting in its best interests?) and, secondly, of how such tissues are to be made available. Again, we are left with our own metaphysical attitude toward the fetus: Can a young fetus (and a dead fetus at that) be said to have "interests"? When all is said and done, allowing the mother to make such a decision is not altogether good, but it may well be the best of a number of objectionable alternatives.

The fear that women may purposely become pregnant so as to utilize their tissues has several versions:

1. Pregnancy and abortion occur so as to sell fetal tissue.
2. Pregnancy and abortion occur so as to benefit oneself (consider, for example, a young lady with diabetes and early renal damage who grows her own pancreatic tissue.

3. Pregnancy and consequent abortion is done to have a father, disabled by Parkinsonism, treated.
4. In another version tissue is sold by a poverty stricken family (say in a "developing" country) which would allow that family to escape hunger, homelessness and perhaps help in providing a future for the other children of that family.

What is needed is neither to forbid the procedure outright nor to allow it to go on uncontrolled; what is needed is an understanding of the different cultural contexts and situations and communal deliberation and action so as to formulate reasonable laws that would guard against what society would define as abuse. Furthermore, if one is to abolish tissue and organ sales in poverty-stricken nations or by poverty-stricken people the reason for their evident need has to be dealt with.

MATERNAL OBLIGATIONS: FORCED C-SECTION, FETAL ABUSE, AND WRONGFUL LIFE

There has been a recent trend to attempt to control the behavior of pregnant women during their pregnancy in order to "safeguard" the fetus. These attempts have ranged from enforced C-sections in which women, by court order, are forced to undergo C-sections to which they object to charges of "fetal abuse" against women who either are felt to have contributed to the damage done to a fetus during pregnancy or refuse to abort a clearly and severely defective fetus, thereby bringing it to term and then, presumably condemning it to a miserable life.

In one such case a woman dying of metastatic cancer and carrying a possibly viable fetus was forced to undergo C-section. This was done on the basis of a court order issued at the hospital's request despite her own, her husband's, and her doctor's vigorous objection. The patient as well as her offspring died.[28] Throughout the country, court-ordered C-sections have become increasingly more frequent.[29] In another case, a pregnant lady delivered a brain-dead infant. She was charged with fetal abuse because she had allegedly not followed her physician's advice: She had taken amphetamines, had intercourse during late pregnancy, and had delayed coming to the hospital when she began to have some bleeding. That case was eventually dismissed on a technicality, but the possibility of bringing other cases of this sort has been very much discussed.[30] Newborns who at birth are already severely damaged because of substance abuse by the mother (especially cocaine, but alcohol and tobacco can likewise be severe problems) are becoming frighteningly frequent. This is especially true in the United States but is currently also on the rise in other nations.[31] What are the moral and legal implications?

First we want to re-state that a given course of action may be considered ethically improper but nevertheless not the kind of issue properly adjudicated by law.

Likewise we must guard against giving too great power to either the medical profession or the state to intrude into our private lives. There are, as in everything else, fads in medicine. The current prohibition against even a single drink in the course of pregnancy is, among others, such a fad.

Obstetricians formerly felt that their professional relationship was with the mother. The child, except perhaps to those physicians who under no circumstances would ever abort the mother, was an important, but not a prime, consideration. Some of these feelings have changed today. Obstetricians often feel as much (and sometimes indeed more) concern for the fetus than they do for the mother. In a questionnaire to the heads of maternal–fetal medicine at 57 institutions, 46% thought that "mothers who endangered the life of the fetus should be detained in hospitals or other facilities so that compliance could be ensured;" 47% thought that the precedent for enforced cesarean sections should be "extended to include other procedures that are potentially life-saving for the fetus;"[29] and 26% advocated "state surveillance of women who stay outside the hospital system in the third trimester." The obstetrical literature reflects the concern of practicing obstetricians. The forceful delivery of patients by operative intervention (or even coercing them to do "as the doctor orders") is becoming more routinely considered and, at times, done. The tendency to consider the mother primarily as an incubator for her developing infant is growing.[30,32–35] In some states (California being one of them) pregnant women lose their right to decline interventions like CPR or blood transfusion—thereby becoming virtually little more than state chattel.

In looking at these issues, there are two problems. The first deals with whether or not it is unethical for a mother who had decided not to terminate a pregnancy but to carry it to term to jeopardize the fetus by doing or neglecting to do certain things. The second deals with the rights of communities to force women to do or not to do things to their own bodies during the course of pregnancy. These are different questions: The first clearly deals with personal morality; the second with the community's right to enforce on its members its own vision of personal morality. There are clearly instances (murder, for example) when communities do, and we feel should, enforce such visions; there are other instances when such visions are not sufficiently clear. Enforcement, under such circumstances, becomes quite a different matter. (See also the discussion of abortion elsewhere in this chapter.)

Women who are pregnant sometimes have the opportunity (but not always: we think here of the many states that refuse to fund Medicaid abortions and the women whose insurance and personal finances make abortion practically not possible) to terminate their pregnancy. When women choose not to terminate their pregnancy, they voluntarily choose to take on at least some responsibility for a developing other.[36] Members of a community have an obligation to help, or at the very least not to harm, other members of community no matter what vision of community they may share. Such obligations are arguably strengthened when certain relationships are freely assumed: those of physician to patient, professor to student, or husband and wife are examples. Pregnant women who choose not to abort would seem to share in such a relationship with their developing infant. That does not mean that every moment of their life or every action of their body must be devoted to this enterprise any more than that healthcare professionals, professors, or spouses must devote every moment of their lives to those to whom they are at least in part responsible. But it does advance the claim that some responsibility—at least responsibility not to harm—exists and that totally ignoring the good of the offspring is basically an irresponsible thing to do. The extent of this responsibility is, to a large degree, a social as well as a personal vision, and one that will show considerable variation from community to community and from individual to individual. But that there is some responsibility is hard to deny.

There is, furthermore, an aspect that indeed concerns the community. As a community we cannot afford (in terms of money as well as in terms of time, devotion, love, etc) to allow the unnecessary creation of persons who will be disabled and reliant on their care upon the community. If a person's "life-style" seriously threatens a developing fetus it is a concern not only of that person but also of the community which will, inevitably, share in caring for it.

When women do have an opportunity to abort an undesired pregnancy but decide to carry it to term, their obligation to a fetus they are carrying is undoubtedly increased. When one has chosen a role (be it that of physician to patient, professor to student, or husband and wife), one's obligations in that freely chosen role are quite different from what they would be if such a role were a forced one (say, if one were drafted into the army without the possibility of conscientiously objecting, or if one had to do an odious job under duress). But that does not mean that persons who did not choose their role, but nevertheless were saddled with it have no obligations at all. At the very least we have an obligation not to bring harm to another person—and the developing fetus of a substance-abusing mother, if pregnancy is allowed to continue, risks becoming a damaged person.

On the other hand, the obligations women have to their offspring are not entirely analogous to those of healthcare professionals to patient or professor to student. Healthcare professionals do not have the obligation (at least we have never heard this argued!) to consistently do things to their bodies (other than to get fatigued, perhaps) in order to promote their patient's welfare. No one would seriously consider it a doctor's or a nurse's duty to donate blood or to undergo a surgical procedure for the patient's good. Healthcare professionals who give blood (or kidneys) to their patients are considered to be doing a supererogatory thing. When women are held to have obligations to their developing offspring, these obligations inevitably must be translated through their own bodies. In the view of some, pregnant women must follow their obstetrician's instructions to the letter even if this means a radical alteration in their lifestyle: They must take, or refrain from taking, certain drugs or foods, must exercise or not exercise, must undergo or not undergo certain procedures, etc. And they must do this not only because it is "good for the baby" but also because their obstetrician has decided that it is, in fact, good for the baby. Under such conditions, the patient's physician for the duration of the pregnancy becomes the patient's master.

Arguing that an obligation toward the developing fetus on the part of the mother exists is not the same as claiming either that this obligation is absolute (that, in other words, women must follow their obstetricians' dictates to the letter) or that such an obligation can or should be legally enforced. The law, under ordinary circumstances, cannot compel persons to do things to their bodies that they do not choose to do.[29] I cannot be forced to go to bed or not to go to bed at certain hours, cannot be forced to undergo bypass surgery or be forced to surrender a kidney or even blood. Communities can force me not to take illicit drugs but the right to privacy that I enjoy modifies their power, even in that respect. Debatably, it may be immoral for me not to take a medication that would enable me to be gainfully employed and by so doing (or not doing) deprive my family of a decent livelihood, but legally forcing me to do this is another matter. Some analogies exist to the issue of "self-causation" when we invoke the patient's responsibility for their own disease.[37] On the other hand, the number of damaged children being born has become an overwhelming social problem and one with which the community must in some way deal. Likewise and not entirely without connection, the abuse and neglect of children appear to be on the rise. Recent attempts have been made to force women who are offenders to choose between alternative courses of action: go to jail (child abuse as well as drug abuse is a criminal act, and unnecessarily giving birth to damaged infants is, it can be argued, a species of child abuse) or have a hormone that prevents conception implanted under their skin.[38,39] Such a procedure does, of course, restrict the personal freedom of the woman and has therefore been strongly contested by groups that tend toward a libertarian type of ethic.[40] Again we are confronted with the problem that actions that threaten to destroy a community (and having a large number of defective children, or abusing one's children, well might) are a threat to the community's prior worth. Well-informed, democratically operating, and interacting societies will have to grapple with such problems and in an ongoing manner evolve appropriate solutions. Safeguarding individual freedom requires a context that enunciates and then safeguards the freedoms it decides to give. Without this context (the community) individual freedom itself cannot long exist.

Thus far we have briefly discussed the issue of maternal obligations toward their developing offspring, offspring who would, were it not for the mother's action or failure to take action, have the opportunity to be "normal," "healthy" persons. If mothers as well as fathers have an obligation to prevent harm that might befall their developing offspring, do they likewise have an obligation to prevent the birth of a severely defective child? The problem (sometimes one, rather quaintly, called "wrongful life") really has two parts. On the one hand, if a person (mother or father) knows that they carry a dominant gene for a disabling illness, should they be morally free to procreate at will? If the gene is not dominant but the chance is large that the offspring will be affected, are their (moral) procreative "rights" limited? On the other hand, if a woman (be it by amniocentesis or otherwise) discovers that her fetus will be born severely defective, is she (morally) compelled or at least advised to abort it?

The underlying proposition—and one that because of space we shall not argue at length—that all humans as far as that is possible should have the best chance of leading happy, productive, and fulfilling lives is not one that many of us would quibble with. That, however, helps very little. Let us take two extreme cases. On the one hand, it is known that the fetus will be born with a condition leaving it totally crippled, immobile, in severe pain, and unable to communicate meaningfully with its environment; on the other hand, it is known that the developing fetus will be affected with a club foot, a poorly functioning left hand, or diabetes. Both cases represent "disabilities" or "abnormalities." The point I am trying to make is obvious: At what point is disability disabling enough to counsel (from an ethical perspective) against procreation or for abortion? At what point is the notion of life being wrongful justifiably to be considered "wrongful"?

A Kantian approach, counseling respect for all who are capable of making rational choice or who have human form, will not do: "Respect" could, one may argue, consist of aborting or not aborting depending on the prior assumptions one makes. A utilitarian approach calculating pleasure and pain (or utility) likewise runs into the very same problem, and the notion of primary worth (while possibly helpful) is likewise in itself insufficient. There are no easy absolute answers to such problems nor are there ready-made formulas or rules that one could with any degree of justification apply. But there are dialogue, education, and a decent respect for personal choice made with rational compassion within a framework of communal values and points of view.

—— INFORMED CONSENT AND CHILDREN ——

Although the validity of parental consent for children has been taken for granted, absolute parental freedom to make final judgments for their children has been found to be increasingly problematic. If, of course, one were to presume that parents invariably choose in their child's best interests, if what constitutes "best interests" in a variety of situations and circumstances were agreed upon, and if, under such circumstances, one were to allow the parents alone to define that interests, no severe problems would exist.

But do parents always have their child's best interests at heart and do they always know what those interests are? What if parents have their own agenda (if, for example, they stand to benefit from a life insurance policy on their child's life or, far more subtly, if they have to make judgments that must balance the interests of a severely impaired child against those of other normal children)? Should the interests of a severely impaired child be allowed to trump those of normal children in the family? What if the parents see it against the child's best interests to be transfused (because in the parents' viewpoint such transfusion spoils the chance of Heaven for their child) or see it concordant with their child's interests to allow the child to become a subject of non-therapeutic experimentation?[41]

Last but not least, is the problem of how to define a "child" and how to adjudicate the child's proper role in deciding what should and should not be done. The legal question ("When is a child a child?") is only feebly connected with the growing autonomy of developing adults. Looking at autonomy in the 3-, the 10-, and the 15-year-old in the same way is patently ridiculous. The "mature minor" rule, a rule that permits medical treatment without parental consent under certain circumstances after age 15,[42] and other statutes covering "emancipated minors" (which vary from jurisdiction to jurisdiction), provides only a partial solution for what is inevitably an arbitrary cutoff point but otherwise begs the question. Maturing, it would seem, is neither an all-or-nothing proposition nor altogether predictable by age. Ethically, the best that we can do is judge individual situations on their own merits using arbitrarily fixed groupings merely as statistical guideposts on our way.

The law in most states provides that a minor who has delivered an infant is therefore "emancipated" and has, among other things, parental control over her infant and over her own and her infant's destiny. Such a law, in our view, affronts all logic and has often led to disaster. What it does is to reward irresponsibility (getting pregnant when one cannot care adequately for the needs of a child) and, backhandedly, to punish those who have acted responsibly by not becoming pregnant (by continuing to treat them as minors).

Rights are virtually never absolute. Property rights, for example, are modified by culture, circumstance, and the nature of the property. While I am at liberty to take an ax to my table, I am not at liberty to take an ax to my dog. Parental rights over their children are likewise far from absolute: Parents are not at liberty to destroy, maim, or neglect their children. The interest of the state in protecting individuals does, at least as far as physical abuse is concerned, act to safeguard children. The question, "Can anyone give proxy consent for another that is not in that other's best interests?" is not easily answered for it leaves the interests of that other necessarily undefined.[43]

While parents, in our society as in most, are not at liberty to destroy their children, their power over such children nevertheless goes very far. Parents, for example, can have their children operated on for religious reasons alone (circumcision, although it also has proper "medical" reasons, is perhaps the most frequent example in our society, and what is called "female circumcision" and often, in fact, is severe mutilation is still practiced in other societies). Parents are free to inculcate their children with rather injurious points of view and teach them rather devastatingly destructive things. For example, parents can bring up their children as flagrant racists or sexists and teach them the gentle art of handling submachine guns without running the risk of community interference. As always, it is a problem of balancing communal and personal obligations. It is, however, a problem with a twist: A third, helpless party and one that is as yet unformed and whose future depends on that shaping is involved.

When it comes to making determinations for infants or for very young children, "interests" as viewed in our community must conform to societal norms. Parents are not at liberty (legally, at least) to abstain from consenting to clearly life-saving procedures. A difference between the right to assent to a procedure clearly for another's benefit (transfusion or appendectomy, for example) and the right to refuse such a procedure for another and risk, or bring about, certain death has been made. Physicians, when confronted with situations in which a child's parents refuse to permit a clearly life-saving procedure in very young children, have historically been able to obtain a court order or, if a pressing emergency exists, to proceed without a court order until such an order can be obtained.[44]

A child's refusal to permit life-saving treatment—especially when that child is of more mature years and has been further matured by experience, when the illness itself is not reversible, or when treatment has only a slim chance or is excessively burdensome—is an agonizing one. In the adolescent, the decision is properly one in which the patient maximally participates;[44-48] in younger children, it is often one in which participation becomes more and more problematic. One must, furthermore, remember that circumstances affect maturity: In the 17th and 18th centuries, children of 16 were quite mature. When children have been chronically ill or when they are living under severely adverse circumstances (in a ghetto or as a Jew under the Nazis), they will mature quite rapidly. Those who fail to mature under such circumstances are, in fact, unlikely to survive. Children with chronic or relatively longstanding illnesses likewise tend to mature faster, and their decisions tend to take on more of the attributes we expect from much older children or from adult persons. Assessing a given patient's maturity, understanding, and decisional capacity is one that health professionals with the advice and counsel of child psychologists and social workers should make. Here, as in all other pediatric issues in which children able to express themselves coherently are involved, the decision is, at the very least, a communal one in which the child is a partner in the communal enterprise of decision making and not an inanimate object to be acted on by others.[46–48]

Experimentation in children is even more problematic. Ramsey[49,50] has argued that "consent as a canon of loyalty" precludes the use of children for experimentation unless, all other means having failed, such experimentation can reasonably be expected to result in direct benefit to the child. Ramsey includes "offensive touching" in the course of experimentation (drawing blood, for example) in this interdiction. Others have taken a more moderate view in which minimally risky and minimally offensive procedures may be permissible, whereas others may not.[51,52]

Experimentation in children, obviously, must first of all conform to the ethically acceptable principles of research.[53,54] Beyond this, and since children, together with prisoners, the mentally defective, and at times the elderly, are particularly vulnerable to abuse, further safeguards are essential.[55–58] Unfortunately, some research, research that is often of the greatest importance to future generations of children, can only be performed on children. Recognizing this, the National Commission for the Protection of Human Subjects came up with specific guidelines in such circumstances.[55] Guidelines included the requirements that (1) risks must be minimal except in circumstances in which the subject itself (having no viable alternative treatment) would have a fair prospect of benefiting; and (2) permission of parents and, where possible, assent of the child is free and informed. Such guidelines fail to answer the moral question in many if not most particular cases. Like all guidelines they are helpful but cannot be conceived as mindless rules to be applied in cookie-cutter fashion, and they do require a further analysis and definition of "risk" as well as of "free consent." If taken as guidelines instead of as substitutes for moral reasoning, they may begin to serve well.

────── **SURROGATE MOTHERHOOD** ──────

Surrogate motherhood—the procedure by which one woman is hired by another to bear children that the first, for whatever reason, was unable to conceive or carry—has become a practical reality and an ethical problem.[59–62] It is a problem because the interests of the surrogate and the mother may strive for the same goal initially but later, due to a great part to the influence of biological, emotional, and social forces, may diverge dramatically. Payment, it has been shown, plays a significant role in most persons' decisions to be surrogates.[63] Inevitably, it is the wealthier who can and do afford to rent the uterus of one who is poorer. It has, furthermore, become a problem not only because to begin with there are enormous associated ethical, cultural, and religious questions at stake, but also because a rather large and profitable industry has grown around it—an industry from which many healthcare institutions, professionals, attorneys, and others profit. This fact makes the problem even more difficult to discuss dispassionately. And yet the fact remains that renting out one's uterus, no matter how one feels about contracts, is not the same as renting out one's garage or even one's time and skill to do a job. A good deal more than that is at stake.

"Surrogate motherhood," at least in one of its manifestations, is, furthermore, a misnomer.[64] "Surrogacy" is used in at least two different ways:

1. The "surrogate" is artificially fertilized with the sperm of the father, carries the pregnancy, and then surrenders the child to the biological father (and, presumably, his spouse). "Surrogacy" here is indubitably a misnomer: The mother, in every sense, is the biological mother.

2. The "surrogate" may have another's ovum artificially or otherwise inseminated by a male implanted in her uterus. She then carries a fetus in whose genes she does not share. In the sense of not having contributed to the genetic makeup of the offspring, strictly speaking, she is not the biological mother; but, because she is carrying the pregnancy and, inevitably, becoming biologically and, perhaps, emotionally involved, she is, in some sense, fulfilling that role.

Here we shall refer to the "surrogate mother" in either instance but caution that the misapplication of the term in the first, and the possible (or at least partial) misapplication of the term in the second instance must be kept in mind. As in many other instances, the language in which we frame problems plays a significant role in their final adjudication.

If one views communities as united merely by duties of refraining and sees in freedom an absolute condition (see Chapter 3), one will have little problem seeing in surrogacy a purely personal concern. On the other hand, if instead of subscribing to a purely autonomy-based justice, communities include beneficence as a necessary ethical condition, such contracts may be more suspect.

Two views of "surrogacy" are then possible. The one bases its justification on the right of consenting adults to control their destinies as long as they do not impinge on others. Such a view holds that surrogacy falls into such a category: Persons have knowingly contracted together; the contract is valid and is no one else's business, and that's the end of it. Contract laws, here to enforce valid contracts, can be invoked. The other view sees surrogacy as involving far more than merely a contract that is no one else's business. While it is a contract between consenting adults who do have a right to control their own destiny, such a right is enmeshed in a social matrix of values and not, therefore, inevitably no one else's concern.

Persons who oppose surrogacy do so because they feel that (1) mothers, during pregnancy, inevitably have a surge of hormones and undergo other changes that more often than not result in their bonding with the child they carry. When such children must then be given up, severe hardship and, therefore, battles likely to be socially disruptive may occur; (2) having one woman (almost inevitably poorer) carry the child of another (almost invariably considerably more affluent) is an act of social condescension likely to be communally disruptive; and (3) the use of resources to create more life instead of taking care of existing life is ill-advised. There are many infants and children in need of adoption who will never be, and perhaps first ought to be, adopted. Many, furthermore, feel that often a racist agenda motivates the desire to try surrogacy instead of adopting a child in need of a home.

Under current laws, women are not allowed to sell their babies or, at least, to do so outright. The restrictions against this are made not only for the good of the infant but, ultimately, to safeguard the community. Those who feel that the sale of infants should be forbidden generally feel that organs, blood, and children are different from cars, houses, and even one's labor and that therefore they ought not simply be for sale. Surrogacy is, perhaps, not quite the same thing. In a community that puts a high value (let alone an absolute value) on personal freedom and choice, especially when it comes to the use of one's own body, making rules that restrict such contracts between well-informed and freely consenting adults is, at the least, problematic. Those who argue for surrogacy largely rely on such an argument. Contracts, on the other hand, exist within the embrace of a social milieu and, at the very least, have to be mindful of it. In communities that put a high value on freedom, making such contracts and enforcing them may not be morally precluded; if, however, such contracts are socially disruptive or if by making such contracts resources that could be otherwise be put to better use, such contracts may be unwise enough to be ethically problematic.

—— ECONOMIC CONSIDERATIONS ——

As with end of life issues, economic considerations cannot be the primary ethical motivating force when it comes to decisions between physicians and patients. In other words: physicians and other health-care providers are not entitled to decide that a given patient "deserves" and another does not deserve a given course of action. Physicians, utilizing those resources made available to them by the community, must use them when they are indicated to further a commonly agreed upon goal. It is the community which must decide what interventions, medications, laboratory or x-ray examination, etc. will and which will not be available for general usage and leave physicians free to decide when and were these are in the patient's best interests and most likely to reach a commonly agreed upon goal.

This statement is historically grounded in our current vision of the physician–patient relationship (see Chapter 6 on the patient–physician relationship). The obligation of health professionals to their patients, however, does not exist outside the social context and cannot be unmindful of it. Economic considerations are necessarily part of the framework in which health professional and patient must interact. At the very least, doing expensive and useless things at a time when resources are sorely needed to accomplish interventions that are desperately needed and rather modestly priced is problematic. (See the section on futility in Chapter 11.) Ways must be found to accommodate the evolving vision of the professional relationship in its social nexus without radically dismantling either one or the other. (See Chapter 8 on macro-allocation.) Physicians together with other healthcare professionals, in today's society, must serve as advisers to the community as well as professionals concerned with individual patients.

There is a further problem beyond this. While we in the industrialized world generally have access to sophisticated, expensive and sometimes exotic ways of dealing with our illnesses, the vast majority of people all over the earth do not even have access to the more primitive and basic forms of medical care. Historically our relative prosperity rests and in part continues to rest on the exploitation of people in what we condescendingly refer to as "the third world" or "the underdeveloped world." It is difficult to deal with this problem—but it is a problem that in decency we must deal with (see Chapter 8).

REFERENCES

1. Durant W. *The Life of Greece.* New York, NY: Simon & Schuster, 1939.
2. Edelstein L. The Hippocratic oath. In: Temkin O, Temkin CL, eds. *Ancient Medicine: Selected Papers of Ludwig Edelstein.* Baltimore, Md: Johns Hopkins Press, 1967.
3. Carrick P. *Medical Ethics in Antiquity.* Boston, Mass.: D. Reidel, 1985.
4. Aristotle. De generatione animalium. In: McKeon R, ed. *The Basic Works of Aristotle.* New York, NY: Random House, 1971.
5. Churchill LR, Simán JJ. Abortion and the rhetoric of individual rights. *Hastings Center Report.* 1982;12(1):9–12.
6. Annas GJ. Righting the wrong of wrongful life. *Hastings Center Report.* 1981; 11(1):8–9.
7. Rachels J. *The End of Life.* New York, NY: Oxford University Press, 1986.
8. Noonan JT. An almost absolute value in history. In: Noonan JT, ed. *The Morality of Abortion: Legal and Historical Perspectives.* Cambridge, Mass.: Harvard University Press, 1970.
9. Thomson JJ. A defense of abortion. *Phil Publ Affairs.* 1971;1(1):47–66.
10. Milby TH. The new biology and the question of personhood: implications for abortion. *Am J Law Med.* 1983;9(1):31–41.
11. Green MB, Winkler D. Brain death and personal identity. *Phil Publ Affairs.* 1980;9(2):104–133.
12. Jones GE. Fetal brain waves and personhood. *J Med Ethics.* 1984;10:216–218.

13. Warren MA. On the moral and legal status of abortion. *Monist*. 1984;10:216–218.
14. Engelhardt HT. *The Foundations of Bioethics*. New York, NY: Oxford University Press, 1986.
15. Gustafson JM. Mongolism, parental desires and the right to life. *Persp Biol Med*. 1973;16:529–559.
16. Duff RS, Campbell GM. Moral and ethical dilemmas in the special-care nursery. *N Engl J Med*. 1973;289(25):890–894.
17. Robertson JA. Dilemma in Danville. *Hastings Center Report*. 1981;11(5):5–8.
18. Campbell AGM. Which infants should not receive intensive care? *Arch Dis Child*. 1982;57:569–575.
19. Arras JD. Toward an ethics of ambiguity. *Hastings Center Report*. 1984;14(2):25–33.
20. Annas GJ. Checkmating the Baby Doe regulations. *Hastings Center Report*. 1986;16(4):29–31.
21. McCormick RA. To save or let die: the dilemma of modern medicine. *JAMA*. 1974;229(8):172–176.
22. Rhoden NK. Treating Baby Doe: the ethics of uncertainty. *Hastings Center Report*. 1986;16(3):34–42.
23. Kopelman LM, Irons TG, Kopelman AE. Neonatologists judge the 'Baby Doe' regulations. *N Engl J Med*. 1988;318(11):677–683.
24. Loewy EH. Waste not, want not: communities and presumed consent. In: Thomasma DC, Monagle JF, eds. *Medical Ethics: A Guide for Health-Professionals*. Rockville, Md: Aspen Publishers, 1988.
25. Task Force for the Determination of Brain Death in Children. Guidelines for the determination of brain death in children. *Ann Neurol*. 1987;21:616–617.
26. Sytsma C. Anencephalic infants as organ donors. *Theor Med*. 1996;17(1):19–32.
27. Loewy EH. Of sentiment, caring and anencephaly: a response to Sytsma. *Theor Med*. 1996.
28. Annas GJ. She's going to die: the case of Angela C. *Hastings Center Report*. 1988;18(1):23–25.
29. Kolder VEB, Gallagher J, Parsons MT. Court-ordered obstetrical intervention. *N Engl J Med*. 1987;316(19):1192–1196.
30. Annas GJ. Pregnant women as fetal containers. *Hastings Center Report*. 1986;16(6):13–14.
31. Slutsker L. Risks associated with cocaine use during pregnancy. *Ob Gyn*. 1992;79(5):778–789.
32. Johnsen D. The creation of fetal rights: conflicts with women's constitutional rights to liberty, privacy and equal protection. *Yale Law Rev*. 1986;95:599–615.
33. Finamore E. Jefferson v Griffin Spalding County Hospital Authority: court-ordered surgery to protect the life of an unborn child. *Am J Law Med*. 1982;9(1):83–101.
34. Jurow R, Paul RH. Cesarean delivery for fetal distress without maternal consent. *Ob Gyn*. 1984;63(4):596–598.

35. Leiberman JR, Mazor M, Chaim W, Cohen A. The fetal right to live. *Ob Gyn.* 1979;53(4):515–517.
36. Shriner TL. Maternal versus fetal rights – a clinical dilemma. *Ob Gyn.* 1979; 53(4):518–519.
37. Loewy EH. Communities, self-causation and the natural lottery. *Soc Sci Med.* 1988;26:1133–1139.
38. Robertson JA. Norplant and irresponsible reproduction. *Hastings Center Report.* 1995;25(1):23–26.
39. Ginzberg JF. Compulsory contraception as a condition of probation: the use and abuse of Norplant. *Brooklyn Law Rev.* 1992;58:979–1019.
40. Engelhardt HT. Current controversies in obstetrics: wrongful life and forced fetal surgical procedures. *Soc Sci Med.* 1988;26:1133–1139.
41. McCormick RA. Proxy consent in the experimentation situation. *Persp Bio Med.* 1974;18(1):2–20.
42. Holder AR. *Legal Issues in Pediatrics and Adolescent Medicine.* New York, NY: John Wiley & Sons, 1977.
43. Langham P. Parental consent: its justification and limitations. *Clin Res.* 1979; 27(5):349–358.
44. Shaw A. Dilemmas of "informed consent" in children. *N Engl J Med.* 1973; 289:885–890.
45. Schowalter JE, Ferholt JB, Mann NM. The adolescent patient's decision to die. *Pediatrics.* 1973;51(1):44–46.
46. Bartholome WG. In defense of a child's right to assent. *Hastings Center Report.* 1982;12(4):44–45.
47. Gaylin WA. Reply to Bartholome. *Hastings Center Report.* 1982;12(4):45.
48. Gaylin WA. The competence of children: no longer all or none. *Hastings Center Report.* 1982;12(2):33–38.
49. Ramsey P. *The Patient as Person.* New Haven, Conn: Yale University Press, 1970.
50. Ramsey P. The enforcement of morals: non-therapeutic research on children. *Hastings Center Report.* 1976;6(4):21–30.
51. McCormick RA. Proxy consent in the experimentation situation. *Persp Bio Med.* 1974;18(1):2–20.
52. O'Donnel TJ. Informed consent. *JAMA.* 1974;227:73–75.
53. Levine RJ, Lebazqz K. Ethical considerations in clinical trials. *Clin Pharm Therap.* 1979;25(2):728–741.
54. Fried C. *Medical Experimentation: Personal Integrity and Social Policy.* New York, NY: Elsevier, 1974.
55. National Commission for the Protection of Human Subjects of Biomedical and Behavioral Research. *Report and Recommendations: Research Involving Children.* Washington, DC: DHEW Pub (77-0004), 1977.
56. Marston RQ. Research on minors, prisoners and the mentally ill. *N Engl J Med.* 1973;288(3):158–159.
57. McCartney JJ. Research on children: national commission says "Yes, if. ..." *Hastings Center Report.* 1978;8(5):26–31.
58. Davies I. contracts to bear children. *J Med Ethics.* 1985;11:61–65.

59. Ethics Committee of the American Fertility Society. Ethical considerations of the new reproductive technologies. *Fertil Steril.* 1986;46(3):suppl 1:1S–81S.

60. Warnock M. Thinking and government polity: The Warnock Commission on human embryology. *Millbank Mem Fund.* 1985;63(3):504–522.

61. Elias S, Annas GJ. Social policy and ethical considerations in noncoital reproduction. *JAMA.* 1986;255(1):62–68.

62. Parker PJ. Motivation of surrogate mothers: initial findings. *Am J Psych.* 1983; 140:117–118.

63. Annas GJ. Death without dignity for commercial surrogacy: the case of Baby M. *Hastings Center Report.* 1988;18(2):21–24.

11

Problems in the Care of the Terminally Ill

Herr, lehre Du mich, daß ein Ende mit mir haben muß, und daß mein Sein ein Sinn
hat eh ich davon muß, eh ich davon muß

Brahms – A German Requiem

Oh Lord, teach me that my life must come to an end
And that, ere I depart, my existence must have had aim and purpose

Translated EHL

—— INTRODUCTION ——

While caring for the terminally ill or those in the last phase of life occupies a lot of the time and thought of health care professionals, clergy-persons and the public at large, it is probably not the main ethical concern that faces those in the heath care professions or even the public at large today. More prominently—even if much less loudly—the concerns of everyday practice and of equitable access to non-end of life health care are, in fact, a much more common problem for many today. While this chapter will be devoted to ethical problems at the end of life, it is essential not to forget that the "end" generally forms only a small part of the whole and unless one is aware of this, the "end" cannot be properly examined or dealt with.

Caring for the terminally ill frequently requires a reshaping of at least some cherished contemporary medical values and a reexamination and reordering of a tacitly accepted hierarchy of medical goals. Concerns for prolonging life will often have to be modified, will lose their rank in the hierarchy of values and goals, and will yield to concerns for bringing comfort or, at the very least, for ameliorating suffering. To cure occasionally, to alleviate sometimes and to comfort always remains an ancient precept of medicine we all too often forget about in an age in which people tend to be viewed predominantly as altered patho-physiological states. In ordinary practice today physicians, while they try to avoid causing pain, do not see the avoidance or amelioration of pain and suffering as their prime concern. The primary mission of

—— 249 ——

medicine, under ordinary circumstances, is to heal patients and to return them to useful and enjoyable life.

Issues at the end of life bear a resemblance to issues at the beginning of life in that we must deal with what we think life is and with what we believe its purposes are. Why, for example, is killing or purposely letting people die wrong? This is a simple question and one that we are apt to wave aside as silly. The answer to the question seems obvious: It is wrong to kill people or to let them die unnecessarily because by so doing such persons end up dead, and few of us would want to be dead. Such an answer, however, is facile and insufficient, for some of us do, in fact, want to be dead, and others, we may feel, are best served by being dead. There must, then, be more to this answer than meets the eye.

As Rachels has so aptly pointed out, killing people (or letting them die unnecessarily) is seen as wrong not because it ends life itself but because it writes "*finis*" to a "biographical life" with its capacity for hopes, aspirations, and social interconnections.[1] It is the difference between "being alive" and "having a life," the former a necessary condition for the latter though not, by itself, sufficient.

The problem is at least in part one of language. The Greeks have two words with rather different meaning for what we in English, French, or German call "life." (This is similar to the Eskimo's many words for snow: a different term denoting a rather different type of snow.) They distinguish between "*zoe*," which is said to denote the biological presence of life (that which would cause a biologist to claim that life exists), and "*bios*," which denotes a biographical condition, one of self-awareness about knowing and acting. The former is what we call "being alive"; the latter what we term "having a life."[2]

When physicians and other healthcare professionals deal with critically ill or dying patients, their own philosophy of life and death tacitly informs many of their actions. When patients deal with their own death, similar concerns inform the patient's hopes, fears, choices and, ultimately, actions. Often the worldviews of physicians and patients differ, and such differences form the basis of many, often poorly understood, conflicts. It is helpful to examine the ways in which death is generally viewed when dealing with death and dying issues: issues dealing with any aspect of death but especially those dealing with suicide, euthanasia, allowing death to occur, or even questions of how we talk about the fact that death may be imminent or that a prognosis is dismal. In examining these ways of looking at death, we are emphasizing Western and largely omitting Eastern points of view, not out of disrespect for Eastern viewpoints but because of a lack of emotive and experiential understanding of Eastern culture.

─────── ATTITUDES TOWARD DYING ───────
AND BEING DEAD

In the Western world in the last two millennia attitudes toward death have undergone great changes. A few examples will make this clear. Until the last few hundred years, persons feared sudden death. Sudden death meant that one was incapable of performing the social and religious rituals of death and dying that were felt necessary to having had a successful life. One could not say one's good-byes, ask for and offer forgiveness, etc. Until the Middle Ages, the catholic church did not provide burial in sacred ground to those who died suddenly—even when they were murdered! The ancient and modern practice of blessing and giving absolution to those going into battle (a strange practice, indeed) may have originated in this. Odd as it seems, persons often seem to have chosen their own time of death, prepared everything for the occasion and then, having apparently done all this in fair health, lay down and died. A person who died suddenly (just as one who was murdered) was suspect and, for a time, burial in hallowed earth was not given to persons who died suddenly or to those who were the victims of violence. Death was a communal affair just as was living; the overwhelming importance of the individual is a rather recent development.[3]

Man has always feared dying alone. Dying alone, without loved ones and without any human companionship, is something most persons have always viewed with horror and, where possible, avoided. When patients died at home, dying alone was far less likely and rarely happened. Obviously, our place of dying has often changed: Depending on our culture, we are apt to die in a hospital or nursing home rather than in the embrace of our family and home. This difference, however, is highly culture dependent: In Austria and the United States most persons die in a hospital or nursing home; in Holland, most die at home. This tradition of dying can significantly affect the issues of medical ethics with which we will be dealing.

Throughout history persons have not only feared a lonely death but have likewise feared being buried alive. In former times (and not so very long ago), the risk of being buried alive was far more real than it is today, and dying alone occurred relatively rarely. Today live burial is almost unthinkable, but in all too many cultures dying alone has virtually become the order of the day. When we die in the hospital (or in the nursing home), we often die surrounded by instruments, connected to tubes and catheters, unable to communicate our feelings and wishes effectively, totally dependant on strangers—and in human terms we often die quite alone. Recently and perhaps in revolt against the stark loneliness and mechanization of institutionalized dying, there has been a movement toward dying at home or in a hospice (a far more homelike and only minimally mechanized setting). The hospice setting has done much to ameliorate this problem of loneliness and the fear of abandonment or isolation. Hospice can be either in the patient's home with hospice workers helping the family care for the patient (external hospice) or in an institution. But hospice, as will be discussed, has its own ethical problems.

The problem of dying can be addressed from a number of vantage points. We can inquire what the criteria for death are, we can examine the dying process, or we can concern ourselves with what we envision the state of death to be. Criteria for death are a largely technical issue in which tests to a state previously philosophically defined are applied ("criteria for" a particular thing are not that thing but only describe how a thing whose essential nature was previously determined can be recognized). Likewise, the dying process deals with a number of technical issues (pain, psychological changes, etc.) and, although these are crucially important and help formulate our final attitudes, there is a prior concern when we consider attitudes toward death.

The prior concern is the way in which we envision the state of death: how we think that being dead is. This often tacit and generally unarticulated prior concern is critical when we deal with dying patients, especially patients whose life we may shorten by failing to treat inter-current illness or pain or patients who request of us the means for ending their own life. Although, as Freud has pointed out, none of us are capable of imagining ourselves as dead, we nevertheless, deep in our bones, know that death occurs and that, ultimately and relentlessly, we too must die.

There are, as Carrick has pointed out, four ways of looking at the state of being dead:[4]

1. The Homeric tradition, also approximated in Egypt, sees death as terrible, unconquerable, and basically as a strange twilight stage in which little of interest occurs. The dead are shadows, morosely wandering through the underworld. The chthonic (from under the earth) religions see immortal persons (not only their souls) as spiritless, pale, and eternal wanderers:

    > Oh shining Odysseus, never try to console me for dying. I would rather follow the plow as a thrall to another man—one with no land allotted to him and not much to live on—than be a king over the perished dead.[4]

2. The Orphic-Pythagorean and later Christian (albeit quite different and varied among Christians) concept is that of immortality of the soul as conceived by Plato and Socrates.
3. The genetic survival ("species survival," as Carrick would have it) concept of Aristotle in which we live on through our offspring and fellow creatures.
4. The personal extinction of the Stoics and Epicureans accepts death as a true end of all existence.[5]

There are, furthermore, variations on these themes. The personal extinction view and a belief in the immortality of the soul are, for example, not inconsistent with a simultaneous belief that our genes or our works survive. Except for the chthonic view, which has largely lost currency, these basic points of view of what it is like to be dead continue to influence us. Understanding the function of these concepts in our everyday lives, and the linguistic and conceptual symbolism of these ideas, is critical if we are to understand not only our own but our patients' attitudes toward the very practical problems of life support, physician assisted suicide, active euthanasia, and suicide itself. Inevitably, as we deal with dying patients, our notions of what it is like to be dead produce feelings ranging from extremes of fear and revulsion to envy.

The Pythagorean and ancient Greek model sees in the continuation of the soul an uninterrupted continuum. The soul, freed of the body, continues on in another existence. Socrates, certain of this in his own mind, drank the hemlock and calmly waited for his own soul to be freed. Christians—at least the early Christians and many to-day—see the immortality of the soul differently. At death, the body and soul die (or the soul lies dormant) and later (on the Day of Judgment) the soul (and in some beliefs also the body) is resurrected.[6]

Depending on the specific sect or church, Christianity today has varied (and, at times, no set) views of immortality. Some conceive of the immortality of the soul as a continuous and uninterrupted process; others affirm that resurrection is a function of God's grace, which may, or, in some views, may not, be mediated through our actions on Earth. Hebrew thought, by no means clear or uniform on this issue, differs in viewing the relationship of man with God as one of covenant in which each must do his or her part. Life here on this Earth, in the traditional and orthodox Jewish viewpoint, is to be valued above all else and therefore to be preserved at all costs. Until relatively recent times (partly arising from the Eastern ghetto experience and its contact with orthodox Christianity), a belief in life after death played no great role in Judaism. Today, Judaism is split into various factions (from Hasidim through Orthodox, Conservative, Reformed and even secular), and the beliefs vary with faction as well as with the individual within that faction. In general, life, not death, is the central theme of Judaism.

Islam has a set belief in the afterlife, and Muslims believe in resurrection and heavenly judgment. Many eastern religions, moreover, subscribe to some form of reincarnation—in some, a reincarnation that continues until perfection is attained. Healthcare professionals (themselves the bearers of their own peculiar belief) are profoundly affected by this mélange of cultural assumptions and must be sensitive to them as they go about dealing with patients from a variety of cultures. Religious or not, the ideology of our culture plays a dominant role in tacitly forming and shaping our individual consciously or subconsciously held beliefs.

Genetic or "species survival"[5] (as Carrick calls it) is the scientific answer to immortality. Our gene pool, our DNA, survives and is passed on as long as there is life. One may extend this and subscribe to a chemical survival in which our building blocks survive as building blocks for other and, perhaps, even different forms of existence or life. While this is, in a sense, a mechanistic and material adaptation of a belief in reincarnation, it is hardly bereft of symbolism and can be of great comfort to the dying as well as to the bereaved. It, however, differs sharply from a belief in a conscious survival.

Personal extinction—the "when I am dead, I am dead" philosophy—rooted in Epicurean and Stoic belief is ostensibly shared by many today. Many of those who enunciate such beliefs will still find it difficult to believe that their own ability to feel and know will be no more. We can, only with difficulty, imagine a world in which we have not existed, but we find it almost impossible to conceive of a continuing world in which those we know and love act without our participation. Indeed, we may find it difficult to believe that those we love (or hate) are distinct from that love (or hate) and not, somehow, contingent upon it. Our deep-seated Cartesian dualism—which allows our body as the *res extensa* to perish—does not allow us to conceive that our knowing self, our *res cogitans*, will vanish.

Both the "species survival" and the "when I am dead, I am dead" view are bereft of personal satisfaction except in the negative sense: if no personal "I" persists, then when that "I" can no longer know or feel, no further evil can befall it. What matters, the Epicureans, Stoics, and those who share their belief today agree, is to live honorably and well. In this belief one does not live honorably and well because one fears punishment or hopes for reward, but because as a human being and a member of a community one shares in a set of values and obligations with others. Life, then, is viewed as the necessary condition of experience rather than as a freestanding value. When the possibility for positive experience is no more and the future holds only negative experience, ending life no longer seems an impossible or immoral choice to make. Indeed, the Stoics (and to some extent the Epicureans) believed that there comes a time of life when suicide is advisable and beneficial.

Such views profoundly influence the way in which healthcare professionals deal with their critically ill or dying patients. Our own unexamined philosophy of such things can become a stumbling block when it runs across the patient's equally unexamined and often quite different philosophy. Misunderstandings are then bound to occur. Physicians and other healthcare professionals are well advised not only to speak with their patients about symptoms, signs, and illness, but also to recognize their humanity by speaking with patients about more personal attitudes and beliefs. If healthcare professionals assume the responsibility to grapple with the "Ought I to do this?" as well as with the "Can I do this?" questions, such conversations become an essential component of such decisions. And they bring relief to both patient and family who often hesitate to approach health care professionals or "to take up their time" (yet such conversation is often more important than prescribing yet another medication!)

We cannot assume that others are just like us: that they share our basic attitudes, beliefs, and values. While especially true as one deals with patients from diverse cultures, such an assumption is also often erroneous when one deals with a neighbor whose culture is similar (but hardly identical) and whose experiences may be radically different from ours. One word of caution: Talking with patients about their beliefs (a "talking" that most fruitfully is really mainly allowing *them* to talk about their beliefs, hopes, and fears—and listening) must not entail proselytizing or in any other way attempting to change another's belief. Under these circumstances, as in any circumstances of the healthcare setting in which healthcare professionals interact with patients, an attempt to proselytize or change a patient's personal belief system is a gross, unprofessional and ethically abhorrent abuse of power. Unfortunately, it still does occur.

STAGES OF DYING

Kübler-Ross has outlined five stages through which patients who do not die suddenly but live with a progressive fatal illness for some time prior to death (say, when they are dying of cancer) generally pass. Briefly put, these five stages are denial, anger, bargaining, depression, and, finally, acceptance.[8] It is helpful to know that such stages exist and that many—but by no means all—patients will pass through them. Such stages must not be (and were not meant to be) used in a cookie cutter fashion. Individuals do not pass through these stages lock step. Some may skip a given stage, pass through it quickly, or undergo a quite different sequence of events. Dying, like living, occurs in a setting of communal, cultural and personal values and expectations and has wide variations. Ultimately, it is something we experience alone. These five stages, then, are not and must not be what "good patients" can be expected to "comply with." At times healthcare professionals get frustrated or even angry when patients fail to behave as expected or skip a stage altogether. Making of death and dying a mechanical and predictable act that all "good patients" (if they are to deserve that name) would be expected to pass through defeats the very intent that motivated the study to begin with: to understand a process better so that patients could be helped more. Understanding these stages can help us deal more effectively with dying patients—but it can also, when misused, get in the way.

During the first stage, patients will often convince themselves that the diagnosis, biopsy, tests, or other critical information are wrong: "It can't be me!" At this stage, they may change physicians, seek an unreasonable number of consultations, and otherwise barricade themselves against the truth. Kübler-Ross states that this stage is especially pronounced in patients who have been informed abruptly or without due regard for their readiness to know. Revealing information is more than merely informing. It takes time, skill, sensitivity, compassion, and proper timing.

The second stage (anger) follows when denial can no longer be maintained: The truth has come home. Why does it have to be me? What have I done to deserve this? Why couldn't it happen to someone else, someone older, someone with fewer responsibilities, someone who has led a less virtuous life, etc., etc.? Often that anger is directed not only towards acquaintances and strangers but also at family members and healthcare professionals. Patients at this stage tend to be hostile, resentful, and difficult to deal with. It is, however, an understandable attitude and not one limited to the knowledge of having a fatal illness. Why did I have to get fired? Why not George? What did I do to deserve becoming ill? (Illness, hearkening back to primordial times, still crops up as a punishment by someone for one's failure.) At this stage, patients are irritated (at times to the point of being infuriated) by those who are healthy or at least not dying. Anger eventuates into resentment. And in religious patients, anger sometimes turns against God.

The third stage, or state of bargaining, follows from the second. Instead of questioning why he or she is being "punished," the patient tries to be rewarded for "good behavior." If I promise something (God, fate, or whatever), do something, or behave in certain ways, my fate will be changed. Such bargaining is not limited to dying. How often in life have all of us, despite all reason or rationality, been tempted to engage in similar activities? The religious person (the nonreligious too, but in different ways) has bargained or tried to bargain to prevail. In a sense a compulsive neurosis ("if I do not step on the cracks in the side-walk everything will be fine," for example) is a quite similar phenomenon.

In the fourth stage, the patient becomes depressed. It is a stage of grieving or loss. Not only those left behind grieve the loss of their loved one; the dying person also grieves for the loss that he or she is experiencing or about to experience. After all, those left behind are losing one person; the dying person is, in a sense, losing everything—or at least everything here on Earth. Often symptoms worsen and more surgery, more treatment, or more medication becomes necessary, and the patient begins to experience and to prepare for loss. Suicide is not apt to occur during true denial or during a time when bargaining is still believed to provide a way out.

The final stage, acceptance, is one in which—often with help in working through the previous stages—patients neither rail against their fate nor are unduly depressed by it. At that stage, patients give up the struggle and bow to the inevitable. Patients make their peace and can, generally, talk about what has happened and what is about to happen.

It is critically important to realize that these five steps are not the steps all patients go through or go through in that order. That does not make them "bad patients" but rather unique persons whose work of art simply does not follow such rules. When these stages are applied mechanically and when we "expect" patients to comply with them ("Now he or she should be in stage # 3!") we violate their innate dignity and their (contingent) right to choose.

At times, giving patients "permission to die" may be an extremely important factor. In our society today, one simply doesn't die—it is considered, in a way, "bad form." Often it is seen as a sign of weakness, sometimes as punishment and often as something that, if one only had the strength, one could resist. Telling a patient bluntly but compassionately that dying is "all right" may, in certain instances, be most helpful—this is true of family members as well as health care professionals. A spouse telling the patient "I shall grieve for you, I shall miss you but I shall be all right" or a physician saying something similar to a long-term patient may bring great relief and result in a more peaceful and accepting death. Our colleague, Dr. Faith Fitzgerald has often related such stories about her patients and first made us aware of this important phenomenon.

Throughout these stages, hope is a common thread. It is, however, hope for different things. At first it may be hope by denial, then it is hope that anger or a clever bargain may prevail over fate, and finally, even in depression, there is some (even if terribly little) hope. Depression is not despair. Ultimately when patients come to accept their fate, it is hope (and often here their basic belief system qualifies that hope) that they may "go to their eternal reward," "meet their loved ones (at the time of death or later)," or be able to live what is left of life with as little pain or suffering as possible. Hope, furthermore, has to be for something remotely realistic. When physicians say that telling the patient the truth "removes hope" they are falling prey to this fallacy. Hope may be re-directed—it may be shifted from cure or prolongation, to not being abandoned, to having their pain controlled, to making the last days still worthwhile, and even, depending upon the belief system, to being helped to die peacefully. Hope is never something that cannot be provided.

In our personal experience, the possibility of committing suicide may present a way of effecting hope, of having the power to choose one's time or to avoid a worse fate. It may be a part of what we shall be calling "orchestrating death."

——————————— ATTITUDES TOWARD ———————————
MAINTAINING LIFE

To the vitalist, all (or at least all human) life is an unconditional good regardless of anything else. It is "sacred"—that is, untouchable or not to be interfered with by (mere) humans. The vitalist, therefore, is committed to supporting life (or at least human life) at all costs and under all and every circumstance. When it comes to saving or prolonging life, vitalists do not concern themselves with whether life is or is not worthwhile. If life is held to be a primary or "intrinsic" good (one that is good in itself and not only as a means to an end), the question of its being or not being worthwhile is incoherent. Being alive—pure biological existence—rather than having a life, is what concerns vitalists.

Those who speak of the sanctity of life sometimes blindly appeal to the vitalist stance. Few persons even among the "right to life" groups, however, truly adhere to such a position. If one holds all life (or at least all human life) to be sacred, then a tissue culture (of any imaginable human tissue one may choose) shares in that untouchable sanctity. After all, such tissue has the right number of chromosomes as well as the right DNA to be human, and has those biological attributes we require to call anything life (this of course, has a great deal of similarity to the stem cell debate, which is discussed elsewhere). Yet, if one is to be consistent, then tissue cultures could not be destroyed. There are, of course, some Eastern sects that do hold all life as sacred. The Jaines of India not only are vegetarian (and generally eat only things like fruits and nuts, which can be taken without destroying life) but often will carefully sweep their paths so as not to step on insects. Although he did not go quite that far, Albert Schweitzer felt a like reverence for all life.

The non-vitalist position, on the other hand, denies that human life itself, just because it is human life, is necessarily "sacred." Instead, it holds that life, far from being a good in itself, is merely an instrumental good: a means toward an end beyond itself. This position either sees life as a condition for experience and life's being or not being worthwhile as dependent upon the nature of the experience or sees life as a bridge to eternity. Attitudes differ from very rigorously supporting life as long as the slightest possibility of again experiencing "good" (or working for salvation) remains to being more ready to abandon support where that possibility seems slim. But life, in the non-vitalist's conception, is a means and not an end in itself.

Many, but by no means all, healthcare professionals today have abandoned (if they ever shared) a strictly vitalist position. Nevertheless, this position is often unthinkingly invoked and, at times, this attitude still lingers when the healthcare team continues to treat in hopeless situations. More frequently the team gets so involved in the complicated business of correcting chemical imbalance, reversing acidosis, or maintaining sinus rhythm that the *goals* of correcting chemical imbalance, reversing acidosis, or maintaining sinus rhythm are forgotten. Technology often becomes its own *raison d'être*: A thing is done because it can be done, and in a sense technology begins to drive itself. This in no way is meant to fault the necessary and careful attention to the patient's biomedical needs nor to underestimate the primary importance of technical competence. It is, once again, to point to the inescapable fact that biomedical competence, if it is to accomplish its mission of serving the true and self-selected good of its patients, must serve a purpose beyond itself. Biomedical competence is the necessary, but quite insufficient, condition of responsible medical practice.

Those who presume that life is a condition for experience and not an end in itself will, of course, have difficulty determining when that experience is, or is not, worthwhile. Concerns with determining whether the nature of the experience is worthwhile, often called "quality of life" issues, are of extreme importance. The threat that the quality of one's life will be determined for us by another is the hidden danger in such situations. Here, as in most other areas where such judgments must be made, every attempt to make such judgments on the patient's and not on the health professional's terms are in order if crass paternalism is to be avoided and a decent respect for persons is to be maintained. Only when patients are clearly incompetent to make such judgments and when their will is unknown and unknowable are substitute judgments acceptable. Health care providers (and even, disturbingly, persons close to the patient) do not have more than a "flip of the coin" chance of making such estimates on the patient's terms. (See Chapter 5 on patient autonomy and Chapter 13 on decision-making.)

Quality of life judgments not made by the person about himself or herself under acceptable conditions are problematic. Inevitably my judgment of another's quality of life is filtered through my own values. It has been shown that when physicians judge their patients' quality of life, they are very likely to be wrong.[9] Judgments of this sort made for another are, as Pellegrino and Thomasma have convincingly argued, thinkable when "life no longer has any possibility of satisfaction to the patient," as in the terminally ill patient in pain. Continuing treatment under such circumstances represents "a kind of therapeutic belligerency."[10] But under all circumstances judgments of this sort need to be carefully hedged. The question of last resort when we truly cannot know is not "what would the patient want?" but rather "what would a person in such circumstances *not* want?"—an easier to answer, because more universally human, question. It is a question we fail to ask often enough.

———————— NO-CODE AND OTHER ————————
NON-TREATMENT DECISIONS

When physicians decide to treat, they do so under two presuppositions: (1) a technical judgment that treatment has a reasonable chance of being technically successful (i.e., that a given intervention is likely to ameliorate the patho-physiological condition for which it is instituted), and (2) an ethical judgment that treatment has a reasonable chance—on the patient's terms and consistent with the patient's self-determined goals—of bringing about a desirable, or at least tolerable state.

Cessation of treatment is done under the converse presuppositions: (1) a technical judgment that treatment does not have a reasonable chance of technical success (e.g., that restarting the heart is useless because it cannot, given current knowledge as reflected in the literature, continue beating), or (2) an ethical judgment that, even if there is technical success, the patient's ability to profit (as defined by the patient or by his or her legitimate surrogate) from this treatment is nonexistent or, perhaps, that prolonging life would lead only to suffering (e.g., that restoring the heartbeat might be successful for a given length of time in a hopelessly ill patient. In such an instance, restoring the heartbeat could either result in intolerable suffering or be irrelevant to a permanently insentient patient. Moreover, it could prolong a life the legitimate decision maker did not wish prolonged).

In the first instance, when treatment is technically almost certain not to be efficacious, treatment is not only unnecessary but is also, in fact, a logical contradiction. Treating the untreatable makes no sense. Doing so is ill advised and, in some institutions and states, obtaining consent to withhold resuscitation under such circumstances does not require consent. While in such instances, writing a "do not resuscitate" (DNR) order does not necessarily involve obtaining consent, healthcare professionals are certainly well advised to inform those involved. (See the section on futility later in this chapter.) Saying to the patient and family, compassionately and sincerely that, unfortunately, no aggressive, cure-oriented treatment is possible and that regretfully, therefore none will be given, is the best that can be done.[11] Writing a DNR order under such circumstances is a judgment made on technical grounds that such treatment is useless. It is not a moral choice that such treatment, while technically possible, is inadvisable in a given patient.

In writing DNR orders physicians need to guard against slipping their own value judgments in under the guise of technical judgments. For example, when a heart could, in fact, be restarted but health professionals believe that doing so would only result in a few days' sentient life, the judgment to write a DNR order is made on ethical and not (as some would argue) technical grounds. The judgment that life is not worth living is a judgment that only the person who has that life (or, if need be, surrogates speaking for the patient) can make. What follows concerns itself with the more frequent ethical quandary of not resuscitating in situations in which resuscitation is technically feasible.

Writing DNR orders has become a daily reality in our hospitals. In former days, most of us died in the midst of our families, friends, and loved ones, or we died alone. Dying, as it were, was a private process. Today, however, deciding to withhold treatment is no longer a matter that concerns merely the patient, the family, and the physician. Dying has become institutionalized and therefore public, and it is consequently supervised and hedged with a multitude of restrictions. Deciding to allow a patient to die peacefully and to withhold treatment or support is far less easy. Under ordinary circumstances, hospitals are properly seen as institutions here to preserve life by all available means, and the burden of proof is, properly, on the person who would refrain from using all possible means to sustain life. Here, as always, proper documentation giving good and coherent reasons for such a judgment—which can be quite brief—must be given.

It must be remembered that cardiac resuscitation was developed in a rather specific and quite recent, setting and circumstance: *i.e.*, to reverse cessation of cardiac action in the face of an acute cardiac event and to give the underlying process a chance to heal. Most orders written in a hospital setting are positive orders directing other members of the healthcare team to do something. Only when "doing something" (such as giving a patient fluids or nourishment) is the norm is an order not to do that thing necessary. After it was shown that in certain instances resuscitation could be effective and could restore some patients to a meaningful life, resuscitating virtually all patients who underwent a sudden cardiopulmonary arrest became routine. It became routine for a quite proper reason: When patients arrest, there is no time to deliberate. If resuscitation is to be successful, literally every second counts. Therefore, a specific order not to resuscitate is needed if the properly conditioned reflex of the hospital team is to be over-ridden.

The primary goal of clinical medicine as conceived today is, at least to the degree possible, the restoration of health and the saving of life. All too often the ancient goals of alleviating suffering (which far preceded the saving of life) are forgotten, neglected or done in a perfunctory manner. Leaving aside the fact that this was not always the case (see Chapter 1), abandoning what has come to be considered the traditional goal of clinical medicine requires justification and the delineation of new goals. When we write a DNR order, we acknowledge either that resuscitation is not technically feasible or that there is consensus between the health care team and the legitimate decision-maker(s) that the circumstances of the case would make the prolongation of life unwise.

The duty to relieve suffering, a much more ancient duty of medicine than the duty to prolong life[12]—and one much more enduring—moves into the foreground of obligation when decisions not to resuscitate are made. Once a DNR decision is made, comfort measures properly become the primary goal.[13] This is an important point: All too often patients who have had a DNR order executed on their chart are avoided by their physicians and other members of the healthcare team. Essentially, this is abandonment. It is ethically improper because it is not really the case that nothing can be done. Rather, the something that can (and must) be done is to bring comfort (to the patient as well as to the loved ones) rather than to sustain life. In the last weeks, days, or hours, human contact and comfort are, perhaps, especially important. In many if not most cultures "touching" is very much a part of this and the tendency to turn care over to machines or other mechanical devices is to be deplored. Physicians have the obligation to make sure that all concerned understand the intent of a DNR order fully: Far from counseling neglect or saying that "nothing can be done," the goal of DNR orders is to shift the goal to comforting and caring, often a much more time-consuming task and one that often requires just as much skill. It may be advisable, in case of doubt, to append to the DNR order a statement clearly emphasizing that all possible comfort measures must be taken. (See the section on orchestrating death later in this chapter.)

In most instances in which resuscitation might be technically successful but ethically and humanly problematic, decisions to write DNR orders should be made jointly. Patients, family members, close friends whom the patient and family have treasured and all members of the team should be involved. The alternative—acting without the consent or knowledge of the patient or of others who are concerned—violates the basic respect due others and is crass paternalism. When patients are still capable of making medical decisions, discussing the matter only with the family merely widens paternalism. Much as wives, husbands, or other loved ones are needed in the process of reaching such decisions, they are not entitled to make decisions for competent patients who have maintained decisional capacity and who wish to be involved. Families properly help and support patients in their decisions, but they should not speak for them unless they have been asked by the patient to do so (see also Chapter 4).[14]

There are, of course, rare cases when conscious patients cannot be consulted. But here the burden of proof (why such patients should not be involved) is on the caregiver who fails to consult the patient. Such a course of action falls on the previously discussed and usually highly challengeable category of "clinical privilege" and, at the very least, requires careful proof and documentation. The belief that discussing such matters is burdensome to the patient has been shown to be a fallacy.[8] Not only are most patients *not* burdened by having such issues discussed, but they generally will also be relieved by the prospect of discussing such matters and by no longer feeling obligated to partake in a complicated game of charades with health professionals and their own families. At this stage of life, a direct and honest approach is generally the most successful and the least cruel. Patients who are ready to hear will hear; those who are not will quite readily signal this to those speaking with them; or, at times, they will simply not hear. (See the section on speaking with patients later in this chapter.)

Often, and preferably, decisions about "how far to go" will have been made long ago when, perhaps, a formal or even informal advance directive had already been given by the patient and discussed with the caregivers. Making decisions beforehand is preferable because it provides an optimal opportunity for the patient to make an autonomous choice (see Chapters 5 and 6) especially, because it allows an early dialogue among healthcare professionals, patient, and family that permits a squaring of moral views. During such a dialogue, healthcare professionals can and must make sure that patients really understand the consequences of their choices, and are given the opportunity to make such choices with as much unbiased information as possible. It is here that a discussion of what it means to be on a ventilator or what it means to be defibrillated is not only in order but essential. In this day and age, laypersons, often conditioned by the media, frequently have warped ideas about such matters. On the one hand they have been sold on the power (in truth, often relatively puny!) of medicine; on the other hand, they have been warned of its tendency (often overblown) to inflict unnecessary suffering. The fear of being on a ventilator, for example, may be grounded on gross misunderstandings—that once this device is used it will be permanent or that it is necessarily associated with great suffering.

When decisions as to how to deal with end of life matters have not been made beforehand, and when circumstances are such that patients might reasonably have doubts about the prolongation of their life, a compassionate discussion about such matters is one of the physician's obligations. It is the physician's obligation because the alternative of substituting his or her own, or the family's, judgment for the patient's is a crass exercise of paternalism and disrespects the basic autonomy of another. In discussing such issues with the patient and the family, the physician should call on the active help of nurses and other members on the team. Beyond this and importantly, hospital chaplains and often also the patient's own clergyperson may be of immense help to all concerned. It is here that, for a religious patient, a clergyperson well known to the patient and not associated with the institution can play an important role—as can a close friend who has no religious association.

When possible, families and close friends should be involved in such critical decisions. The role of the family or of close friends is properly to advise patients or to act as their sounding board. Families or friends may support or disagree with a viewpoint or decision; family members and friends may counsel, advise, cajole, and argue. But normally the family and friends act through and not for the patient. Unless patients are unconscious and previous choices in such circumstances are unknown, or unless patients are clearly incompetent, the patients must make the final decision. When patients clearly lack decisional capacity, are unconscious, and when no prior wishes are known, the family is involved in such decisions because it is assumed that they are most familiar with the patient's world view and that they truly have the patient's best interest (defined largely on the patient's terms) at heart. "Family" here is used in the broadest sense: not merely kinship but any others closely associated with the patient and, therefore, likely to have the patient's best interest at heart and to be most familiar with his or her worldview. Judgments as to what ethically (not legally) constitutes "family" in that sense are judgments that physicians and other healthcare professionals will find most troublesome to make but that, nevertheless, they will have to make to the best of their ability. The decision made and the reasons for it again need to be briefly but clearly documented.

These days resuscitating patients has, on the one hand, lost much of its original appeal and, on the other, been mindlessly applied even to cases in which doing so has not the slightest chance of success.[11] Often our enthusiasm to restore the heartbeat or to support respiration even in cases in which success is possible is tempered by the fear of "creating a vegetable." This unfortunate fear is largely based on essentially anecdotal experience. At the very least and, yet, most importantly, resuscitation "buys time" for deliberation. Part of this fear of "creating a vegetable" is grounded in a belief that once a heart has been successfully restarted, health professionals will be compelled to continue treatment even when the patient has no chance of regaining consciousness. The belief that stopping something is different from not initiating it—and that the law is in essential agreement—is one of many fallacies that are difficult to eradicate. Meisel has called these beliefs "legal myths."[15] Well-thought-out DNR orders, including the failure to treat inter-current illness—are a necessary part of medical practice. But failing to resuscitate a patient who might have lived a few more years of a full life because of a fear that doing so might result in an irreversible vegetative state is indeed, an unnecessary tragedy. First of all, as has been amply shown in the literature, creating such vegetative states is rare.[16] Secondly, having tried and been unsuccessful does not compel one to continue. The decision not to continue treatment can, in such cases, be appropriately made (see the section on acognitive states in this chapter). Here, as is so often the case, failure in some cases is the price paid for success in others.

—— RELIEVING PAIN AND SUFFERING ——

On the whole the professions' willingness and capacity to treat pain in terminal illness (especially in cancer) has greatly improved in the last decade. It is rare that one hears the paltry excuse that using appropriately large amount of narcotics will "addict" the dying patient or that doing so will shorten life. In fact, it has been amply shown that the use of adequate narcotics will, if anything, prolong life—something not necessarily desired but something which one accepts as the price of relief. And yet, cases of pain that can be alleviated continue to occur even in the setting of known terminal illness. With today's means (pumps, sustained action medications, etc.) managing severe pain is often a "technical fix" which is not altogether hard to accomplish. There continue to be a small percentage of patients in whom relieving pain remains impossible.

Nevertheless and parenthetically (for these are not dying patients) the management of pain in chronic as well as in acute situations continues to be generally ignored or badly done. Often pain—because no obvious physical cause is readily visible—is simply not believed or, even worse, shrugged off. Dr. Hüseboe—one of Europe's leading palliative care specialists—has repeatedly made the statement that "pain is what the patient says it is": It is a subjective phenomenon and not one an outsider can readily assess or quantify.[17] Malingering is quite another matter but one, which an experienced health care team should be ready to deal with—for malingering, too, is not a "state of health!" Patients with highly painful conditions—rheumatoid arthritis is only one example—are inadequately treated with narcotics: despite the fact that proper pain control will not only not decrease but, in fact, will increase their capacity to function and to lead a full life and will rarely result in addiction. Such patients are frequently (to their face and, perhaps worse, behind their back) accused of showing "drug seeking behavior"—as though it is inappropriate for a patient in pain to seek relief! In a way it is like accusing a parched person of showing "liquid seeking" behavior or a hungry person of showing "food seeking" behavior! This deplorable state of affairs is in part—and only in part—the "fault" of the physician; more often than not it is a result of inappropriate federal and state regulation that forces physicians to fill out special forms and subsequently often be questioned (and indeed harassed) by agents of the government. But here too, unless physicians and other health care providers play their role in altering such archaic mechanisms, they fail their patients. It is not that addiction may not be—and often is—a severe medical as well as social problem, but most of the time it does not originate in the offices of legitimate health-care professionals but on the street. And it is usually motivated not by pain but either by seeking some form of escape from a socially undesirable environment or by a bizarre form of "thrill seeking" behavior that incidentally eventuates in addiction.

One must keep in mind that pain and suffering—even though they often are causally related—are not the same thing (see previous chapters). The relief of pain by most physicians (and in complicated cases by experts in that field) can be generally accomplished—it is, indeed, generally a mere "technical fix." The relief of suffering—which may or often may not be related to physical pain—is a far more complex, multidisciplinary and often virtually impossible matter (see section dealing with euthanasia). And yet it is at least equally important.

——— DECISIONS NOT TO LEAVE HOME ———

Decisions not to leave home are made more and more frequently today.[18,19] Under proper conditions, these decisions represent a reasoned judgment made by patients in concert with their families and caregivers not to seek hospital care but to remain and die at home. More is involved here than in other cases in which patients ask not to be resuscitated. Since patients who decide against institutionalization remain at home, their families and loved ones are intimately and critically involved. It is the patient who chooses to remain at home, but it is the family and other caregivers who provide the necessary conditions by which remaining at home is made possible. Decisions to remain at home are therefore invariably communal decisions involving a number of people joined in and working towards a common enterprise. A patient's wish to die at home is, given today's conditions of living, not always one that can be implemented; but when it is the patient's wish and when it can without serious disruption to others be implemented, healthcare professionals ought to be willing to support and carry out such a choice.

Decisions to remain at home and die are more complex than decisions not to resuscitate for yet another reason: Although in both of these decisions technical intervention that would not serve the best interest of patients (as determined by patients themselves) are ruled out, decisions not to leave home are made in a quite different milieu. DNR decisions are normally made in a *milieu* in which such a decision is part and parcel of everyday life and in which the implementation of such decisions is by persons who are socially accepted as being responsible and accountable for such decisions.[19] When patients remain at home, the family, who must ultimately implement the decisions (by, for example, not calling the emergency medical squad when an emergency occurs) have no professional credentials to act in such a manner. Families fear that by not calling for help, even when decisions of this sort are made beforehand, they will "not be doing the right thing." They have doubts, feelings of guilt, and, at times, a fear that social disapproval by neighbors or other members of the family may follow.[19] Helping with such decisions, therefore, requires even more support and understanding from the health care team than does the decision not to resuscitate. It is one of the circumstances in which external (*i.e.*, home-based) hospice may help a great deal.

In addition, things are not necessarily quite this cut-and-dried. While in many instances the treatment *milieu* of the home may have a lot more to offer to the patient than does the hospital setting,[20] unforeseen circumstances can occur. Patients and caregivers may, at the time the decision is made, feel that homecare would be best. Yet patients with advanced cancer or who have for some other reason made such a decision may be confronted by an unexpected complication. They may, for example, suddenly develop florid pulmonary edema, break a leg, or have a painful abscess associated with sepsis at a time when removal to a hospital to alleviate the severe discomfort of such an inter-current condition seems advisable. Such a sudden event may, in addition, cloud the patient's judgment and force caregivers into making decisions for them. If we are to "re-humanize" dying and make what Stollerman calls "lovable decisions,"[21] temporary removal to the hospital may, in fact, be necessary.[20] Decisions of this sort may be guided by considerations similar to decisions to limit treatment (see the following section).

—— HOSPICE—PROMISES AND DANGERS ——

Hospice, a relatively old concept in new clothes, has been heralded as a solution to the "cure-oriented" preoccupation of modern medicine. Some of the hype that has accompanied the rise of hospice in this country is undoubtedly correct and well deserved. However, as with all effective movements as they become organized on a larger scale, the delivery of hospice care has not proven an unmitigated good—there are drawbacks as well as benefits. In this section, we will, *very* briefly, sketch, in broad outline, the history of the modern hospice movement, how it currently functions here in the US, and discuss some of its strengths and weaknesses.

The word, "hospice," shares the same Anglo-Saxon roots with such words as "hotel," "host," "hospital," "hospitality" and "hostel." From early medieval times, the word "hostel" has denoted a house of sojourn, a place of refuge for weary travelers. By the 11th century, C.E., a wealthy and powerful religious military order had been dedicated to the rather curious combined task of tending to the sick and poor and waging war on the Muslims. The former activity, tending to the sick and poor, earned them the name, "hospitaliers." By the 17th century, C.E., the activities of hospital and hospice began to become more distinct: increasingly hospitals focused on caring for patients and, if possible, curing them, while hospices ministered the poor, travelers and the homeless sick, offering food and shelter though not the services of physicians.

The modern incarnation of hospice began in the mid-1960's with the work of Dame Cicely Saunders, a London physician, who created an inpatient facility, St. Christopher's, which devoted itself to the complex needs of persons terminally ill and close to death. The growing awareness of Dr. Saunder's work and the nearly simultaneous publication of Kübler-Ross' *Death and Dying*[22] combined to bring to the public's attention many important findings about the dying person's experiences and perspectives. As a result of this cross-fertilization, the rise of the hospice movement has been rapid and widespread. As of the year 2000, in the US, Puerto Rico and Guam today, there are over 3200 such programs, though unlike its European counterparts, 96% of all hospice care hours here are outpatient services (in most other countries the predominant form of hospice remains inpatient, though that does seem to be changing). *Per diem* re-imbursements vary according to locale and source of payment but, since 79% of all patients admitted to hospice claimed hospice medicare and 5% claimed Medicaid, the Health Care Financing Administration's *per diem* re-imbursements have, in essence, set the standard for re-imbursement. As of 2000, these rates are approximately $110/day for routine (outpatient) home care, $645/day for continuous home care, $120/day for respite care, and $490/day for general inpatient care.[23]

Since approximately 84% of hospice services here are currently reimbursed by Medicare and Medicaid programs, the remainder are covered by private charity and, increasingly, by HMO's and third-party insurance providers.[23] However, reimbursement is tied to very specific criteria that limit care to palliative measures specific to a patient's terminal or chronic-declining-to-terminal illness (*i.e.*, progressive end-stage diseases or disability such as CHF, ALS, end-stage renal disease, *etc.*). Under Medicare and Medicaid, patients must choose a pre-set hospice benefit *instead* of standard benefits for the terminal illness that, unfortunately, can result in a denial of coverage for some patient-specific needs over and above the standard co-payment arrangements that seem to be endemic to all of medicine these days.

The preliminary summary of the strengths of the hospice phenomenon might best be articulated by the National Hospice and Palliative Care Organization (NHPCO), originally called the National Hospice Organization (NHO): "'[H]ospice' refers to a steadily growing concept of humane and compassionate care which can be implemented in a variety of settings—in patients' homes, hospitals, nursing homes or freestanding inpatient facilities."

However, there are a numbers of difficulties—both conceptual and practical—with these claims. While ideally hospice is supposed to be not a place, but a frame of mind it is, as we described earlier, predominately an out-patient experience in the US and patients too often find themselves forced to choose between hospice which, because of the structure of the system and it's accompanying economic constraints, is forced to deal exclusively with palliative measures, and traditional medicine which also, in part, due to reimbursement constraints, is too often aggressively "cure-oriented," emphasizing a commitment to eradicating the disease or disability rather than being singularly committed to the needs of the whole patient. Unfortunately, NHO has done little to improve the situation with its much publicized advertisements of how hospice allegedly differs from traditional medicine. According to NHO, hospice:

- Offers palliative, *rather than* curative treatment
- Treats the patient, *not* the disease
- Emphasizes quality, *rather than* length of life
- Considers the entire family, not just the patient, the "unit of care"
- Offers help and support to the patient and family on a 24-hour-a-day, seven-days-a-week basis [italics ours][24]

Traditional medicine—the recent preoccupation with cure notwithstanding—has always had an implicit broad commitment to caring for patients. The most recent articulation of this age-old goal can be found in this excerpt from the 1984 edition of the American College of Medicine's Ethics Manual: "The primary goals of the physician are to relieve suffering, prevent untimely death, and to improve the health of the patient while maintaining the dignity of the person." Indeed, the traditional goals of medicine have always included cure—though never exclusively to prolong biological life, but they also have always included:

- Relief of pain, symptoms and/or suffering
- Minimization of dysfunction; maintenance or improvement of function
- Avoidance of harm and unnecessary risk to patients
- Promotion of health *via* preventive education
- Bearing witness for and/or advocating for patients (especially when any or all of the above are no longer possible)

To insist on claiming the existence of such an arbitrary, unnatural division of labor between hospice and "traditional" medicine is disadvantageous not only to patients, but should be quite worrisome to health care professionals and society as well. For one thing, it puts potential patients in the US (and that means *all of* us!) in the rather peculiar position of having to choose between curative and palliative therapies so that, even if one were able privately to fund both simultaneously—quite unlikely in today's economic climate—truly holistic care would still be fragmented because the goals of hospice and the goals of the "cure-oriented" tradition of which the patient's original primary care physician is a part can only be reconciled with the greatest difficulty, if at all. As a result, the original physician of record can begin to feel like an intruder and the patient may feel 'abandoned' to a group of strangers by both the 'traditional' system and a physician with whom they may have developed a quite deep and trusting relationship. Defenders of the *status quo* have made the argument that transferring physicians can always continue to make *pro forma* visits; he or she will simply not be able to be reimbursed. Our response is that this is an unrealistic—and, perhaps, even cynical—expectation of many physicians given, for example, the tightly controlled economic environment of an HMO practice. Secondly, because of the structure of reimbursement, third party payers—whether public or private—will ultimately determine (by willingness to pay) when patients must move from a curative to a palliative mode. Thirdly and, perhaps, most troubling, is that with the rise of this particular form of hospice, we have created a whole sub-set of health care professionals who are showing signs of becoming narrowly focused on and dedicated to the dying *process* instead of the dying *patient*—ironically threatening to replicate in hospice care what was originally so roundly criticized in "traditional" medicine!

    Undoubtedly, it is an unfortunate, but unavoidable tendency of all systems to become regimented instead of organized. With some difficulty, this tendency can be overcome; but the more serious problem lies deeper than this. The more insidious threat is that, in the attempt to offer good, effective and efficient patient care—*i.e.*, offering patients expert biomedical means so that they may continue to pursue their overall goals (which include, but are not limited to, health), health care professionals can easily fall prey to a certain idealized view of what we call the "good patient," in which they unwittingly superimpose their own view of the patient's goals and how best to reach them—a *pendant* of what hospice rightly has criticized in cure-oriented practice. Instead of assuming, as do many cure-oriented clinicians, that the patient's goals are identical to their own (*i.e.*, the aggressive pursuit of cure), and that the patient's demeanor is (and rightly ought to be) one of passive acquiescence and grateful acceptance of all the health care team offers, the hospice team assumes the patient's goals are identical to their own (*i.e.*, merely palliation), and that the patient's demeanor should rightly be one of passive acquiescence and grateful acceptance of all the hospice team offers—this is clearly an ethically unjustifiable paternalistic tendency, whether cure-oriented or not! Simply because health care professionals have had more experience and, therefore, developed a substantial degree of expertise in certain areas so that they can articulate the range of biomedical possibilities for a given patient does not give their view of the patient's good priority. Nor does it give them the right to label a patient who is brave enough to object to such covert paternalism "difficult," "non-compliant," or simply "not a 'good' patient," as so often occurs.

In the final analysis, there will always be some patients who will be most grateful to accept health care professionals' assistance in revising their long-term life plans and goals as well as in choosing the best means by which to accomplish those revisions. However, some patients will never make peace with their impending deaths and may feel coerced by and resentful of health care professionals who expect them to live up to some idealized notion of how the "good patient" ought to behave. The best way for clinicians to avoid this sort of covert paternalism is to strive to establish a collegial relationship with patients that stresses the importance of interdisciplinary collaboration between the patient and the various members of the health care team; to be cognizant of the fact that, in the final analysis and whenever possible, the decision-maker ought to be the person most relevantly affected—the patient (and if not the patient, compelling reasons why not!)—and to be ever sensitive to the danger of making uncritical assumptions about our own or others' values and goals.

— DEMENTIA AND LIMITING TREATMENT —

Severe and permanent dementia (in which all reversible causes have been excluded, in which no reasonable doubt as to diagnosis or prognosis remains, and in which there is an evident and progressive lack of meaningful thought processes) is a very troublesome problem. A sometimes long road separates its early stages from the ultimate, lingering non-being of the vegetative state. Along this road, various problems and options will inevitably present themselves and will need to be dealt with. Alzheimer's disease is one of the ways in which dementia most frequently presents itself, but it is not the only one. It can be an inborn state of affairs in which a severely retarded person can neither speak nor socially interconnect in any way; it can occur due to accident or illness. Whatever its mechanism, dealing with permanent dementia is one of the more frequent and more troubling ethical problems in medicine. For caregivers—no matter how loving—it often represents a sometimes infuriating challenge which, in turn, causes them to have severe feelings of guilt. Explaining that such anger and frustration are not an unusual manifestation of callousness but are, indeed, the common lot of mankind may help (even if not alleviate) this problem.

Dementia may, but usually is not, unforeseen or sudden. Its onset is not something most patients are not fully aware of. And it is a matter which must be openly, frankly and fully (even if painfully) discussed. Patients' wishes must be ascertained and these must (when still possible) be re-examined with the patient and the family.

If advance directives were not executed (as they should have been) long before, they need to be strongly encouraged now and when they have been previously executed they must periodically be reviewed.

When patients are mildly demented, are still able to communicate fairly well, and still have sufficient decisional capacity to make their own choices, no real problem (other than the problem of determining decisional capacity, a terrible problem in itself at times) exists. Patients capable of deciding their own good on their own terms must be given the respect due all persons. Likewise, patients who are no longer capable of making such choices but who have expressed their wish explicitly (by means of some form of "advance directive") or by a tacit or explicit but not necessarily duly executed understanding with healthcare professionals and family over the years do not pose an insurmountable problem: Their expressed will, accepted by all concerned, forms the framework of their care.

When healthcare professionals face such problems in patients for whom no framework of care has been previously established, they are often inclined to obfuscate and to evade facing the issue. Against their better judgment, health care professionals treat, but often they treat reluctantly and with half measures.[25] Establishing a framework for the care of such patients is at best difficult, and it inevitably leaves (and probably ought to leave) a considerable residuum of doubt. Ambiguity is not something we like to deal with and is often something we deal with badly, but ambiguity is something anyone in the profession must learn to accept and deal with. Often relatives, by helping physicians understand patients more fully, aid by acting as acceptable advisers and surrogates.

At other times, relatives may be absent, disinterested, or materially prejudiced and may, consciously or not, do other than seek the patient's "best interests." A substitute judgment, at best difficult, may then have to be made. Further, and perhaps more frequently, physicians are asked (and, we believe, asked properly) to guide and advise on treatment decisions—not only to offer an array of available means but also to share in formulating appropriate ends or goals. It is helpful if prior to being asked to guide and advise, healthcare professionals have explored their own feelings and have begun jointly to discuss the problems, their options, and their willingness to pursue the various possibilities. Such dialogue is something that healthcare professionals among themselves should actively pursue at all times.

Often one must deal with cases in which it seems appropriate to limit treatment but inappropriate to forego treatment altogether. Examples of this are not limited to adult dementia: the newborn whose illness limits life span to a few months or years (say, children with Werdnig-Hoffman or Tay–Sachs disease) or the severely mentally retarded of any age who will never be able to communicate or evince more than the ability to perceive discomfort, to name but two. The grounds for determining what treatments are, and what are not, appropriate are often not well delineated, and decisions often therefore tend to be *ad hoc*. At times, regrettable decisions are made.

In enunciating grounds for making the decision to limit treatment, our obligation to refrain from causing suffering and to prevent harm (and "harm" cannot merely be defined as allowing death to occur or allowing a disease to go untreated but must be "harm" in its broader sense) as well as our obligation to sustain life must be considered. In deciding to treat or to forego treating, at least four considerations are appropriate: (1) the immediacy of the threat, (2) the relievability of the suffering caused by the disease, (3) the suffering entailed in the treatment, and (4) the patient's ability for sustained understanding of and cooperation with treatment.[26]

In general and when no provisions to the contrary have been previously established, an acute threat to life is appropriately met by an attempt to reverse the threat. Among many other reasons this is done to preserve the opportunity for deliberate choice. Doing otherwise denies that option. In order to minimize suffering, diseases that impose a serious burden of suffering on the patient and whose treatment burden is less than the burden of the disease itself must be treated. The goal here is primarily the relief of suffering; prolonging life is now, if it is a consideration at all, a secondary consideration. When treatment itself is severely burdensome, physicians must re-examine their projected goals and determine whether the burden imposed seems, all things considered, to be justifiable.

Such decisions must be made in the context of the specific situation. Patients no longer capable of sustained understanding and, therefore, incapable of co-operation are easily frightened. Often, in order to accomplish treatment, they must be tied down or otherwise restrained. Such patients may easily forgive or forget the imposition of brief and relatively slight discomfort, but may have their last days darkened by fear and terror when the reason for imposing severe or prolonged discomfort is not understood.[26]

What then, in a demented or otherwise mentally severely impaired but yet conscious patient, is the difference between, say, treating pneumonia or operating for appendicitis and starting dialysis or doing coronary bypass surgery? Using the criteria mentioned in this section may help. Treating a severely senile or severely mentally defective patient for pneumonia (or taking out his or her appendix) may be justified because pneumonia (or appendicitis) (1) poses an immediate threat at a time when the patient still has a reasonable short-term ability to profit from life, (2) entails relievable suffering and because treatment (3) involves little prolonged suffering and (4) does not require enlisting the patient's sustained cooperation or understanding to accomplish the goals of therapy. Starting patients on long-term dialysis (or doing bypass surgery), on the other hand, may not be justified because (1) the threat may or may not be immediate, (2) suffering is relievable in other ways, (3) the proposed treatment entails considerable and/or prolonged suffering that, furthermore, the patient cannot understand, and (4) the patient's cooperation is necessary for success.

Asking ourselves such questions and seeking to work out answers in community with others concerned for and with the patient may be troublesome. The alternative (not asking these questions or seeking these answers or appealing to predetermined principles rather than using these principles as guidelines along the road to decision making) has the appeal of simplicity but yields unsatisfactory results if one assumes that a sense of humanity and compassion is a necessary part of moral medical practice. In specific situations, specific judgments will depend on an understanding of the patient, the context, and the community in which such decisions are made.

ACOGNITIVE STATES

The "acognitive state" is defined as a state of biological existence in which the ability to think, know, and feel is permanently absent. This definition would include brain death, irreversible coma, and the vegetative state (see Figure 1). A brief technical description follows.

The cerebral hemispheres, necessary for perception, volitional motor function, and thought, depend on the reticular activating system located in the rostral brain stem. The reticular activating system provides wakefulness and therefore, when higher centers are intact, allows cognitive function. Without it, coma ensues. The caudal portion of the brain stem is necessary for respiration. A patient whose entire brain including the brain stem is destroyed is, therefore, comatose and unable to breathe without mechanical assistance. According to our current definition, such patients (whose central nervous system from the cerebral hemispheres to the spinal cord is permanently destroyed) are termed "brain dead," a condition that in all states and in most (but not all) of the Western world is today equated legally with death.

A person whose hemispheres are destroyed but whose reticular activating system is preserved will have periods of wakefulness but will lack awareness, perception, and cognition. Such patients are said to be in a vegetative state. Patients with a preserved brain stem but a destroyed reticular activating system will be able to control their breathing but will be permanently unconscious. They are called permanently comatose.[27–30]

Taking proper care of patients—especially in the notoriously often poorly run nursing homes as they exist in the United States today—has become an almost impossible task. Our geriatric centers—in which physicians often see patients only at monthly intervals—tend to warehouse rather than care for people. In some other nations, for example, the development of decubital ulcers (in or out of a hospital or nursing home) is not merely a medical matter but must be reported to the district attorney who in turn must investigate the matter and where needed take appropriate legal action.

When we speak of "brain death," death of the whole brain is implied. This, according to accepted standards today, is legally considered to be the equivalent of death in most of the Western world. Without respirator support cessation of all other vital functions will occur in a very brief time. And at that time, organismic death will ensue. Some persons have had philosophical problems with this definition, especially since this redefinition of death (formerly defined as permanent cessation of cardiac action) was done for convenience: It was found that such patients were the most suitable as "cadaver" donors of organs.[31–34] Denmark is unique in not accepting this definition. After considerable debate carried on through their "ethics councils" it was decided not to equate death of the whole brain with death. Rather, with proper instruments of donation, organ retrieval is allowed. Since such persons still have measurable vital signs, metabolize nutrients and excrete waste products they can hardly be called "dead." The Danish approach seems more honest and makes a lot of actual and conceptual sense.

Attempts to redefine death more broadly so as to include other permanently

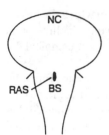

NC: Neocortical Structures
Cognition
Thought
Feeling
Integration
Suffering

RAS: Reticular Activating System
Walking, sleeping
Blinking
Swallowing

BS: Brain Stem
Vegetative functions
Breathing

BRAIN DEATH

Definition: Permanently nonfunctioning neural tissue from the spinal cord up. It includes the brain stem.

Criteria:
1. Reflexes may continue
2. No spontaneous respiration (ventilator dependent)
3. Frequent sympathetic disturbances
4. Without ventilator, death ensues
5. Lack of perfusion (radioactive)
6. Flat EEG: caution-false positives and negatives

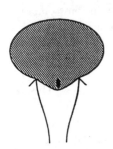

PERMANENT COMA

Definition: Neocortical structures irrevocably non-functioning (dead); RAS permanently nonfunctioning but brain stem otherwise generally intact.

Criteria:
1. Clinically determined
2. Reflexes continued
3. Spontaneous respiration
4. Frequent inter-current infections, needed for artificial nourishment, catheter and bowel care, decubiti, etc.

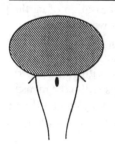

PERMANENT VEGETATIVE STATE

Definition: Nonfunctioning (dead) higher neural centers with RAS and brain stem preserved.

Criteria:
1. Clinically determined
2. Reflexes continued
3. Spontaneous respiration
4. Presence of waking–sleeping cycles, etc.
5. "On timers"

Figure 1.

acognitive states are even more conceptually problematic. Setting criteria for something (death, for example) implies that something has been previously satisfactorily defined. Such a definition, a philosophical task, should, if we are to respect persons, be made in an informed and democratic way and should be acceptable to a broad consensus of those concerned. As Hans Jonas has so aptly noted, we can either manipulate facts so as to redefine death when, in fact, death has not occurred, or we can "squarely face the issue of the rightness of continuing solely by our artifice what may still be called life."[32] The attempt to squarely face the issue is one that optimally we must try to reach through dialogue and consensus. In explaining the problem and in grappling with the process, ethicists as well as healthcare professionals can do their share to help lead; they cannot—and should not—by themselves, resolve the issue.

The vegetative state, consisting of an intact reticular activating system but without cortical function, is one of the most difficult emotional as well as ethical issues to deal with. Patients here look "so alive": They frown, grimace, swallow, and blink. The state is analogous to one's placing radio, lights, and window shades on various timers and then leaving the house. Observers outside the house would conclude from the changing lights, shades, and music that someone is at home. But such is not the case. In the vegetative state, also, the appearance, but not the fact, of "being at home" is preserved. Emotively this presents a terrible burden for the family as well as for healthcare professionals associated with the care of such patients. It is easy to see how wishful thinking together with the obvious fact that such patients do frown, grimace, swallow, and blink can lead the family as well as healthcare givers to entertain false hopes. The permanent vegetative state is, therefore, extremely difficult to deal with. Hearkening back to what was said about methodology, we first and foremost must successfully resolve the first question: Many cannot come to terms with "where we are" and therefore cannot rationally deal with the question of where we are going or how we might get there (see Chapter 13 for discussion of the importance of these three questions).

When the reticular activating system no longer functions but the brainstem is preserved, acognitive patients do not, unless other conditions are also present, require a ventilator. They are evidently unconscious without signs of wakefulness or volitional response. Such patients are said to be permanently comatose. The appearance of such patients, however, is decidedly one of coma rather than one, as in the vegetative state, of sleeping. The emotional trauma of all concerned, therefore, tends to be somewhat less severe.

When we consider (permanent) acognitive states other than brain death, we must come to grips with why it is that life is being maintained. The difference between being alive (discussion, beginning of this chapter) is critically important in thinking about this issue and must be compassionately explained to family and friends. If we are to eschew the vitalist position—if, in other words, we are to affirm that life serves as a *condition* for experience and not as an *end* or a "good" in itself—we will see no ethical imperative to maintain such an existence. On the other hand, neither will we discover an ethical imperative to discontinue all life support. Such patients, since they now lack primary moral worth (since, in other words, they are incapable of being harmed or benefited on their own terms) are no longer the primary objects of concern. Their relatives and the healthcare team, for whom they are of secondary as well as of symbolic worth, and the community (as representing a symbol of humanity) for whom they are of symbolic worth now move to the forefront of decision-making. Patients of this sort, patients who may be said to have lost primary moral worth, may have two other sorts of secondary worth: (1) They may have positive secondary worth to another in vital need for life or function of transplantable organs; or (2) they may, by consuming a vast amount of resources, have negative secondary worth.

It is often essential to allow time to pass. Families and loved ones must come to terms with a situation they had generally not anticipated or refused to think about. Often anecdotal stories of patients having regained full function after being in what was considered to be a permanent vegetative state enter into the equation. While these are often due either to misreporting or to a diagnosis too hastily made they cannot simply be shrugged off. Here, obtaining further consultation, allowing time and affecting compromise ("let us wait another week and then re-assess") may help affect an equitable, honest, ethically proper and acceptable solution. The fear, of course, is that the pressure of cost and the institutional rules of MCO's may rear their ugly head.

THE LOCKED-IN STATE

Rarely can a combination of lesions produce a state in which patients are awake, retain the apparent ability to think, but do not retain the ability to communicate in any way. It is not just that communication is merely difficult; it is quite impossible. In the usual difficulty, a speech therapist can be of immense help in communicating with patients: Pointing to letters or nodding and shaking one's head or blinking an eye and painfully spelling out words is still communicating, and a patient with decisional capacity (and to ascertain that may be in itself a difficult task but one that when doubt exists must be undertaken) has the right to make their own informed decisions. But in the truly "locked-in state" no capacity for communicating with one another exists. In the locked-in state patients have intact hemispheres but lesions of the motor pathways in the pons or midbrain prevent them from speaking, controlling their facial muscles, or otherwise expressing themselves. They are totally paralyzed; their mentation is totally disconnected from their ability to act or to make their desires known. Consciousness is maintained since the reticular activating system is spared. These unfortunate persons can think, feel, and know but can in no way express themselves. It is an extreme example of "freedom of the will" entirely and hopelessly without "freedom of action." Such patients pose incredibly difficult ethical problems. This is especially and poignantly the case when their previous wishes relating to life support are not known.

It is difficult for any of us to think ourselves into such a situation. In a way thinking about the locked-in state is similar to thinking about being buried alive, of which it is the modern and extreme analogue. We cannot know how anyone would feel once they became permanently locked-in a place from which there is no return. However, there are certain things we can cautiously and tentatively assume: Such persons do not want to suffer. If they maintain sanity and do not withdraw into a schizoid state, one would think that they would, at the very least, want to have their pain and suffering minimized. Beyond this, most of us would not want such an existence artificially prolonged.

——— WITHHOLDING FLUIDS AND ——— NUTRITION

Even when physicians are willing to write DNR orders or to limit treatment in the hopelessly ill, controversies about withholding fluids and nutrition continues to exist. In part, this is because a fundamental (and frequently artificial) distinction between "caring" and "treating" has been made when, in fact, "treating" is a form of "caring." Health professionals may often be willing to stop treatment but find themselves poorly equipped emotionally to abandon what they perceive to be care. The symbolism of caring[34–39] to many denotes an activity that is more human than it is professional: All humans care, or, at least, ought to, but only a few "treat."

Intuitively, traditionally and linguistically, we differentiate between caring (a human and not necessarily active sustaining) and treating (a professional and generally active task). Professionals willing to stop treatment are unwilling to divest themselves of their basic humanity by, as they see it, stopping to care. Caring, however, can and should never stop. We "care" for people in different ways and in different ways at different times:

1. When patients can and wish to be healed or to have their life prolonged, we "care" for them by skilled technical intervention tempered by using such skill in as humane a fashion as possible (my surgeon repairing my hernia and seeing to my being as "pain free" as possible afterwards "cares" and "cares" in a way appropriate to the situation)

2. When patients no longer wish to be healed or have their life prolonged, we care for them by our concern, by our making certain that their wish is truly their wish, and ultimately by our understanding.

3. We care for people no matter what their condition when we provide comfort (that is, when they are able to be benefited or harmed in themselves and are, therefore, of primary moral worth).

4. At times, we care for persons whom we can no longer benefit or harm because by so doing we can bring comfort to others to whom such persons are of value (i.e., the patient, now no longer of primary moral worth, retains secondary and symbolic worth, and caring for him or her therefore remains an obligation).

When we care for patients, however, we must be very clear as to what we are about.

Keeping patients clean, taking care of excreta, washing and dressing them, and preventing or treating decubital ulcers all seem essential.[30] It is not that insentient patients can know or feel the difference, not that they can benefit or suffer. Other considerations demand such care. Respect for the family, as well as for the patient, enjoins that the body of their loved one be kept presentable and free from the appearance of suffering. The hygiene of hospital wards or nursing homes requires meticulous standards of cleanliness so that infection does not spread. Symbolic value and aesthetics insist that the human form not be neglected and allowed to become offensive. Above all, perhaps, respect for oneself exacts this task: We cannot allow ourselves the inhumanity of dealing disrespectfully with a symbolic representation of ourselves. But caring includes the notion of concern for the welfare of others and protection of their interests. It is not a blind obeisance merely for the sake of form.

Nutrition and fluids are certainly essential for biological survival. When patients, no matter how ill or demented, derive comfort from such measures or are made uncomfortable by their lack, no one can doubt that the administration of nutrition and fluids is a moral necessity. However, when patients are acognitive and unable to derive comfort from such measures, the issue is less clear. There is recent good evidence in the literature that the use of PEG tubes and other methods of artificial nutrition are not only ethically problematic but in fact do not even serve to prolong life itself,[40–42] and may, at times, even be more burdensome than beneficial.

There are some who appeal to an inchoate and atavistic sense of humanity when they state that under such circumstances these measures are still ethically needed.[36–38] They feel that supplying food and water to our fellow creatures is a basic expression of humanity and love and one that we can ill afford to lose. Furthermore, there are those who fear that allowing healthcare facilities and healthcare professionals to withhold fluid and nutrition from their patients would be socially disruptive. It would distort the existing relationship between health professionals and healthcare institutions and their patients and be contrary to "the integrity of the medical profession as a learned and ethical" one. Such people feel that "in medicine the ultimate value is life," and life must be supported.[38] Such persons are essentially of the vitalist persuasion, often if not always religiously and in fact fundamentally motivated. Their belief falls under the rubric of personal morality rather than ethics.

The way the question is framed and understood makes a great deal of difference to our ultimate decision in the matter. If we look at fluid and nutrition given by nasogastric, gastrostomy, or enterostomy tubes as "supplying" basic elements of life, and if we frame the question as "denying" this, we may obtain a far different answer than if we frame the question as not "forcing" such feedings. Offering the basic elements of life to persons who may or may not avail themselves of them is a far different matter from forcing helpless patients to accept our ministrations. No one has proposed that fluids be "denied;" what has been proposed, and what is up for debate, is the propriety of not artificially or forcefully administering fluids and nutrition by artificial means to the permanently unconscious or to those terminally ill patients who no longer wish to receive fluids or nutrition.

Not infrequently patients will refuse nutrition (and sometimes fluids). There is good evidence that in the last stages of serious illness patients do not feel hunger or thirst the way normal or less ill patients do.[43] But that is not always the problem. Food and drink do have a symbolic function—all societies share food and drink in this manner. Some patients refuse to eat (or sometimes to drink) because they are forced to do so by themselves. I vividly remember an elderly lady in a nursing home who refused all food until one of us suggested that the nurse share her lunch-time meal with her, whereupon she promptly began to eat and continued to do so even when (as inevitably happened at times) the nurse did not have the time to do so and explained this to her.

There is, however, another and more serious problem. Some patients retain some capacity to experience, but their world is constricted until suffering seems to be their only experience. Such patients are often called "anhedonic"—that is, apparently quite unable to experience pleasure (which unfortunately does not entail that they cannot feel pain). The elderly patient totally demented but yet capable of feeling pain, the cachectic terminal cancer patient whose movement is restricted by intravenous lines and feeding tubes, and the terminally ill patient who simply says, "Leave me alone, I want to die," are examples. Some such patients lie in their beds in a fetal position and their only response to stimulation appears to be a groan of pain. Inserting devices to feed such patients usually necessitates restraining them and evidently causes considerable discomfort. When our attempts to supply fluids and nutrients burden the patient's last days, such attempts no longer, in truth, are ways of caring.[30,39,44] Instead, such attempts seem to pervert "caring" into officious and unwarranted assaults. At the very least, the notion of caring should encompass the notion of not causing pointless suffering. When our "caring" produces undesired and/or meaningless pain and suffering, one wonders at whom such "caring" is aimed: at ourselves and our own feelings, or at the patient whose interests we should (even when it hurts us) be willing to serve?

Providing nourishment and fluids to the dying can be a human and a humane duty, a useless obeisance, or an officious assault. When patients are acognitive and permanently incapable of feeling pleasure or pain, speaking of "maintaining their comfort" is nonsense—in truth, what we are really doing is maintaining our own! Some feel that stopping fluids and nutrition under such conditions is morally permissible.[45-49] However, even if ethically permissible, it is certainly not ethically or for that matter legally required. Such patients, now no longer of primary moral worth, maintain a high degree of secondary as well as symbolic worth to their loved ones as well as to the hospital context and to the community. When failure to supply fluid and nutrients seems morally, humanly, or aesthetically offensive to the family, friends, or members of the healthcare team, there is no reason why support cannot continue until, perhaps in the fullness of time, opinions slowly change.

Things stand differently, however, when terminally ill patients are burdened by our attempts to continue feeding them or infusing fluids. In terminal patients, our overriding duty to comfort and, above all, to refrain from causing pain would suggest that forcing nutrition and fluids under these circumstances is difficult to defend. There is no reason why, when support seems cruel or offensive, it cannot be stopped. The nature of the case and the context in which it is embodied will decide our choices in specific situations.[49]

In the United States and although hedged by a variety of legal precautions the nature of which depends upon the several states, the withdrawal or not starting of artificial fluid and nutrition is, according to the Supreme Court in *Cruzan*, viewed legally no differently from the withdrawing or stopping any other form of therapy.[16] Not starting or withdrawing fluid and nutrition in patients who are either burdened by or who refuse such intervention or not starting or withdrawing fluids and nutrition in the permanently comatose or vegetative may affront our aesthetic, religious, or moral sensibilities; but legally speaking—and with the presupposition that the legal conditions prevailing where it is stopped are adhered to—stopping or withdrawing fluid and nutrition is legally permissible.

——————— VOLUNTARY DEATH ———————

We want to make certain linguistic distinctions, emphasize some differences, cast doubt on others, and then briefly discuss ethical problems that may be associated with each of the following:

1. Suicide is understood as the killing of oneself without assistance by anyone else (it may be rational, irrational or capricious).
2. Withholding possibly or probably effective treatment for a pathophysiological state for reasons other than the patient's direct refusal (this will only be alluded to here but discussed in another section of this chapter).
3. Physician assisted suicide will be understood as help rendered by a member of the health care professions (usually by providing a prescription which the druggist then knowingly fills) and which patients may or may not use.
4. "Active euthanasia" in which someone (usually the physician but at times a close friend or family member) administers a substance or uses another modality to bring about a patient's death with the sole intent (rightly or wrongly) of serving that patient's good.

——————————— Suicide ———————————

Historically, attitudes toward both suicide and euthanasia have varied. In ancient Egypt suicide was clearly proscribed. Those who committed suicide were felt to have cut themselves off from the gods.[50] Yet physicians were not expected to treat the incurably ill. Indeed, morally speaking, treating the incurably ill was felt to be morally ill advised. Greek attitudes were complex. In the Platonic dialogues there is an evolution in thinking about suicide and euthanasia until, in the *Republic*, qualified approval is given in cases of lingering and hopeless illness. The proscription against administering poisons in the Hippocratic corpus may have been as much directed toward preventing physicians from participating in political intrigues as it was against the killing of patients.[5] Most physicians, moreover, were not Hippocratic physicians, and there is good evidence that the use of poisons to assist consenting patients to end their lives was frequent.[5] The Pythagoreans (a sect living largely in Crete, with great influence on Greek philosophy, Mathematics and Ethics) in many respects subscribed to what might today be called a "right to life stance"—things like suicide, euthanasia or abortion were beyond the limits of their religious morality. Aristotle, in the *Nichomachean Ethics*, likewise rejects suicide altogether, while the Epicureans and Stoics not only decidedly approved but recommended suicide and active euthanasia in cases of hopeless or painful illness or when "one's powers were waning."

Greek physicians were seen as obligated to (1) ameliorate disease, (2) provide comfort, and (3) desist from treating the hopelessly ill.[13] These obligations neither precluded nor included helping patients to die. The Church firmly opposed suicide and euthanasia. In the medieval era, when most physicians were priests and when a physician's first obligation was to see that patients had been "shriven" (given last rites) so that their souls were safe, suicide and active euthanasia were, of course, strictly forbidden. The first departure from this point of view came with Sir Thomas More's *Utopia*, in which this very pious Catholic scholar envisioned a system in which patients afflicted with conditions entailing hopeless and severe suffering would not only be permitted but actually encouraged and helped to commit suicide.

In more recent times, attitudes toward suicide and euthanasia have begun to change. John Locke, himself a physician, still felt that suicide and active euthanasia were wrong, basing this on the "inalienability" of certain rights. Unlike other property, the right to which could be waived, life belonged to God and could therefore not be taken. On the other hand, Hume—likewise a physician as well as a philosopher—argued eloquently for the propriety of suicide. Kant categorically opposed suicide. His rationale—not, one suspects, entirely untainted by his pietistic upbringing—was that to do so was to give up the possibility of making further rational and autonomous choices. Furthermore, not committing suicide could, like not killing or lying, be argued to be an obligatory duty (one can always refrain from doing these) rather than being an optional obligation (like benevolence, which one cannot refrain from altogether but has a choice of when and when not to discharge). In Kant's system obligatory duties invariably trump the more optional ones—one can always refrain from assisting with suicide or performing active euthanasia but even the most benevolent intention to relieve suffering would be trumped by the absolute duty not to kill.

Act utilitarians' approval of suicide, PAS and euthanasia is parasitic on their worldview, which approves actions that bring about the greatest good or, at least, actions that, all things considered, do the least harm. Newer "interest utilitarians" like Peter Singer would have the propriety of such an action commensurate with the person's true interests. What such an interest would, in truth, look like is, of course, debatable.

Rule Utilitarians, different from act utilitarians, would have to look upon the shorter and longer-term consequences that instituting or failing to institute a general rule might have. In other words, whether legalizing PAS or active euthanasia would be likely to have positive or negative consequences in a given social setting would be the deciding factor. In general rule utilitarians, have not been opposed.

In medicine, the change in attitudes has encountered two opposing forces. On the one hand, medicine since the 16th century has added the obligation to save life to its other obligations. Preserving life at all costs has become more and more a modern obsession.[13] On the other hand, as religious thinking has played a decreasing role in everyday life, as considerations of autonomy have moved more and more to the fore, and as technology has made things possible heretofore impossible questions about preserving life under all circumstances have been repeatedly raised and the right of persons to take their own lives has lost much of its previous stigma. Curiously enough, some states still make suicide illegal—most no longer do. Punishments, except for confining persons in psychiatric institutes are generally not meted out— presumably capital punishment would be inappropriate!

If humans are to be granted an autonomous will and the right to carry out those actions in accordance with it that do not negatively impinge on others, then taking one's life, at least under some circumstances, cannot easily be seen to be morally wrong. The wrongness of suicide, then, must inhere in the harm that one's suicide brings to another. In fact, the harm done to others is generally how arguments in opposition to suicide have usually been couched. The others at issue may be relatives, friends, society at large, those others (students or patients, for example) who depend on us or, ultimately, God.[51]

It is not rare that health care professionals are confronted by actively suicidal patients or by patients who have attempted to kill themselves and are then brought to the emergency department. At times a person threatening suicide may be desperately seeking help. There are few persons who wand to die—most simply cannot bear "to live that way" and see no other way out. In many cases compassionate and active intervention by a skilled health care team can do much to help patients return to a satisfactory life.

Sometimes patients are dead on arrival at the emergency room and the physician must decide whether to call or not to call a case "suicide." In many cultures and municipalities calling something a suicide casts "shame" on the family. Suicide victims about whom the ER physician knows nothing are brought to the ER for treatment. At other times, terminally ill patients seeking to end their lives are brought to the ER and will ask physicians not to interfere or at times even to facilitate their failed attempt.

Historically our attitudes toward suicide and toward voluntary active euthanasia have had much in common. In suicide, persons, for whatever reasons, decide voluntarily to end their own life; in voluntary euthanasia, another assists in carrying out the person's wish. Common to both is that the person who wishes to be dead and who, if successful or if their wishes are complied with, ends up dead is the one who makes the decision.

━━━ STOPPING OR NOT STARTING ━━━ TREATMENT IN PAS AND ACTIVE EUTHANASIA: Differences and Experiences

━━━━━━━━━ Introduction ━━━━━━━━━

Differentiating between PAS and active euthanasia is by no means clear-cut. In PAS physicians prescribe appropriate drugs, pharmacists fill the prescription and other health care professionals and institutions are frequently involved. In active euthanasia physicians (or it could be anyone else knowledgeable to do so) inject or otherwise administer such a drug (usually but not necessarily the same) for patients unable to swallow or take the medication themselves. When physicians deliberately omit or fail to initiate a treatment liable to reverse a life-threatening patho-physiological condition they are, whatever they may call it, aiming at the very same end. The subterfuge that, in such cases, it was the illness and not the medical team that caused the death is a slim reed to lean on if not indeed truly disingenuous. At times this may be a dangerous delusion—calling things by euphemisms leads to self-delusion, is more readily and less thoughtfully done and ultimately hinders our learning process as much as it does our growth as moral agents. There is no doubt that, to the health care team, not starting, abandoning, directly aiding in death by PAS or actually performing active euthanasia "feels" quite different, and, since health-professionals are also very much human, this is hardly trivial. But in its ultimate consequences as well as in its intention the difference tends to vanish.

━━━━━ The Doctrine of the Double Effect ━━━━━

In many cases in which treatment is stopped or some otherwise ethically problematic way of going about things is adopted, the doctrine of the double effect has been invoked. That doctrine, called the Doctrine of the Double Effect (DDE) briefly stated, provides that

1. The act itself must be good or at least neutral.
2. The intention of the actor must solely be to intend the good effect
3. If the good effect could be brought about solely without being associated with the (non-intended) bad effect that route would have to be chosen.
4. The good effect is not the result of the bad action.
5. There are compelling reasons to accept the bad effect.

Historically this doctrine originated with the Catholic Church in the middle ages and has been invoked ever since. It would, for example, justify the removal of a pregnant and cancerous uterus—the death of the fetus was "not intended." Although this doctrine has been most comforting to many members of the health-care team, we find little to justify it. Like claiming that the disease (which could easily have been treated) and not the failure to treat were responsible for the death of the patient, this doctrine is, at its roots, highly challengeable. Even in the law (in the EU as well as in the United States) persons are held culpable not merely for the consequences of their action but likewise for the consequences they could have readily foreseen.

We have neither time nor space to examine this doctrine (which we consider rather peripheral to the discussion since its use has been so very much cast in doubt); however, we shall summarize the objections briefly:

1. It is difficult to define the action itself—almost all acting can be isolated and subdivided into its component parts. ("All I did was write—or fill, or administer—a prescription", for example)
2. Intentions are difficult (even for oneself) to be clear about. Rarely if ever does an effect have only one cause or is brought about by merely one intention, something we have labeled as the "fallacy of uni-causality or uni-intentionality"
3. What the "good effect" really is can often be a matter of debate.
4. What counts as compelling reasons can invariably be questioned and is highly dependent upon the values, goals and worldviews of all involved.

Health professionals are quite relieved by invoking the DDE—it, so to speak, "lets them off the hook." This, however, is the very danger. It permits ongoing self-delusion that, like every other self-delusion is an ethically as well as pragmatically dangerous habit to adopt. The conviction that the health-care team never does a bad thing (even when it is the least bad available thing) and that, whatever else, their "ethical virginity" has been preserved is a dangerous delusion.

If, however, we acknowledge responsibility not only for what we directly do but equally for what we can reasonably be expected to foresee then the ultimate responsibility remains where indeed it belongs—to the person who ultimately decides and acts. In this case, it is with the physician in charge.

——— Not Initiating or Discontinuing Therapy ———

A difference has been ethically sought between knowingly failing to initiate possibly "effective" therapy in a moribund patient and discontinuing therapy under such circumstances. In the United States the matter seems to have been legally (but hardly ethically) settled by the decision of the Supreme Court in the Cruzan case: given the same documented circumstances there is no legal difference; withholding as well as not initiating are, under proper circumstances, legally proper. First and above all we must be clear that failure to act (or decide) when one could have done so is in fact in no way different.

But the debate continues—physicians often fail to initiate CPR or other interventions because they fear that after creating an insentient state they are bound to continue therapy. The fact that this is not the case seems—probably for emotive reasons—almost ineradicable.

The ethical difference should be dissected out by questions we have previously raised:

1. Do we—and above all does the patient, whose existential circumstance it is—have a higher value for life at the price of suffering or for suffering at the price of longer life?

2. Does the fact that, by not treating, we might feel emotively less involved in the patient's death than we would be by treating have any ethical significance?—again do we really believe that we can keep ourselves in a state of greater "moral chastity?"

3. Just what do we mean by "refraining to treat"?—not defibrillating? not treating a complicated uro-sepsis? or failing to perform by-pass surgery?

4. Where is the physician's responsibility to omit "treatment" when patients with decisional capacity and severe suffering ask to have their lives ended?—for that too is an omission of treatment! And it may be an omission of the physician's obligation to ameliorate or prevent interminable suffering.

The question of acting or refraining to act is by no means uncomplicated. It is a question that, above all, cannot be decided outside a careful assessment of the context of the specific existential situation in which the patient, the friends and family as well as the health-care team are intimately and inevitably intertwined. And it is a question that cannot be addressed purely by a resort to reason alone; it is one in which emotion and even aesthetics will inevitably play their role. Faced with such decisions that ought to have been, but more often than not have not been previously discussed, extensive consultation among all concerned (and in the religious and those open to it in the presence of the person's clergy-person) should be sought—preferably with all together. At times an ethics committee or consultant may help think through—but should never decide—the issue.

Physician Assisted Suicide and Active Euthanasia

Many if not most physicians have come to accept that there are circumstances when no further treatment to prolong life is defensible. Under some circumstances when life is only a heavy burden (a fact that except under most exceptional circumstances is one that must accord with the patient's and not with the healthcare professional's feelings about the matter) and no hope for cure, amelioration, or return of function remains, to continue aggressive treatment to prolong life is to practice what Pellegrino and Thomasma have so aptly termed a "kind of therapeutic belligerence."[10]

When we speak of passive euthanasia, we speak of the deliberate omission of an act that, under normal and expected circumstances, would reverse an inter-current condition and prolong life. It is frequently spoken of as "letting nature take its course." Allegedly, "the disease, not the doctor, has killed the patient." Refraining from treating septic shock in a patient hopelessly riddled with metastatic cancer would be a rather classic example, but there are many, many more. Often what we are doing is unclear. When we refrain from resuscitating someone who we think could (technically) be resuscitated, disconnect a ventilator from a patient whom we know to be ventilator dependent, or stop supplying fluids and nutrition to a vegetative person, we may or may not call such an act, or such a failure to act, passive euthanasia. However, what is clear is that, except for our acting (when it would have been easy to refrain) or not acting (when it would have been easy to act), the patient might well be alive. To claim that it is "the disease" and not we that were involved in and, at least in part, caused death is to leave out a crucial link in the causal chain that ended in someone's death— it is, to say the least disingenuous.

Active euthanasia is another matter. As used here, active euthanasia must be deliberate, the stopping of a life that, without such an action, would have continued. It must be done under circumstances in which the death of the individual is reasonably felt to be inevitable and shortly at hand, in persons whose future offers no choice except (on the patient's terms) one filled with relentless suffering and a longer existence or one in which suffering is abrogated but length of life reduced. Furthermore, it must be the repeated request of the patient (or, perhaps, upon request of a legitimately designated and reliable surrogate). Above all, active euthanasia, to be called euthanasia, must be done in the interest of the person being killed and not in the interest of the state or of anyone else in mind. The way the term, euthanasia, is used here (or anywhere else in the civilized world) is not equivalent to what the Nazis called "euthanasia" and which was, in fact, murder. Persons were killed not so as to serve such persons' own (assumed) interest but to serve the interest of the state. The use of the word was one of a series of euphemisms and hypocrisies that graced that regrettable period and that has until today deformed our language and affected our ethics.[52] Hypocrisy— something we have consistently felt to be one of the prime ethical evils—leads to the delusion of others as well as oneself and is one of the real dangers to be guarded against. It compounds an already evil action.

When active euthanasia is discussed, three different questions must be answered. The first (and the basic ethical question) deals with the ethical permissibility of any killing of humans, including their active euthanasia: Is killing a person ever a morally allowable act? If that question cannot, at least in some circumstances, be answered in the affirmative, then all other questions are moot. The next (what we have called "the professional" question) question deals with the probity of involving health professionals in such an act: Is there something in the concept "health professional" as used in our society today that would make it impermissible for them to participate? The third question (which is really a social-cultural and legal one) deals with legalities: Is it advisable to legalize such a step. These are separate (even if not entirely separable) questions.[53]

As for the first question, it is clear that we countenance the preventable death of persons. We set speed limits at certain speeds (knowing full well that setting them lower would save lives), mine coal and build skyscrapers, wage war and (regrettably, in our view) execute people. Moreover, we know that poverty is associated with a higher disease and death rate, that many children (about one-third even within our own country) are hungry, and that many persons go without even the most basic of medical care. To claim that, as a society, we are life affirming is simply a false claim. In speaking about euthanasia, we limit the discussion to persons who have met the criteria set forth earlier in this chapter. Technically there is a difference: Direct voluntary active euthanasia (in which the patient speaks for himself or herself) is not quite the same thing as indirect voluntary active euthanasia (in which responsible surrogates are allowed to speak) and quite a different thing from involuntary euthanasia, a form of euthanasia where some member of the healthcare team, without consultation with patient or legitimate decision-maker, acts by himself or herself.

Most health professionals have come to agree that in hopeless circumstances, treating merely to delay what is clearly and imminently inevitable is often not justifiable. Treatment decisions of this sort should properly accord with the patient's wishes and, preferably, are decisions mutually agreed upon previously among all concerned. Only when no such arrangements have been made should surrogate or substitute decision-making be considered. Decisions not to treat merely to prolong the life of a patient who is terminally ill and in pain may or may not be justified. The decision here is largely a personal one depending on the patient's values, plans, and goals. A patient near the end of life and in pain may, nevertheless, choose to prolong life for personal reasons: say, to see a daughter graduate, a book published, or a son get married. Prolonging life here may be a definite part of what I [EHL] have called "orchestrating death." Under a similar medical condition, another patient may choose to forego all treatment. This decision, too, may properly be seen as a part of orchestrating death (see section of same title later in this chapter). Physicians and their colleagues cannot properly make these decisions based on the medical condition alone without ascertaining their patient's wishes. At times, healthcare professionals will be asked their advice and should, under ordinary circumstances, give it as part of their ongoing obligation. At other times, patients may choose not to make a choice, leaving such a choice up to the healthcare professionals. While physicians and other healthcare professionals dealing with such patients may feel that this imposes a heavy burden that they are loath to assume, such a decision, when properly made by a patient, nevertheless needs to be, perhaps with a heavy heart, respected.

Serious questions have been raised about the distinction between active and passive euthanasia, between what has been called "killing" and "letting die." Under most circumstances, physicians will not be faced with the issue of permitting intolerable pain to persist or ending their patient's life. In almost, but not all, circumstances today, pain can be successfully obtunded, and obtunding pain, even if doing so hastens but does not directly bring about death, is considered generally acceptable, even within the Catholic church.[54] But not all circumstances are circumstances of intolerable pain, and not all pain is entirely relievable. Persons whose pain is moderately well controlled but who as a result are nearly obtunded, unable to think, and only poorly able to communicate may feel that such a life is a life of suffering and is not, on their terms, worth living. The nagging question remains: "Is it always and under all circumstances wrong to kill?"

Those who claim that there is always a difference between killing and letting die base this claim on two things. First of all they claim that killing is quite different from allowing death to occur because one can always refrain from a voluntary action (killing) but one cannot prevent all death. Secondly, they claim that simply allowing something to happen does not hamper others from preventing it while causing it to happen is a more surefire thing.[55] Both of these arguments, as Rachels has pointed out, are severely flawed.[56] In dealing with active *vis-à-vis* passive euthanasia, the point is not that one cannot prevent all death but that one has deliberately chosen not to prevent a specific and presumably preventable one. The question, furthermore, is not that others may do what one has neglected to do (for example, institute a treatment that has deliberately been omitted) but that one has deliberately chosen not to intervene in a situation which, under different circumstances, might oblige one to do so. By not acting to prevent a death that could have been (and could perhaps have been easily) prevented one has, whether one wishes to admit it or not, become a necessary and therefore critical and inescapable link in the causal chain of death.

Those who affirm that there is a clear moral difference between killing and letting die regardless of context will find that dealing with practical considerations may be troublesome. Consider, for instance, John, hopelessly trapped under a burning vehicle and pleading with his friend, who has a gun, to shoot him. Would his friend be acting more responsibly by allowing John's agonizing death or would shooting his friend be a more morally acceptable option? The answer that killing is, under all circumstances, wrong, is an absolutist answer ignoring interests, intention and context. If a resolution is sought by trying to find a morally acceptable ("praiseworthy") course of action, none will be found; if, on the other hand, one seeks an answer that, in the circumstances thrust upon us, tries to chart the least objectionable (least "blameworthy") course of action, a beginning may be made (see also Chapter 4).

Often the way we linguistically and conceptually "stack the deck" determines whether we call a given act killing or letting die. When we turn off Mrs. Smith's ventilator or refrain from treating a vegetative patient's pneumonia, the distinction is far from completely clear. Consider a patient who under anesthesia is entirely dependent on ventilator support: If I were surreptitiously to disconnect the ventilator, I doubt that a judge would have much patience with my plea that it was not I but the anesthesia that caused the patient's death. When we turn off a ventilator connected to a patient we know to be ventilator dependent, our claim that it was not we but his illness that caused his death is disingenuous: Both facts (the fact that illness caused ventilator dependence and the fact of our turning off the ventilator) were causally needed to produce death, and one without the other would not have done the trick. A *prima facie* duty not to cause death remains. Killing patients, under all circumstances, is a wrong even if, at times, it seems like a lesser wrong…however, in the human condition, it is a wrong we cannot always evade.

Moreover, the way we frame a question often goes a long way toward determining the answer. If we frame the problem as one of "killing," we may be inclined to give a quite different answer than if we frame it as "relieving suffering." And yet refraining from treating intercurrent disease in a terminal and agonizing condition, thus allowing death to occur, or deliberately giving too much of a drug and killing such a patient can be viewed in either of these two ways. Euthanasia, performed under the circumstances stated, is both "killing" and "relieving suffering"—but the latter would not have been possible without the former. It is a situation in which being alive supports a life that has ceased to have any meaning for a subject who has, in truth, only the option to live a bit longer and suffer more or to live a bit less long and suffer less.

Another linguistic maneuver frequently resorted to is framing the question in terms of "harm," *i.e.*, that healthcare professionals must "never harm" their patients. Not harming (or harming as little as possible consistent with achieving a mutually agreed goal of treatment) is part of the implicit social contract between healer, the one seeking to be healed, and the community, which ultimately underwrites the enterprise. No one will quibble with that. But arguing that, therefore, allowing to die or at the request of the patient actually killing them is of necessity "harming" merely begs the question by defining harm and killing or letting die in terms of one another. What constitutes harm is context-specific; thus harming must be defined on individual terms: Depending on circumstances, values, and worldview, what is a harm to me may be a benefit to another. To make the claim that keeping a patient who is suffering (and has intact decisional capacity) alive against her expressed will is not inflicting "harm" whereas acceding to her wish and killing her is inflicting harm defines harm in a most peculiar way—it is especially onerous because it excludes the patient's own assessment and evaluation of what for her, in these particular circumstances, constitutes being harmed.

While refraining from treating is, at least in some circumstances, acceptable to most health professionals, many health professionals find participation in active euthanasia to be reprehensible. The idea of killing under all circumstances affronts their sensibilities. The issue, as was put in a recent paper, "touches medicine at its moral center."[57] The feeling that "if physicians become killers, or are merely licensed to kill, the profession will never again be worthy of trust and respect as healer and comforter and protector of life in all its frailty" is a cogent and powerful one. Physicians and other health professionals are dedicated to life and to its preservation. Society in its contract with medicine presumes this obligation. But physicians and their colleagues are likewise and with equal force dedicated to relieving suffering and to helping patients (whose freedom of the will may be intact while their freedom of action is severely damaged) shape what remains of their life. Those who oppose abortion (and generally—albeit not always—basically do so on religious grounds) will use arguments against the practice that have great similarity to those used against euthanasia. Since they argue that abortion is likewise a form of active killing, they feel that the issue likewise "touches medicine at its moral center." The difference, among others, is that the one is and the other is not legalized and that the one is and the other has not as yet been officially accepted by medical bodies. But neither of these facts speaks to the ethical questions of the issue.

Not rarely, patients who have terminal cancer or other terminal conditions ask their physicians for the means of suicide. Physician-assisted suicide (where a patient still capable of carrying out the act is provided by a physician with the means—be they specific directions or a specific prescription) has become a rather widely discussed issue. Some states have attempted to make physician-assisted suicide illegal— that is, physicians could be punished if found out. In states in which suicide itself is illegal, this is, at least, not irrational. But to make assisted suicide illegal in states where committing suicide is legal, however, makes little sense. Logically speaking, how can it be illegal to assist someone in performing a legal act?

Patient requests to their physicians for help in committing suicide are unfortunately often met by physicians and other healthcare providers with a brusque refusal or, at times, by an appeal to the law. These are issues that ought to have been carefully discussed long before the event precipitating the request. It may well be that personal moralities conflict here and that referral early on to a physician who shares the patient's worldview was indicated—early on, because a later refusal may readily be interpreted by the patient as abandonment. Physicians who believe that suicide or helping another commit suicide is immoral cannot in good conscience comply with such a request; but neither can they hold their patients hostage to their own idiosyncratic beliefs. A compassionate approach to such patients (including, perhaps, referral to another physician whose worldview does not preclude assisting suicide) is necessary if physicians are to act from more than merely subjective "morality" (*i.e.*, based on their own personal belief system.

At times nurses, pharmacists or other healthcare professionals may be involved. They may be explicitly asked to participate (by filling a prescription or by providing or helping to provide information) or they may, against their will, become involved through the actions of their employers or co-workers. When some find assisting suicide to be (for them) morally unacceptable, they cannot explicitly or tacitly, directly or indirectly, become or be expected to become involved. In helping orchestrate death (see the following section) the issue of assisted suicide is an important issue for all healthcare professionals to feel secure about. Realizing that at least some patients will request such help, healthcare professionals must be clear about their own feelings and possible courses of action. This is why it is crucial for every healthcare professional to think about and come to terms with this issue and to let their patients understand their personal moral point of view well before they inadvertently place themselves in a position or specialty that has a good chance of conflicting with or compromising the strongly held beliefs of their own personal morality.

There is one other point connected to allowing patients to have access to an acceptable means of suicide: It empowers them. Not rarely (and both of our personal experiences bear this out), empowering patients in this way helps them to persevere a bit longer than they often otherwise would. Here as elsewhere, allowing patients as much control over their own destiny as possible is not only ethically proper; but by giving them responsibility for their fate, it tends to motivate patients to cooperate.

Laws concerning PAS and active euthanasia differ from state to state and from country to country. For example, in Germany PAS is not punishable except for one quaint circumstance derived from another law. In most countries in the EU failure to render assistance to one in distress is a punishable offensive—one cannot simply walk by an assaulted or stricken person. When patients following planned PAS lose consciousness no one unprepared to give aid can be present and fail to render assistance without becoming legally liable. This state of affairs (which we consider an absurdity much in need of correction) needs badly to be re-evaluated and, in our view, abolished. In Austria PAS as well as active euthanasia are strictly forbidden; in France the laws against both tend not to be enforced (albeit most physicians would not participate) and in the Netherlands PAS and active euthanasia are legal, provided certain stringent criteria are met and retrospective review is conducted. Active euthanasia is forbidden in all states of the United States and only in Oregon is PAS (but not active euthanasia) legally permitted—though there are strict criteria which must be met.

Although PAS and active euthanasia have been only recently been permissible in the Netherlands, a system which allowed the district attorney, providing strict criteria were met, not to bring charges has been in place for some years now. The problem in the Netherlands is that proponents and opponents of this system have never agreed on the validity of the data and it is, therefore, difficult to come to any empirical decision.

Oregon, for the last four years, has had data on which both proponents and opponents substantially agree.[58-60] In effect, the numbers of those seeking PAS and receiving an appropriate prescription has not changed over the years, remaining at about 9 out of 10,000 deaths in the entire state—the claim by opponents that hordes of people would seek this way out has, at least to date, not been borne out. Most people who requested PAS were given substantial doses of short acting Barbiturates (sometimes with anti-emetics), had few if any deleterious side effects, were asleep in a few brief minutes and dead shortly thereafter. Pain was not—just as it was not in the Washington study[61]—the chief reason for seeking death. Once again the fear of the loss of self-determination and the suffering brought on by the inability to translate one's will into action, to care for oneself or to engage in any of the activities the patient enjoyed led the list of suffering. Not without interest is the fact that over 70% of patients were registered with hospice at the time of their death—a fact which either means that hospices were "doing a bad job" (which they were not) or that good hospice care did not suffice to eliminate some of the pain and/or suffering. It is also of interest that by no means all of the patients given prescriptions used them—in fact, many had had them for some time before their death. Once again—empowering patients (and, by that very act, making them responsible for it) is of extreme importance.

———————— ORCHESTRATING DEATH ————————

As is sometimes the case, our language in a sense predetermines or at least seriously affects our answers. When, as we have mentioned before, patients can no longer be effectively treated so as to restore function, ameliorate illness, or "cure," we have a tendency to say that "there is nothing more we can do." Attempting to do what "cannot be done" is seen either as foolish or, even worse, as a waste of time that could be devoted to something useful. Abandonment easily follows.

In general, when we make the statement that "there is nothing left to do," we mean that we can do nothing more to effect cure, restore function, or ameliorate disease. But medicine has other and at least equally critical obligations: Ameliorating suffering is, as has been repeatedly pointed out, an historically enduring obligation. It is an obligation that under the usual circumstances of medical practice is of the greatest importance. Generally, however, it is a value that is less important than cure, restoration of function, or amelioration of illness. Even though good health professionals will do their best to keep pain to a minimum and to obtund it where possible, health professionals will not hesitate to inflict some pain so as to attain what is seen to be a more important goal. When patients are beyond achieving cure, however, or restoration of function or amelioration of their illness, the question of dealing with suffering assumes primacy. Such dealing with suffering is very much "doing something." It is what can be called "orchestrating death."

Orchestrating death is, in fact, focusing on life.[62] Dying is not somehow a state set apart from living: It is part of life itself and like the rest of life can be lived more or less well. The emphasis of all concerned with patients should not be on their inevitable death (a fact of which patients are generally well aware and that should not be hidden from them), not really on "preparing them for death," but rather on enabling them to live their lives as fully and as richly as their illness and circumstances permit. It is a time of life when "having a life" can greatly be facilitated by what healthcare workers do or refrain from doing. Orchestration is a complicated business. It involves directing the various parts of the orchestra, calling forth the strings here, muting or strengthening the woodwinds there. Like a good conductor or orchestrator, the leader of the team (generally the physician in concert with others—and especially the patient and the family) must be ready to call on various backgrounds and skills to facilitate the task at hand. Orchestrating consists of doing or calling in help to do a variety of things: pharmacological, physical, psychological, spiritual (if the patient is so inclined), emotional, and social. By working together, by understanding a patient's biography, values, needs, hopes, and fears, life can still be meaningful, pleasurable, and rich. Indeed, seeing that this is done properly is a difficult, important, and often rewarding task. It is very much "doing something."

An understanding of what death means in the patient's life, what hopes it forecloses, what plans it shatters, and what possibilities for the realizing of at least a part of such plans in the time left, is essential if one is to successfully "orchestrate." Emphasis throughout should be not on the inevitable death, which we can at best (or, sometimes at worst) forestall; rather, emphasis should be on making what is left of life richer and more satisfying. Understanding the stages of death that Kübler-Ross described while not "expecting" patients to necessarily behave accordingly can be of great help in accomplishing this task.

If PAS or active euthanasia may, in some rare cases, fit in with orchestration will be debated. The fact remains that (1) suffering and pain are not the same and relieving pain, as important as it is, does not by any means necessarily remove suffering, and (2) most patients in a variety of studies are not motivated to seek help in dying because of pain but because of factors such as loss of the capacity to care for themselves and to translate their willing into action, incontinence, the inability to engage in activities that they previously enjoyed, etc. Pain invariably is down low on the list. The use of large doses of narcotics (while often and, at the patient's wishes, indicated) may, in fact, cause even greater suffering since it tends to grossly interfere with the patient's freedom of action.

Orchestrating death—if it is to be well done—is a complicated skill and one that must be learned and, therefore and systematically taught. It is a multidisciplinary skill, has its theoretical underpinnings and a multitude of variable ways of application. In fact there is not a single college of medicine or nursing specifically teaching this skill to all students.

———————— TALKING WITH PATIENTS ————————

Physicians and their colleagues in other health professions find talking to patients whose prognosis is poor or for whom no further therapeutic intervention is possible very difficult. Such persons, at times consciously but more often unconsciously, avoid dying patients and, when they must be with them, avoid the issue by resorting to idle chatter or pointless jocularity (a thing quite distinct from humor). Likewise, the body language often sends a clear message of distancing to the patient. At the very time that patients need communication most, communication is cut off. This is not a problem unique to health professionals. It is a problem that involves the family, who, likewise, are loathe to address the problem head-on and who "to be kind" hide behind optimistic platitudes, half-truths, self-delusion, or outright lies. And yet studies have shown time and again that patients generally want to be involved, want to share, and want, above all, to communicate about this issue.[8]

Healthcare professionals often do not involve patients in critical matters. Often they fail to tell them the truth about their prognosis, and they frequently fail to involve patients in critical life-and-death decisions.[8,63–65] When this happens, the professional–patient relationship becomes a game of charades. Often such a failure to communicate demoralizes all members of the healthcare team. Nurses, asked by patients who find communication with their doctors cut off, find themselves in a terrible quandary: They can tell the patient the truth, knowing that this may bring about the physician's wrath as well as perhaps otherwise "get them into trouble;" they can join in the game of charades; or they can simply obfuscate and lie to the patient. Physicians, as the acknowledged heads of the healthcare team, owe an obligation not only to the patient but likewise to the other members of the team. Open communication and discussion among all members of the team is of utmost importance.

On the other hand, while patients, because of the respect humans owe one another, ought not to be lied to, telling the truth comes in many shades and gradations. As one paper has put it so aptly, these conversations are an elaborate *pas des deux* in which both the two parties feel each other out.[63] Patients can only be told what patients are ready to hear; they will turn a deaf ear and literally deny having been told if told at a time when they are not receptive. I [EHL] well remember a lady who was told about her diagnosis of cancer at least four times by her physician, and each day would ask again and simply deny having been told. Sick people are not, as Eric Cassel put it so beautifully, simply well people carrying "the knapsack of disease."[64] They are people in whom a whole host of changes and adaptations are and have been taking place. Simply assaulting patients with the truth is hardly the proper thing to do.

Patients must be gently led to receive bad news. The manner in which this is done varies with the personality of each patient, each healthcare professional and his or her role, each situation, and the peculiarities of the specific relationship. It depends on the assessment of the situation (an assessment that can receive invaluable help from the other members of the healthcare team as well as from the family) and defies a stereotypic approach. Human understanding, rather than technical knowledge, is what is needed, and humor, as always, has its place. Physicians have to be as sensitive to the patient's implied wishes as they are to the "letter of the [moral or legal] law." Shoving the facts down the throat of a helpless patient who may derive some lasting comfort from deception is "moral" in no more than a very aseptic sense. The physician has to "size up the patient" (a comment made by Eric Cassel to Professor Jonas some time ago[32]) and deliver a judgment as to the patient's desire and capacity for truth.[65] Talking to patients who must be given the news that they are hopelessly ill truly requires compassionate rationality: compassion, so as to be able and willing to share humanity with one another, and rationality, so that sentimentality does not swamp the situation.

Often if not invariably it is proper for physicians or other members of the health-care team to sit on the patient's bedside: there is little as intimidating as to have a person in a white coat loom over a prostrate body often tied to tubes. We would strongly suggest that during such conversations a cup of coffee, another beverage or even food be shared. The sharing of food and drink has tremendous significance in all civilizations and it's importance today—albeit sometimes differently expressed—is as strong as ever. The main issue is to have such talks as far as that is possible as equals trying to understand each other's goals and gaining confidence and trust in one another. The evident and inevitable difference in power and the patient's understandable desire to please the health-care team makes such conversations difficult.

Under most circumstances, most patients must eventually, if they desire to hear it, be told the full truth. Just as patients have a "right" to be informed, they also have a "right" to choose not to be informed. Patients, in other words, must have the right to choose and to have their choice respected. Respect for persons demands as much. In some cultures patients are traditionally not generally told bad news, a tradition that, however, seems to be rapidly changing. But even here the physician with the active help of other healthcare professionals must try to ascertain the patient's wishes and act accordingly (see Chapter 5).

When the truth is to be told, however, it has to be gently given. Physicians and other healthcare professionals involved must allow ample time to share what they perceive to be the truth, as well as what they know to be their ignorance and uncertainty, with the patient. A hurried approach to patients under these circumstances violates the duty of respect as much as does not telling the truth at all. Patients who are willing to hear must be told the truth, but they must not necessarily be informed the moment the "facts" become known. Physicians are well advised to defer such conversations to a time when they are not pressed for time, not "due at the office in 5 minutes," and not visibly harassed by conflicting demands. Likewise, other healthcare professionals need to choose their time to speak with patients, if necessary telling them that they will return when they have more time and then returning and spending what time is needed. Sitting at the patient's bedside, perhaps sharing a cup of coffee, perhaps touching a shoulder or an arm, all are appropriate maneuvers that can convert the process from one of mutual pain to one of a sharing of mutual mortality. It is here that a start can be made at orchestrating the rest of the patient's life so that it is as full as possible.

Patients may ask not to be told. This does not occur frequently, but it does occur. If it occurs, and if it seems to be a truly autonomous decision, the request must be respected. An ample opportunity for patients to change their minds must be given, but the patient who in effect "leaves everything up to the doctor" and does so knowingly has made a deliberate choice. Physicians who cannot live with such a choice (and some may not be able to) are well advised to communicate this to their patients and to reach with them a shared agreement of how to proceed. It may be that the patient delegates a family member; it may be that the physician may have to obtain consultation or turn the case entirely over to another physician.

On the other hand, relatives may beseech physicians not to tell the truth to their patients. A wife may "forbid" the physician to tell her husband of his metastatic cancer or even threaten suit (the fact that patients threaten suit does not mean that they will sue, that they, in fact, can find cause to sue, or that, ultimately, they can hope to prevail in such a suit). They may plead that they know their husband well and that he "couldn't stand to know." In general, physicians are ill advised to follow such counsel. First of all and in most circumstances, who will and who will not be told should be explicitly agreed upon long before the need arises. Secondly, when circumstances have made this impossible, the physician's first obligation is to the patient: Unless physicians know in individual cases that something about a specific patient makes the administration of truth ill advised or that the patient has specifically stated that he or she does not wish to be informed, they are obligated to tell it. Streptococcal disease is treated with penicillin except in the very unusual case when penicillin is contraindicated because of some special condition (allergy, for example) peculiar to the particular patient. Truth telling, in the moral sphere, has a somewhat similar standing: Unless overwhelming reasons to the contrary can be given, the truth must be told. In general, it is not *if* but *how* the truth should be told that is the issue. Physicians must "size up their patients" (guided, perhaps, by their family but certainly not dictated to by them) and reach a decision about how much, how quickly, and above all how to share bad news.

Most of the time, when they are speaking with patients, healthcare professionals do not have to deal with such vexing and emotionally disturbing matters. It is in the daily conversation with patients that patients and the professionals caring for them can feel each other out and get to know each other. And it is these daily conversations and this mutual understanding that forms the basis of later, far more difficult conversations. When healthcare professionals and their patients have come to know each other over time, such difficult conversations become far easier: patients and healthcare professionals have tacitly or explicitly come to know, understand, and hopefully respect each other's world view. But that is the ideal. Often healthcare professionals and their patients do not know each other well when illness strikes. Even here, however, the less emotionally wrenching parts of the conversation precede and it is here, even if the time is brief, that mutual understanding and trust can be achieved.

Critical to all of this is not so much "talking to patients" as listening to them and observing as they interact with their families and other healthcare professionals. The good listener not only must listen but also must gently try to steer the conversation so that a real conversation rather than (as my father used to put it) "a mere exchange of meteorological data" results. They must learn what the patient and their loved ones know, fear, or hope about their illness and about its course and consequences. In other words, they must try to understand what the disease or illness means in the context and totality of that patient's life. The good listener (not the bored or the uninvolved listener who "simply listens" and fails to engage himself or herself) is the one who will be able to accomplish the most for their patient when it comes to dealing with ethical issues or confronting emotionally trying problems.

GRIEF

In dealing with end of life issues the problem of grief becomes an essential component. Not only after the patient dies, but during his last few months, relatives, friends and (hopefully and to some extent) the health-care team are beset with grief. But the patient, also—and perhaps to the greatest extent—grieves about many things but especially about the fact that while relatives and friends lose one person (as central as that person may be to their lives) the person who dies loses everything: friends, relatives, the world. He or she ceases to have a life. One must be attentive to this grief and by body as well as verbal language make clear that the subject is neither a disagreeable nor a taboo one. Grief cannot be handled in a stereotyped way, but in a manner very much authentic for those involved. Re-assurances that "everything will be all right" are pointless and often very much resented clichés for they preclude all further communication. Likewise, during the last few months we need to make relatives aware that one of their feelings will be guilt.

We well remember a lady in her high 80's who had been a controlling and thoroughly obnoxious person. When she obviously had not much longer to live we had a long talk with the daughter-in-law and quite bluntly told her that she should first of all not send in her young child to wake up grandma and, most importantly, upon finding her mother-in-law dead would experience an initial sense of relief followed by a feeling of guilt at that relief. When a terrible problem or burden is removed (no matter how) there is a part of us that, at least initially, feels relieved. The guilt that follows can be destructive. The expected happened—the lady died, the daughter-in-law felt relieved and was able to deal with guilt involved. In fact, she returned for the sole purpose of saying "thanks."

Relatives and friends taking care of patients in the hospice setting and (sometimes likewise in hospital) may likewise feel this sense of relief—inevitably followed by guilt. Their job (especially in home hospice care) is difficult and burdensome to the rest of their lives. It behooves the team to prepare relatives for this event.

After the death of the patient—who may, at that stage of the game, not be the only one treated and, in fact, my be no longer at center stage—it is ethically problematic not to follow through with relatives and friends, assuring them of continued interest and of, when needed, support. One must be careful—many persons grieve alone and consider obtrusive support to be disturbing rather than helpful.

———— ECONOMIC CONSIDERATIONS ————

Ordinarily, and within our current vision of the physician–patient relationship physicians, when dealing directly with individual patients, cannot put the cost or inconvenience of a given procedure above their individual patient's good. The fact that such a statement must somehow be integrated into the totality of a community whose resources are not infinite is likewise beyond rational dispute. We address this troubling issue concerning the possible ways of integrating these two obligations in Chapter 8.

Because we are such a crisis-oriented society vast resources are often expended at the end of life. The very same patients who had no access to routine medical care, who could not afford immunization or often even food and warmth, suddenly become the object of vast, often heroic, efforts when they are found to be critically ill or on the verge of dying from starvation or exposure. Uninsured patients in most areas, for example, will find it difficult to have increasingly severe benign prostatic hypertrophy with partial urinary obstruction treated by a relatively simple surgical procedure. The same patient who has, in consequence of this neglect, become uremic will find that nothing will now be spared to save his life: hospitalization, dialysis and even, at times, transplantation.

This is not the place to speak about limiting exotic interventions that may marginally benefit individual patients. Largely these are issues of macro-allocation that must and will be addressed elsewhere. There is, however, what most of us would feel to be a great waste of resources occurring in all of our hospitals. This waste consumes large amounts of time, effort, and money: It is the use of resources for patients known to be beyond hope. We speak here of the brain dead patient kept on a ventilator, of the irreversibly vegetative or comatose, of the infant with extensive and diffuse grade 4 intracranial hemorrhage who will never be able to lead a sentient existence, and of many more tragic states. The problem is not, as has so often been charged, a problem of the elderly. It involves all age groups. I do not speak of those who, at whatever level, retain a capacity to enjoy existence but rather of those beyond hope who are incapable of being benefitted although, at times, they retain the capacity for being harmed. Such patients consume a vast amount of resources in their tenure on earth. These resources cannot be measured only in money, material, or space. Much more is involved. Resources include the energy, hope, and love of relatives and friends as well as those of the healthcare team. Once all hope is gone, it seems morally indefensible to consume such resources in the empty discharge of an empty duty while others starve or go without even the most basic health care.

Hardin compares modern civilization to herdsmen sharing a commons.[66] As conditions become more crowded, herdsmen seeking only their private good by increasing their herd jeopardize the commons. "Freedom in a commons brings ruin to all."[66] In our commons, consuming limited resources beyond the hope of reasonable gain seems only questionably justified. Hiatt voices concern for three types of demands on medical resources:[67] (1) those that pose a conflict between the society and the individual, (2) those for potentially preventable conditions, and (3) those of no value. Using up material as well as other resources to maintain patients at the end of life who are insentient (or who are beset by unrelievable and unending pain and who do not wish their life prolonged) is, at the least, ill advised and to the detriment of all concerned.

PROBLEM OF FUTILITY

While economic considerations lead us directly to considerations of "futility," ethically speaking, they ought not to become the driving force behind what is considered to be futile and what is not. That is not to say that economic factors may not preclude making a given treatment modality generally available. One could certainly argue that a medication costing several million dollars a day, which had to be given over a long period of time and which had a slim chance of prolonging meaningful life, could not be made as readily available as aspirin or penicillin. We should be honest about this: Such a modality should not necessarily be considered to be futile. It might, however, be considered to be too expensive and therefore not be made available for general use.

The problem of futility, for reasons quite aside from economy, has become an important issue in medical ethics today. There are several reasons for this. First of all, unless they are unexpectedly found dead in bed, few patients who die in the hospital (or even often at home) today die precisely when they would "naturally" have done so. Almost always, there is something that could be done to prolong life for a few minutes, hours, days, or even weeks. Even when a "code" is called off, a decision that "enough is enough" and that to proceed would be "futile" has been made. Patients, moreover, are increasingly afraid of being insentient captives of a heartless medical system that, in their view, has allowed technology to drive it. In part this fear is realistic; in part it results from the often-contradictory messages sent by the media. On the one hand a message of unwarranted hope (the "miracles" of modern medicine can cure "everything"); on the other hand the often-unjustified fears (the "terrible things" done to patients in hospitals, the terrors of being on a ventilator, and so forth). Time and again the charge that physicians do things that are futile is heard. In a sense yesterday's fear of being buried alive is alive and well in the fear of having one's body sustained with one's mind gone (or worse yet, having one's painful dying needlessly prolonged). And while economic factors cannot be allowed to be the driver in defining futility, there is no doubt that this prolonged dying has consumed an enormous amount of resources at a time when these resources have been found to be quite limited.

The problem with using a concept like "futility" is that there is a total lack of unanimity in its definition. To get on with dealing with the problem, we will either have to reach consensus in definition or seek for another language in which to couch a very real problem. Several attempts at definition deserve mention. Nancy Jecker and Lawrence Schneiderman (two of the pioneer workers in this field) have defined "futility" in the medical setting as (1) "a treatment which has not worked in the last 100 cases," and (2) a treatment that "merely preserves permanent unconsciousness or cannot end dependence on intensive medical care."[68-70]

The first of these two definitions (futility in the quantitative sense, as it has also been called) is based on probability: It hasn't worked in the last 100 cases and, therefore, is unlikely (but not impossible) to work in the 101st. Its validity is statistically determined and what statistics allow us to presume depends on prior choices as to precisely what we consider to be "statistical validity." The second of these definitions may be based on one or both of two separate judgments. The first is that continuing one's existence in a permanently unconscious state or permanently dependent on intensive medical care is not something anyone would desire (also called the definition of futility in the qualitative sense). The second is (and perhaps with more force) that one of the goals of medicine historically or today is not exemplified by supporting such a state. The first of these assumptions is an assumption about how most people confronted with such a situation would feel about it; the second involves a judgment as to what the legitimate goals of medicine in fact are. As such, both of these are inevitably subjectively interpreted by the parties involved.[71-73]

Beyond this, Jecker and Schneiderman (1) distinguish between "a treatment effect which merely alters some part of the patient's body" and "a treatment benefit which can be appreciated by the patient and enables the patient to escape total dependence on intensive medical care" and (2) consider a treatment that merely alters some part of the body (or its function) without being appreciated by the patient to be likewise "futile." These further assumptions, moreover, are based on essentially similar and "qualitative" judgments: (1) that life lived like this would be unacceptable to most, if not all persons and (2) that supporting such an existence is not among the current legitimate goals of medicine

As regards treatment considered futile by such definitions, Jecker and Schneiderman pose three increasingly more restrictive points of view. Such treatments (or, by inference, diagnostic procedures) could be considered to be (1) ethically neutral—doing things that are futile is neither ethically proper nor ethically improper; (2) ethically inadvisable—doing things that are futile is, first of all, foolish and at least ethically questionable, and physicians should be counseled against doing futile things; and (3) ethically inadmissible—to knowingly providing therapy that is futile is ethically wrong. In their papers they argue for the last point of view.

Futility and its concepts, definitions, and ethical implications have been an area of increasing concern in the literature. Definitions have varied from the very narrow (i.e., treatment would fail to reverse a physiological disturbance) to the very broad. In *Care of the Dying* published by the Hastings' Center, the feeling that in order for the concept of futility to be applied it should be narrowly defined is implicit.[75] Unless a patient's death was inevitable and imminent, physicians ought not make a unilateral decision to forgo treatment. Broader considerations have variably included concerns for the suffering of the patient and considerations of other aspects of the quality of life as well as considering economic and other social factors.

The Oxford English Dictionary defines futility as (1) incapable of producing any results, (2) failing utterly of the desired end through intrinsic defect, (3) useless, (4) ineffective, or (5) vain. Common to all of these definitions is that some goal or end for which one strives is, by the method used at least, beyond achieving. In the practice of medicine, then, a futile procedure would be one incapable of achieving what is considered to be a legitimate goal or end of medicine. Therefore, there are first of all and before we can go on, really two questions to be addressed: (1) What are and who defines the (legitimate) goals of medicine? and (2) How in practical terms, once we are a bit more clear on the first question, do we define what is futile? Beyond this there are other questions. Are the goals of medicine changeable and if so, how, by whom, and by what process are they changed? Are healthcare professionals entitled to use their knowledge and skill to pursue what are not (today) considered to be within the framework of "legitimate" goals?

Goals are not determined for all time: The ends of medicine are not immutable but are changeable and must adapt to technical as well as social and historical conditions. The goals of medicine, while in part a product of the work people involved in medicine do, are not set independently by or in the field. They are ultimately rooted in their society just as are the persons (physicians, nurses, or other healthcare workers) who perform the everyday tasks. Medicine can no more evolve its goals independent of the community in which it functions than can the community evolve these goals without the advice (and ultimately consent) of those who must perform the tasks comprising the concept. In a sense this dialogue between healthcare professionals as "experts" (and "experts" who ultimately must do the work) and their community is a form of negotiation. Clearly, speaking about medical futility presupposes a clear notion not only of our own or even of "reasonable" values, but also of the social as well as technical goals and possibilities of medicine itself.

When asked to discuss the goals of medicine, most physicians or students will name three: dealing with illness so as to restore health or at least bring about better physiological functioning in the face of disease; ameliorating suffering; and preventing illness. Medicine, however, is a social art: What is and what is not accepted as a proper part of medicine not only depends on the judgment of health professionals themselves but also reflects the values of society at large. Physicians today are expected to have—and expect themselves to have—far richer and more complex goals than in the past. After all, physicians work as plastic surgeons improving the appearance of patients (not truly includable among the goals noted before: my ugly nose is ugly but is hardly a "disease"!), prescribe birth control pills and provide sterilizations (fertility is a state of health and pregnancy not a disease), and feel obligated to help the physiologically and psychologically well families of their patients and not only the physiologically or psychologically ill patients themselves. Many physicians, furthermore, are deeply involved in counseling; serve in relation to insurance, job, or other applications; fill out disability forms; and do many tasks that cannot be subsumed under the usual mantra. Even beyond this, and highlighted by the smoldering debate about euthanasia in many parts of the world, physicians are (in some countries at least) involved in performing active voluntary euthanasia that, although it can be argued to be in accordance with the goal of alleviating suffering, is certainly an alleviation of a very special kind. Moreover, some have written, and there is in our view a great deal of merit in this, that one of the physician's most important tasks is the restoring of patient autonomy, hardly something intuitively evident from medicine's traditional, and traditionally paternalistically conceived, obligations.[64]

When we speak of treatment as "futile" we need to be most mindful of the complexity and evolving nature of contemporary goals. We agree with the formal statement that physicians cannot be forced (or even expected) to give "futile" treatment to their patients. But to make such a statement contentful, the goals of medicine as well as the goals of specific encounters within medicine need to be understood and fleshed out.

As always in ethical analysis, it may be more fruitful to address this issue by dealing with the areas that are clear and then working toward the grayer zones. The generally held first three goals of medicine (eliminating or at least ameliorating disease and suffering consequent on disease as well as preventing both) form the one end of the area—an area that we would generally—dependent upon internal definition (what is "disease," what "amelioration," what "suffering," and what "preventing?")—accept as legitimate goals. The other side (those goals we would reject out of hand as not being legitimate) would be such goals as helping the state torture or interrogate patients, performing procedures merely to enrich oneself, or addicting patients to drugs we can then sell to them. These are not proper goals, and their not being proper goals would need little further discussion. Such a way of proceeding may be of help by delineating the two ends of the field in which the further discussion takes place, but such a way of proceeding can only begin to initiate the discussion. Is, for example, prolonging the act of dying on one side or the other? And to which side is it closest?

If we can agree on anything concerning the goals of medicine, it is that the health professions are among the "helping professions:" that is, that they are constituted and socially recognized precisely because their function is, in some sense, seen as using their special expertise to help their clients. Such "help" (the obverse of harm) was and remains (just as does "harm") ill defined. Certainly physicians and other health professionals can be expected to help (or to do no harm) only within certain spheres of competence. Helping their patients win a case in court, invest their money, or find a job is not part of the kind of helping health professionals, at least today, are expected to do. Doing no harm to their patients when patients are competing with them for electoral office or when they are engaged in a boxing contest with their patient is not a thing from which health professionals *qua* health professionals are expected to refrain. This is the case because physicians and others in the health professions do not pretend ("profess") special competence in these areas, do not use the tools of medicine or the information they have derived in the course of doing things "medical" to accomplish their task, and because when patients seek out their physician they do not ordinarily have expectations that encompass such activities. The skills a profession or occupation claims to have, as well as the expectations society has for the function of a given profession or occupation, do not define but help shape the obligations professions in fact have. The areas in which physicians are to "help" (or the areas in which physicians are expected to do no "harm") are largely socially defined. And these areas at any one time constitute the proper goals of medicine.

In some sense, "helping" means assisting others to attain their goals. Such others need not be individuals: The term "helping nature along" implies that it is nature that is being helped to attain its telos or that which we hold to be its *telos*. Helping, as usually defined in the medical setting, means assisting patients to attain their particular goals within the framework of the skills taught healthcare practitioners. In selecting these goals as "proper," both societal and personal values come into play. In general, personal values are narrower than are those of the community at large. In speaking of "community" here, we really have two interactive communities in mind: the medical community and the larger community in which the medical community is embedded.

It is from an interplay between these two communities that the legitimate goals and values of medicine properly emerge. A given healthcare practitioner may find that doing, say, abortions is something they cannot personally be involved with. Since, however, communal values are broader, healthcare practitioners cannot advance the claim that doing abortions is not one of the many goals of medicine. (When the physicians personal morality clashes with general ethics this should—where appropriate be discussed with the patient beforehand and proper referral made.)

Likewise, in judging what is and what is not futile, the medical profession must define what is and what is not "futile" within the context of the larger community and its values. Idiosyncratic interpretations then become just that: values that a given practitioner may not be able to act upon but values that still remain "legitimate" in a social sense.

Medicine and the society in which it operates have largely failed to grapple with the problem of futility. Furthermore, as medicine begins to grapple with the concept it has been apt to conceive it narrowly and in a quite individualistic manner. What is considered to be futile seems in general to be centered merely on one patient (or decided upon by one group of healthcare practitioners) at a particular instance in time rather than striving for a wider perspective. Is treatment that was necessary only because the condition for which it is being given was not prevented "futile" in a social sense? We do not wish to suggest that treatment for persons who find themselves ill with an easily preventable condition should not be helped and helped to the fullest. But we do suggest that the concept has dimensions that transcend the purely individualistic one of helping a particular patient at a particular time. It may not be futile to treat Ms. Jones' severe complications of pregnancy that are due to the fact that she was unable to receive proper prenatal care because of social and economic circumstances; it may not be "futile" to treat Mr. Smith's lung cancer (which most probably is related to his smoking two packs of cigarettes per day, a habit at the very least stimulated by and in his social setting). But it may be truly futile to expend tremendous amounts of energy and resources repetitively giving last-ditch treatments for such patients while ignoring and neglecting the very factors that caused them to be in need of such treatment.

Medicine's frequent claim that these issues are separable simply does not ring true: They may be separate issues when it comes to treating Mr. Smith or Ms. Jones, but in being causally connected they are by no means entirely separable. Treating Mr. Smith's lung cancer (or Ms. Jones' eclampsia) needs to be done: By then it is too late to invoke the concept of futility, and it certainly is not futile as far as Ms. Jones and Mr. Smith are concerned. But it is futile to expend enormous resources on treating such patients when nothing to prevent the problem is done (or worse, when the very production of a harmful substance is subsidized by the government!). Treating individuals afflicted with these conditions is not futile, but expending great effort on treating conditions whose causes we ignore may, in the long run, bear some resemblance to such a problem.

In general, we feel uncomfortable about allowing concerns of cost to enter into the discussion when it comes to speaking about "futility." We fear that society may yield to the temptation of labeling as "futile" things that are merely expensive. Even if we were to agree on a definition of futility (say, we were to agree that intervening in intercurrent illnesses or providing artificial nutrition for permanently vegetative or comatose persons is "futile"), allowing costs to enter the equation might be dangerous: Might we not be tempted to "stretch" such a concept by applying the label of "permanently vegetative or comatose" earlier than we otherwise would? Labeling something "futile" that is not futile by social definition merely to save money is a hypocritical way of evading the responsibility of answering the hard social question: When and under what circumstances is treatment that might be effective simply too expensive to apply? None of us like the question and many of us might, at first, be tempted to answer "never." Inevitably, however, and since we live in a real and not in an ideal world, costs will have to enter into our decisions. It is safer to make such decisions explicitly and forthrightly because of costs rather than to use the concept of "futility" as a euphemism for cost-saving. A given procedure, treatment, or diagnostic modality may simply be too expensive to allow its general use: It is not "futile" (it might help someone or might make a diagnosis) but it is, at least today, simply too expensive. Using it would take funds from something else judged more important. Communities through their representatives, with advice from their health professionals and preferably with sufficient dialogue in the populace at large, not only are entitled to make such decisions but inevitably must make them if they are committed to an orderly and just system in which all have equal entitlement.

Once goals are defined by the general and by the medical community, the general community must be aware that a broad definition does imply enormous costs. Should the community decide that keeping permanently unconscious or vegetative persons alive at all costs is a proper goal of medicine, it must be aware that by doing so they must find the additional monies somewhere or other social as well as medical goals cannot be met. It must be ready to define what is truly futile as well as what is simply too expensive and be honest and clear about it. There is another aspect where costs and futility inevitably come together. When no real track record exists but when a given procedure, treatment, or diagnostic modality seems worth exploring, experimentation becomes necessary. One must differentiate carefully between experimental and therapeutic procedures, treatments, or diagnostic modalities.

Such a process (a social agreement that, however, must allow considerable flexibility in individual cases) will be challenged because of its "looseness." Rules, however, are made for general situations: They are not straitjackets but guidelines within which individual problems must be resolved. As a society we must continue to grapple with an ever-evolving definition.

When it comes to what we shall call "futility in a technical sense" (that is, a treatment that has zero chance of prolonging a patient's life), some guidelines may help. In the intensive care setting (and applicable to other settings) there are guidelines called by the acronyms APACHE (for adults) or PRIZM (for children) that have worked out the statistical chances of recovery under a variety of strictly delineated circumstances. Of late the validity of some of these data have been challenged. As long as the rules are not "bent" (i.e., as long as these figures are applied strictly the way they were meant to be), such information is most helpful in dealing with clinical issues at the bedside.[76,77]

What are some practical conclusions? Are healthcare professionals obligated to use treatments judged to be futile? Are they obligated to obtain the permission of patient or family to withhold such therapy? Once a procedure, treatment, or diagnostic modality has been labeled as "futile" (i.e., once it has been decided that using it would not meet what has been socially defined as the ends of medicine), healthcare professionals should feel no compunction in denying its use to their patients.[78] Indeed, if definitions have been socially made, doing so could be argued to violate an implicit contract between healthcare providers and their community. To ask whether a procedure, treatment, or diagnostic modality judged futile by such means may be withheld is arguably ethically inadmissible because it offers an unrealistic choice where, in fact, there is none, gives hope only to have hope dashed, and suggests that the social contract between healthcare givers and their communities is open for individual bargaining. Individual variations will have to be considered: The social agreement between healthcare giver and society should not be so narrowly conceived as not to allow individual variation to accommodate (for a short time) the feelings of patients, relatives, or friends. Keeping a vegetative patient alive for another three days or applying intensive care measures for a short time so that a patient can see a child born, must be accommodated within such a definition.[72–74]

SUMMARY

In summary, there are certain critical considerations this chapter has tried to sketch. Among these are cultural and historical differences in thinking about death and the stages of dying; the difference between life as a biological and life as a cognitive, self-realizing, and self-actualizing state and the ethical implications such considerations may have; various forms of acognitive states; problems of limiting therapy and of artificial feeding; suicide and euthanasia; and problems of futility. Central to all of these issues is the issue of orchestrating death and in so doing communicating with patients and with their families. Healthcare professionals—no matter what their role or area of practice—should seek to have an ongoing dialogue with their patients and with their patients' families. If they do this, if they share their fallibility, their humanity, and their agony with the patient and with their patient's families (instead of acting like remote gods), many of these issues will be much easier to deal with than they often are today.

———— REFERENCES ————

1. Rachels J. *The End of Life*. New York, NY: Oxford University Press, 1986.
2. Kushner T. Having a life versus being alive. *J Med Ethics*. 1984;1:5–8.
3. Ariès P. *L'Homme Devant la Mort*. Paris; Édition du Seuil, 1977.
4. Lattimore R (trans). *The Odyssey of Homer*. New York, NY: Harper & Row, 1967.
5. Carrick P. *Medical Ethics in Antiquity*. Dordrecht, the Netherlands: D. Reidel, 1985.
6. Plato; Rouse WH, trans. In: Warmington EH, Rouse PS, eds. *Great Dialogues of Plato*. New York, NY: American Library, 1956;460–521.
7. Cullman O. Immortality of the soul or resurrection of the dead. In: Stendhal K, ed. *Immortality and Resurrection*. New York, NY: Macmillan, 1965;132–164.
8. Kübler-Ross E. *On Death and Dying*. New York, NY: Macmillan, 1969.
9. Pearlman R, Speer J Jr. Quality of life considerations in geriatric care. *J Am Geriatr Soc*. 1983;31(2):113–130.
10. Pellegrino ED, Thomasma DC. *For the Patient's Good: The Restoration of Beneficence in Health Care*. New York, NY: Oxford University Press, 1988.
11. Blackhall LJ. Must we always use CPR? *N Eng J Med*. 1987;20:1281–1284.
12. Amundsen DW. The physician's obligation to prolong life: a medical duty without classical roots. *Hastings Center Report*. 1978;8(4):23–31.
13. Pariser JJ. Comfort measures only for DNR orders. *Conn Med*. 1982;46(4):195–199.
14. Loewy EH. Patient, family, physician: agreement, disagreement, and resolution. *Family Med*. 1986;18(6):375–378.
15. Meisel A. Legal myths about terminating life support. *Arch Intern Med*. 1991;151:1497–1501.
16. Bedell SE, Delbanco TL, Cook EF, et al. Survival after cardiopulmonary resuscitation in the hospital. *N Engl J Med*. 1983;309:569–576.
17. Husebö S, Klascjik E. *Palliativmedizin: praktische Einführung in Symptomkontrolle, Ethik und Kommunikation*. Berlin, Germany: Springer Verlag, 2001.
18. Stollerman GH. Decisions to leave home. *J Am Geriatr Soc*. 1988;36:375–376.
19. Loewy EH. Decisions not to leave home: and what will the neighbors say? *J Am Geriatr Soc*. 1988;36:1143–1146.

20. Rossman I. The geriatrician and the homebound patient. *J Am Geriatr Soc.* 1988;36:348–354.

21. Stollerman GH. Lovable decisions: re-humanizing dying. *J Am Geriatr Soc.* 1986;34:172.

22. Kübler-Ross, E. *On Death and Dying.* NY: Macmillan Publishing, 1969.

23. National Hospice and Palliative Care Organization (formerly the National Hospice Organization) web site, facts and figures section.

24. Brochure written and made available by the NHO *circa* 1997.

25. Hilfiker D. Allowing the debilitated to die: facing our ethical choices. *N Engl J Med.* 1983;308:716–719.

26. Loewy EH. Treatment decisions in the mentally impaired: limiting but not abandoning treatment. *N Engl J Med.* 1987;317:1465–1469.

27. Daugherty JH, Rawlinson DG, Levy DE, et al. Hypoxic-ischemic brain injury and the vegetative state: clinical and neuropathological correlation. *Neurology* 1981;31:991–997.

28. Plum F, Posner J. *The Diagnosis of Stupor and Coma,* 3rd ed. Philadelphia, Pa: F.A. Davis, 1983.

29. Plum F, Posner JB. Disturbances of consciousness and arousal. In: Wyngaarden JB, Smith JB, eds. *Cecil's Textbook of Medicine,* 20th ed. Philadelphia, Pa: W.B. Saunders, 1993;2061–2076.

30. President's Commission for the Study of Ethical Problems in Medicine and Biomedical and Behavioral Research. *Defining Death: A Report on the Medical, Legal, and Social Issues in the Determination of Death.* Washington, DC: U.S. Government Printing Office, 1982.

31. Jonas H. *Das Prinzip Verantwortung;* and Jonas H. *Technik, Medizin und Ethik.* Frankfurt: Suhrkamp, 1987.

32. Jonas H. Against the stream. In: Jonas H, ed. *Philosophical Essays: From Ancient Creed to Technological Man.* Englewood Cliffs, NJ: Prentice Hall, 1974.

33. Jonas H. The right to die. *Hastings Center Report.* 1978;8(4):31–36.

34. Ramsey P. *The Patient as Person.* New Haven, Conn: Yale University Press, 1970.

35. Capron AM. Ironics and tensions in feeding the dying. *Hastings Center Report.* 1984;14(5):32–35. Also see footnote 38.

36. Meilander G. Removing food and water: against the stream. *Hastings Center Report.* 1984;14(6):11–13.

37. Callahan D. On feeding the dying. *Hastings Center Report.* 1984;14(6):22.

38. Derr PG. Why food and fluids can never be denied. *Hastings Center.* 1984;16(1): 28–30. Also see footnote 36.

39. Lynn J, Childress JF. Must patients always be given food and water? *Hastings Center Report.* 1983;13(5):17–21.

40. Sanders DS, et al. Survival analysis in PEG: a worse outcome in patients with dementia. *Am J Gastroenterology* 2,000; 95(6): 1472–1475.

41. Callahan CM, et al. Outcomes of PEG among adults in a community setting. *J Am Geriat Soc.*

42. Sanders, DS et al. PEG: a prospective audit of the impact of guidelines in two district hospital. *Am J Gastroenterology* 2002, 97(9): 2239 – 2250.

43. Winter SM. Terminal Nutrition. *Am J Med.* 2000, 109(9): 740–741.

44. Wanzer SH, Adelstein SJ, Cranford RE. The physician's responsibility toward hopelessly ill patients. *N Engl J Med.* 1984;310(15):955–959.

45. Towers B. Irreversible coma and withdrawal of life support: is it murder if the IV line is disconnected? *J Med Ethics.* 1982;8(4):203–205.

46. Steinbock B. The removal of Mr. Herbert's feeding tube. *Hastings Center Report.* 1983;13(5):13–16.

47. Green W. Setting boundaries for artificial feeding. *Hastings Center Report.* 1984;14(6):8–10.

48. Annas GJ. Do feeding tubes have more rights than patients? *Hastings Center Report.* 1986;16(1):26–28.

49. Paris JJ. When burdens of feeding outweigh benefits. *Hastings Center Report.* 1986;16(1):30–32.

50. Amundsen DW. History of medical ethics: ancient Near-East. In: Reich WT, ed. *Encyclopedia of Bioethics.* New York, NY: Macmillan, 1978.

51. Battin M. *Ethical Issues in Suicide.* Philadelphia, PA: Prentice Hall, 1994.

52. Klemperer V. *Lingua Tertii Imperii: Notizbuch eines Philologen.* Leipzig, Deutschland, Reclam, 1975.

53. Loewy EH. Healing and killing, harming and not harming: physician participation in euthanasia and capital punishment. *J Clin Ethics.* 1992;3(1):29–34.

54. Pius XII. The prolongation of life. *Pope Speaks.* 1958;4(4):393–398.

55. Trammel R. Saving life and taking life. *J Phil.* 1975;72:131–137.

56. Rachels J. Active and passive euthanasia. *N Engl J Med.* 1975;292(2):78–80.

57. Gaylin W, Kass LR, Pellegrino ED, et al. Doctors must not kill. *JAMA.* 1988;259: 2139–2140.

58. Oregon: www.chd.or.us./chs/pas.htm

59. Steinbrook R. Physician assisted suicide in Oregon—an uncertain future. NEJM 2002; 346(6):460–464.

60. Hedberg K, et al. Legalized physician assisted suicide in Oregon. 2001, 2002;*N Engl J Med.* 346:450–452.

61. Back AL, et al. Physician assisted suicide in Washington State: Patient Requests and Physician Responses. *JAMA* 1996; 275(12):919–923.

62. Loewy EH, Loewy RS. *The Ethics of Terminal Care: Orchestrating the End of Life.* New York, NY: Plenum/Kluwer; 2001.

63. Freedman B. Offering truth. *Arch Intern Med.* 1993;153:572–576.

64. Cassel E. The function of medicine. *Hastings Center Report.* 1977;7(6):16–19.

65. Good MD. The practice of biomedicine and the discourse on hope. *Anthropol Med.* 1991;7:121–135; and Pellegrino ED. Is truth telling to patients a cultural artifact? *JAMA* 1992;268(13):1734–1735.

66. Hardin G. The tragedy of the commons. *Science* 1968;162:1243–1248.

67. Hiatt HHL. Protecting the medical commons: Who is responsible? *N Engl J Med.* 1975;293(5):235–241.

68. Schneiderman LJ, Jecker NS, Jonsen AR. Medical futility: its meaning and ethical implications. *Ann Int Med.* 1990;1112:949–954.

69. Jecker NS. Knowing when to stop: the limits of medicine. *Hastings Center Report*. 1991;19(2):5–10.

70. Schneiderman LJ, Jecker N. Futility in practice. *Arch Int Med*. 1993;153:437–441.

71. Loewy EH, Carlson RA. Futility and its wider implications: a concept in need of further examination. *Arch Intern Med*. 1993;153:429–431.

72. Loewy EH. Futility and the goals of medicine. *Eur Philos Med Health Care*. 1993;1(2):15–29.

73. Nelson LJ, Nelson RM. Ethics and the provision of futile, harmful or burdensome treatment to children. *Crit Care Med*. 1992;20(3):427–433.

74. Loewy EH, Carlson RA. Futility and its wider implications: a concept in need of further examination. *Arch Intern Med*. 1993;153:429–431.

75. Hastings Center. *Guidelines on the Termination of Therapy and the Care of the Dying*. Briarcliff Manor, NY: Hastings Center, 1987.

76. Faber-Langendoen K. Resuscitation of patients with metastatic cancer. *Arch Intern Med*. 1991;151:235–239.

77. Gray WA, Capone RJ, Most AS. Unsuccessful emergency medical resuscitation—are continued efforts in the emergency department justified? *N Engl J Med*. 1991;325:1393–1398

78. Hackler JC, Hiller FC. Family consent to orders not to resuscitate. *JAMA*. 1990; 264(1):1281–1283.

12

Common Problems in Everyday Practice

The ethical problems of "trivial" or of chronic illness that the physician sees every day in his or her office are discussed much less than the more esoteric but much more spectacular and flamboyant ones dealing with euthanasia, abortion or genetic manipulation. In truth, they are hardly trivial and certainly not trivial to the patient afflicted with the problem. Perhaps, in the sense of being a daily occurrence they would better be referred to as "mundane."

And yet these "mundane" problems are the far more frequent and often more vexing ethical problems that practicing physicians face. They range from truth telling on insurance or disability forms, to the unwillingness of a managed care organization to allow a procedure considered necessary by the physician, to dealing with inquiries about patients by persons the physician does not know, or to being frank with a patient who has been grossly mistreated by a colleague and now presents in your office with something which should have been remedied long before. The problems are legion, the contexts in which they appear countless and there are no "good" answers to most of these questions. They are problems that are often—but by no means always—problems inherent in the system, a fact that does not imply that they can be shrugged off with that excuse and without a serious and united attempt to affect systemic changes that spawn such problems. There are several rules of thumb: the first and perhaps most essential is that, according to society's understanding, physicians and other health-care professionals are, above all (unless there are highly unusual circumstances to the contrary), committed to their patient's good. This is not only society's understanding but is also a claim repeatedly made by the profession itself. And that claim has, at least until recently, generally been justified and has formed one of the pillars of the patient-physician relationship.

That health-care professionals bear primary loyalty to their patients is the case for a number of reasons: (1) the societal understanding of the role of physicians from ancient times to the present; (2) the words and actions of physicians who claim to be primarily interested in the patient's good; (3) an inevitable imbalance in power which, under almost all situations, places physicians as the more powerful partner and, therefore, as the partner obliged to use that power with due care. In this sense it is a fiduciary relationship in which I, the patient, knowingly put my trust in the physician.

The long-term relationship which so very often allowed physicians and patients know one another is becoming nearly extinct. One of the assumptions made in much of the ethics literature—that patient and physician have had an enduring relationship over years—is not true today. Having only sporadic interactions between health-care providers and their patients is already reflected in the language. Those seeking help who were once called "patients" are now referred to as "clients," or "consumers" and by MCO's and HMO's as "lives" or "units of care." Correspondingly, physicians, nurses and other allied health care professionals have become "providers." This use of language is not trivial—it conditions our feelings and, therefore, our approach to the matter at hand and in the case of health-care tends to remove or attenuate the warmth, intimacy and trust that a healing relationship entails.

DEALING WITH INSURANCE, MCO'S AND HMO'S

Physicians who are seeking to use an investigative or therapeutic modality which they feel is essential for their patient's good but which do not quite meet the criteria set up by a particular HMO or MCO are often tempted to "bend the truth" so as to get what they feel is necessary for their patient.[1] To put it bluntly: they are put into the position of either lying and then being able to do what they believe is indicated or telling the truth and not living up to doing what, in their professional opinion, would be best for their patient. A neglected colonoscopy may result in the late detection of a cancer that the physician who knows the patient suspects on grounds that do not quite meet the accepted criteria. These criteria are, of course, a gross misuse of the "average" case applied to an individual. Inventing blood in the stool (when none has been present) may get permission for the desired test but puts the physician into the position of lying and undermines the habit of truthfulness. The habit of truthfulness is not peculiar to health-care ethics but rather is a general and generally accepted principle of ethical behaviour—no matter what system of ethics to which one may appeal. Being between a rock and a hard place physicians must choose—and will often choose in the best interests of the patient. This, while not being a praiseworthy act (lying never is), may under such circumstances be less blameworthy than neglecting a relationship of unequal power based on trust. But such practice forms a habit that runs counter to the prior habit of telling—to the best of one's ability—the truth. It is unwise—and in the long run counter-productive—to weaken one of the basic foundations of all ethical behaviour. That physicians increasingly feel obliged to violate the basic ethical principle of truth telling in order to safeguard their patients ought to be a wake-up call for any social system.

When it comes to filling out a disability form and back-dating it so that patients can receive disability benefits for a condition they allege to have had but one which was not seen by the physician, it is another matter. Patients who are ill enough and who know that they will require such illness to be certified are arguably obliged to communicate with their physician and get his or her recommendation to stay at home. Barring such contact a physician cannot be expected to lie. Here the imperative of truth telling is undoubtedly a higher mandate than is that of helping one's patient— the mandate of doing things in the best interests of the remiss patient does not ordinarily excuse a lie.

In order to get life insurance persons often sign a waiver permitting their physician to release information to the insurance company. It is not rare that patients will afterwards beseech their physician to withhold certain information. Physicians are first of all committed to their patient's good—that is generally understood as being committed to their good when it comes to matters directly associated with their sickness or health. They are not committed to their patient's financial or other non-medical good. Omitting or altering information is, in effect, a form of lying. That does not mean that physicians should volunteer that for which they are not asked. But it does mean that direct questions—whether the patient likes it or not—will have to be truthfully answered.

WHEN PATIENTS HAVE BEEN IMPROPERLY TREATED

It is not rare—especially in tertiary care centers—that patients referred there have not been properly investigated or that they are found to have been treated wrongly—usually in a way that does not reflect "state of the art" practice. Such patients raise a disagreeable problem for physicians and the lack of proper investigation or treatment too often goes unmentioned. If medicine considers itself to be a profession—and it most certainly (and generally rightfully) does consider itself to be—it assumes the duty of policing itself which, among other things, implies not sweeping mistakes under the rug. When confronted with serious mistakes made by physicians, we would argue, that the profession has at least two obligations: (1) to inform the patient and (2) to inform one's colleague in an attempt to teach and prevent similar occurrences in the future. In rare and crass circumstances informing the medical supervisory board may likewise be necessary. These obligations rest first of all on the assumed trust within the fiduciary physician–patient relationship and secondly on the obligation physicians have to help educate themselves and their colleagues. Informing the patient is often a very difficult thing to do—physicians, and rightly so, feel a sense of loyalty towards their colleagues and are loathe being a part of the cause that might foment a malpractice suit. On the other hand—and, we fear, equally if not indeed more importantly— physicians and institutions hesitate to anger colleagues who refer patients to them, thereby potentially losing referrals and money. The fear of losing referrals is a very real and a very powerful one—but when all is said and done, the physicians (or institutions) financial good must be weighed against truth telling and, above all, against the good of future patients who are again likely to be the victims of the same error by the same physician. The question is not so much whether to tell but how to tell in such a way that the least possible damage is done.

Consultations could and should be a marvelous means of mutual education. Both the referring physician and the specialist have the opportunity to learn from one another. It is, we would argue, incumbent upon the consultant to call the referring physician and explain where things have gone wrong. This is not a pleasant task but it is a far more pleasant one than allowing misinformation to stand that ought to be corrected. And at times it may well be that it is the consultant who has missed a critical aspect of the case. Conversing with one another as colleagues and keeping in mind that consultations and referrals have mutual education as one of their primary goals can make such exchanges fruitful and not necessarily unpleasant. Indeed, if handled right, such conversations may lead to a relationship far more fruitful than the sterile interchange of (usually stereotypic) formal letters.

COMMUNICATION AMONG REFERRING PHYSICIANS, CONSULTANTS AND PATIENTS

In medicine today no single physician can know everything, albeit one ought to hope that at least all share a basic understanding of patho-physiology. Good ethics relies on good facts—and the good specialist must understand more than merely the function of a particular organ or organ system to effectively treat the patient in whom that organ system is located. In other words—referrals often have to be made and the proper evaluation by the specialist requires good communication. Referring physicians must —if they are to practice effectively and, therefore, ethically—speak extensively with the referring physician and learn not only the "medical" but in many cases the psychosocial "facts" as well. It is astonishing—and in our view frightening—how rarely this is the case; it is even more frightening—and in our experience not uncommon— how rarely the primary care physician really "knows" the patient beyond his or her disease history. Good communication—among physicians as well as between physicians and patients—is one of the essential ingredients of good praxis. Yet time and again the authors have observed that physicians presenting ethically problematic cases often know little about the patient's past history, have not troubled to look at readily available past records and, far more often than not, have failed to communicate with the patient's primary care or referring physician.

Consultations at the beginning of the last century still implied a face-to-face discussion between referring and consulting physician. Today this has become a rarity. In the hospital the consultant leaves a note (often cryptic!) and possibly a few orders that are indicated from his or her point of view. This problem has magnified greatly. Most seriously ill patients are seen by a variety of consultants who communicate only in writing. In ambulatory practice this is likewise the case: a short consultation note is sent to the referring physician. In most instances it is desirable that referring and consulting physicians discuss the case with one another. This is true not only of ethical decisions but of "medical" decisions as well. One consultant's recommendation may be inappropriate in the context of a particular case. Those who argue that such "consultation" is too time consuming in today's world are really arguing that practicing good medicine has become too time consuming! Undoubtedly much of this lack of communication is one produced by the "system." But systems can be changed and one of the ethical obligations of health-care providers is to see that the system within which they work allows sufficient space for practicing good and ethically proper medicine. In today's world most of us are connected by e-mail conversations. Even if hardly as satisfactory as face to face conversation, it may be one way of re-introducing true "consultations" once again.

One should not forget that in a medical interaction it is not only the physician or other health care provider who examines the patient but that, equally important, the patient examines the physician. Out of such communication an atmosphere of trust develops or fails to develop—a trust that is essential if cooperation is expected and healing is to be effected. It is difficult to develop a relationship with a person one sees rarely and who, instead of looking at the patient, looks at a computer screen. Communication between patient and physician is a continuing process that ought not to be relegated to an auxiliary member or—worse yet—done by means of a checklist. What patients do not say—and how they say what they do say—is at least as important as what is being said. In today's world such conversations often do not take place in a consultation room in which the ambience invites the sharing of information or, worse yet, it occurs in a small examining room in which the physician sits in front of a computer, types in the information the patient has given and rarely, if ever, truly communicates with the patient. It is no wonder that a growing number of patients seek out practitioners of alternative medicine who at least listen to their patients as one human being listens to another.

In most countries of the world the physician who treats the patient in an outpatient setting is not the same physician who admits him or her to the hospital. Both of these systems have evident advantages and disadvantages. The main disadvantage to the system in which the same physician treats in and outside the hospital is that many community hospitals are without well-trained physicians much of the 24-hour period. In such a setting and without a team approach mistakes are more apt to occur. On the other hand, when the physician treating a patient in the hospital is different from the physician who may have known the patient for years, the importance of good and extensive communication is evident and a new patient–physician relationship and trust has to be established. The reality is that the ideal picture of a physician having known a patient for years is vanishing, thanks to our system—in which the insured has little choice but take what his employer has to offer. So called "hospitalists" are becoming more and more frequent and it may well be that this represents a trend for the future. What we want to emphasize here is that, once again, health-care ethics and its problems are, to a great degree, system-dependent.[2]

Medicine is a task in which continued learning and "keeping up" is vital. That, of course, can be said of virtually all occupations. No physician who fails to "keep up" in his or her field can be said to be practicing ethical medicine. Consultations, therefore, do not merely have the purpose of helping the attending physician deal with a thorny problem—at least as importantly, they are vehicles of education in which the attending learns from the consultant and, as often as not, the consultant learns from the attending. The attending physician asking a colleague for a consultation should, therefore, not only expect a series of recommendations aimed at a particular patient but should rightfully also learn more about the problem itself. In today's medical practice the use of consultants as teachers is all too often neglected or forgotten.

The optimal time, by the way, to start a dialogue about advance directives is when patient, physician and the entire team are getting to know one another. Suggesting that an advance directive be thought about and discussed with family, friends and the health-care providers may save much time and may permit ethically far more informed and appropriate action when a patient becomes critically ill and can suddenly no longer make decisions for him or herself. It is during such conversations that matters like organ donation or autopsy can also be discussed, thought about more dispassionately and eventually executed and documented.

——— FRIENDS AND UNCOOPERATIVE, ——— ABUSIVE OR HOSTILE PATIENTS

In everyday life there are a few people we genuinely like, some few that we genuinely dislike and the mass of people towards whom we generally feel indifferent. In every-day life we can generally simply avoid the latter two categories. In medical practice we have obligations towards patients that transcend the obligations we have towards other casually encountered minimal ethical obligations.[3] Sometimes there is psychopathology on one or the other side.[4,5] To deny that there are patients we dislike is either self-delusion or simply not true. To claim to be friends to and with everyone is not only absurd but devalues the meaning of "friendship."[6] In one sense believing that all people are "our friends" means that we do not truly have any friends. The habit of calling every casual acquaintance by their first name is pathognomonic of this state of affairs. Medical practice is no different: physicians like a few patients very much, cannot (for good reasons or bad) stand a very few others and feel neutral about the vast majority. Patients whom health-care providers like very much have, over time and fueled by shared interests and world-views, become friends. Taking care of members of one's family or one's friends is a matter of temperament. Some physicians may be able to separate these two things (being a friend or family member from being their physician); some may not. If physicians truly cannot separate these two, then referral to a competent physician or other health-care provider may be the wisest choice. It is also not purely the physician's reaction—a true friend may, and for that very reason, choose to have a physician with whom he or she has a less personal relationship.

Physicians dealing with close family or friends professionally have a difficult decision to make: treat such a person as they would any other patient or help select a competent physician and support what he or she is doing. Neither of these two choices is ideal—once again we must choose between unsavory courses of action. That does not mean that health-care providers do not chat a bit longer with their friends or pay more attention to their non-medical wants. But it does mean that when it comes to every-day medical care that may influence the outcome, friend and foe are treated alike.

The vast majority of patients are people towards whom we are essentially indifferent. The often-stated belief that friendship will provide better care is absurd. As has been pointed out it may, in fact, be inferior. If there is one thing handed down through the millennia, it is that physicians are obligated to do their best for each patient. Deviating from that principle would imply that physicians are doing less than the best for the majority of their patients. We do not mean that physicians will as a matter of course chat a bit longer with their friends than they would with the patient with whom they are not true friends. It does, however, denote that they will order the same tests and give the same type of therapy to their friends as they would a stranger. The statement that "I wouldn't want to do a flexible sigmoidoscopy on Charlie because he is my friend" in a case in which the physician would do so on other patients serves as a not infrequent example. The moral tradition of medicine and the expectation of society is that physicians will do what is in the best interest of their patients. Any deviation from "the best" (and that because the patient is a friend) is, by definition, inferior.

Occasionally a particular patient may be one who is disliked (for good, bad or indifferent reasons) by the physician or by other members of the team. In an emergency, liking or disliking a particular patient is beside the point. But in long-term care, physicians are wise to refer such a patient to an equally competent—and willing! — health-care professional. Healing is more than doing the right test, performing the proper procedure or prescribing the proper medication. It involves a human relationship that cannot be reduced to its technical aspects. Physicians, nurses or therapists are, in fact, very important therapeutic "tools." Their very presence and caring can be the most important aspect of a patient's relief or even cure.

Very rarely physicians will encounter patients whom they not only dislike, but whom, for good or bad reasons, they hate. In an emergency there can be no doubt that physicians are obligated to do the best they can; in a non-emergency it is wise to refer such patients to someone equally competent. One of the authors (EHL) was trained as a cardiologist. Soon after establishing his practice he was called to the bedside of a former SS guard at one of the concentration camps in which (incidentally) his relatives had perished. Since his blood group was tattooed, the author knew at once with whom he was dealing and it did not escape the patient's attention who his physician was. Nevertheless, we were stuck with one another—the patient had a leaking abdominal aneurysm and a severe cardiac arrhythmia. The physician monitored the patient throughout surgery and throughout the patient's hospital course. On the second post-hospital visit when things were stabilized and going well the patient was told to seek another and equally competent physician's care.

It is not rare that patients fail to take prescribed medication, follow their diet or fail to cooperate with the physician's recommendations in other ways. This problem often starts linguistically—"the non-compliant patient" is a problem about which much has been written. The language of "compliance" denotes an athmosphere of "obedience" and tends to set the tone. Patient's lack of cooperation—a term that lends itself far more appropriately to such situations than does the term of compliance or lack thereof—needs to be explored in depth. Often patients may be unable on their income to buy their medications or follow their diets. And often they are ashamed to "admit" this. Many patients—and especially elderly patients—are "too proud" to admit to their poverty and would rather forgo necessary treatment than admit to inability to pay. This is something that physicians and other health care professionals must deal with with great tact. Sometimes approaching the problem obliquely by a casual story can help "open up" a patient and reveal the real source of the problem. At times drug companies can be talked into supplying such patients' needs; at other times social workers or associations of elderly or retired people can be of immense help. But simply prescribing and then labeling as "non-compliant" the patient who, forced to choose between eating and buying the medicine, chooses food is, in our opinion, not fully discharging one's professional obligation towards one's patients.

Sometimes patients have failed to understand (or have not been given) an adequate understanding of their illness and the function that medication or diet has. Sometimes (and this is rare but it does happen) patients simply lack either the rudimentary education or basic intelligence to understand. A brief story will illustrate the point. One of us (EHL) had a patient in his late 50's are 60's referred with a rapidly expanding (and perhaps leaking) aneurysm. At the end of the consultation the patient was taken into the consultation room and on hand of a blackboard and coloured chart half an hour was spent trying to help him understand. Analogies like leaking inner tubes, etc. were used and the discourse was frequently interrupted to ask for questions. There were none. At the end of this long explanation the patient was again told that he surely must have some questions. At that point he said "yes, one doc: what's that thing pounding in my belly?" At that point he was admitted, asked to have his family contact us and told that surgery would take place within two hours. This is an example of soft paternalism: trying to prevent imminent harm from befalling a person who simply did not understand.

There are, of course, steps to be taken before a patient is dismissed from one's practice and in a managed care type of setting this may be difficult. It is, we think, well to try to find out what the patient is "saying" by his or her non-cooperation. Understanding "where the patient is coming from" may be a most important tool in handling such problems. Sometimes a blunt question as to "why are you doing this?" is called for. Some patients may—and sometimes for deep psychiatric reasons—be self-destructive. If possible we should try to understand the motives. Understanding does not denote approval nor does it always solve the problem—but it helps. At times non-cooperation denotes a patient's fear of losing control. In the medical setting understanding may help us develop a strategy that allows the patient to feel in control while cooperating with a therapeutic regimen.

Physicians are humans—not automatons. Dealing with uncooperative patients is time-consuming and nerve-racking. It inevitably plays a part in the way the next patient will be approached. Physicians and other members of the health care team must not forget that there is an "opportunity cost" (not in terms of money but in terms of actual patience and understanding in dealing with the next patient)—time spent with recalcitrant patients is time not spent with another patient who may very well be a cooperative one. There comes a time when patients must be made aware of this. Physicians are obligated to conserve their resources and eventually may have to dismiss such patients from their office. Dealing with the uncooperative patient is an unnerving experience and may very well detract not only from the care of such a patient him or herself but may also "set the tone" for an encounter with the next patient upon whom the physicians wrath is visited. Physicians and their coworkers are certainly obliged to do what is reasonably possible—but dealing with the recalcitrant uncooperative patient (who cannot or will not be helped) steals time better expended in taking care of patients only too eager to cooperate. In a sense this is a profoundly ethical problem for it robs patients who could benefit by spending time and more time battling with a patient who has no inclination to cooperate

There is a small group of patients who are incapable of forming a lasting relationship with their physician or anyone else. These are most challenging patients and these are the very ones who may "run out of physicians" willing to accept them as patients and be forced to utilize episodic emergency room care. Nevertheless, physicians and other health-care providers are not obligated (except in an emergency) to provide care.

——— PATIENTS DEMANDING ——— INAPPROPRIATE TREATMENT

Unfortunately it is not rare today that patients present themselves to their physician—not with a set of symptoms and a vague (and sometimes quite correct) idea of what is wrong, but come convinced of a diagnosis and demanding a particular type of treatment. At times they are right; often they are not. It is not rare that—after taking a proper history, doing a thorough physical and ordering appropriate laboratory or other examinations, informing them of the diagnosis and prescribing a treatment different from that which the patient wants—the patient demurs and insists on the medication he wants. The frequent example is the request (nay, it is often a demand) for an antibiotic for a viral infection. A careful explanation very often will not convince the patient. Oftentimes physicians sigh and give in to the patient's demands—this is done to save time as well as to avoid losing patients. The general excuse is that "if I do not do it, someone else will." While that is undoubtedly true and while it may serve as an explanation, it most certainly does not serve as an excuse. It is improper medicine and good ethics starts with good medical practice.

Patients also frequently want to use over-the-counter medications or herbal concoctions about which the physician knows little and, in the case of herbal concoctions, will generally have little luck in finding accurate information. Some of these may, in fact, be harmful. Others may not and may, in fact, be quite useful. What is lacking is a disciplined study of these preparations. Physicians cannot stop patients from taking these along with the prescribed treatment but they are certainly well advised to tell their patients that they cannot be responsible for adverse effects and should record this in the chart.

When physicians are asked to perform procedures (abortions, for example) or prescribe medication (birth control or "morning after pills" for example) which are perfectly legal but which happen to run counter to the physician's personal morality, physicians have an obligation to inform patients of this and to suggest that they consult another equally competent physician. Many physicians will object even to referral claiming (not without justification) that referral helps the patient in what they consider to be an immoral act. As is so often the case in ethics: there is no perfectly good choice. When it comes to holding a patient hostage to the physician's personal morality or referring her to someone else it seems obvious that holding the patient hostage is ethically the far more unacceptable choice. Indeed, when physicians have such views, it is morally incumbent to inform the patient of this at an early visit—or at the earliest opportunity, which ever comes first.

OVERWORKED PHYSICIANS AND CME—AN ETHICAL PROBLEM

Physicians in general work grueling hours. They are socialized from medical school forward not only to expect this but also to feel that they are not doing their duty if they are not in truth overworked. Very few persons would chose to have a pilot who just piloted a plane across the ocean pilot their plane back again. The decisions physicians make after continuously working for 12, 16 and sometimes 36 hours are not necessarily "bad" (although they well may be) but they are not as apt to be as good as they would be were they well rested. This is not only a problem of residency. It is a problem that has afflicted medicine in residency and practice (at least in the west) for well over a century.

Overworking people not only makes them apt to make decisions that are not as good as they might be it also tends to make them less sensitive to the "non-medical," social or ethical problems their patients present. After all, when a person is called out of bed after a half hour's rest when they had previously been 20 hours on their feet, their interest is to take care of the immediate problem and return to bed. Their patience with exploring values, getting to know the sick person as a person and not just as a disease is, at that point, limited. Fatigue does not make people kinder—it makes them, and quite understandably, anxious to get rest. In practice this is often not very different from residency. A physician may be able to sustain working the entire day and then being on call at night but he or she will neither do the best they can nor be encouraged to then spend their spare time reading in the current literature.

It is a commonplace that physicians must keep up with the current literature in their field, attend continuing education sessions and remain "up to date." Peculiarly in many places continuing education is to be done in the physician's vacation or other "off" time. This is unfair—learning new material is at least as taxing as treating patients—and leads to undervaluing the process and to reducing it to something of a sham. Many post-graduate courses take place in the setting of ski trips, cruises and so forth. Two or three hours are spent in actual lectures and presentations and the rest are devoted to whatever a particular resort has to offer. And often the post-graduate sessions are skipped altogether. If we are to have a good health-care system then we must begin to recognize that CME is as necessary and as difficult a task as patient care. We must compensate physicians adequately—just as we do other professionals—for attending proper CME courses and make sure that the courses offer sufficient content and that they are in fact attended.

Good ethical practice starts with competence in one's chosen field. Periodic re-examination of some type would help accomplish such a task. But before we attempt to make re-certification into a requirement for licensure or specialty practice we need to carefully assess the questions or methods used to assess "competence." In some areas of the world most physicians spend some determined time practicing in some other setting under strict observation. Thus the family doctor may have to practice in a University type setting for a few weeks every few years and the Professor of Medicine may be required to work with and under the supervision of a family physician. This may yield a better understanding of each other's problems and create a feeling of collegiality among physicians. Making sure that physicians are and remain competent and that they have an understanding of and for each other's problems is the basis for ethical practice. If the United States ever creates a just health-care system it must also acknowledge the need for CME—and it must do so by reimbursing physicians for post-graduate CME and not, as is now often the case, penalizing them by taking such time out of their vacation time and expecting physicians to do it at their own expense.

————— THE OBLIGATION OF PATIENTS —————

Throughout this book we have, in one-way or another, emphasized the obligations healthcare professionals or society at large owe to their patients. We have emphasized the importance of a team approach and suggested that where possible patients should be in a partnership relation with health-care providers. Much of the literature of health-care ethics has emphasized the same thing. Strangely enough, the obligations that patients owe to the health-care team and others have been severely neglected.

In the little that has been written obligations such as truthfulness, adherence to a prescribed regimen, etc. are emphasized; however, all such obligations are really self-serving since proper healing depends on them. We would argue that since proper healing depends on adherence to these "duties" they are obligations owed to oneself—a peculiar concept (see Chapter 3, on Theory).

Patients, by virtue of their illness, have not been absolved of all obligations but merely of obligations their illness may make difficult or impossible to perform. Fathers, for example, have an obligation to support their offspring—but may not be said to have this obligation when they are truly incapacitated. They certainly are not absolved of the shared obligations of courtesy or considerateness of others which all humans share. Patients, by virtue of being patients and expecting "state of the art" medicine have obligations which, should they be capable, should be clear to themselves as well as to their caregivers. Patients—by virtue of being patients—have their autonomy reduced and at times are loathe to make decisions deferring rather to the wishes of their healthcare providers. They are not—as Eric Cassell wrote many years ago—the same person as when well except for the knapsack of illness carried on their back. But just because their obligations may be attenuated or changed, in no way implies that they are relieved of all obligation—they will be relieved of some but will also acquire a different set befitting their role as patients.[7]

Patients *qua* patients can be argued to have obligations other than self-serving ones. They have certain universally accepted expectations of their healers and they have benefited from the education and experience that their healers have enjoyed. They expect that their children likewise will get competent and ethically appropriate care. Patients play a crucial role in the education of physicians. For a patient to refuse to be "used" by medical colleges to take a history or do a physical examination in the course of their illness denotes that they want the benefits of rigorously trained professional health care workers for themselves and their children but are unwilling to contribute to their learning so as to benefit future others. Likewise in the setting of a teaching hospital patients will be examined and treated by a hierarchy of residents and fellows. Patients have every right to insist that such learning takes place under the strict observation (and, if need be, correction) of those who have become experts in their field. Patients who refuse to give a history, have a physical examination and, perhaps, ultimately have a procedure done by a closely supervised student or resident are either saying that they do not value experts for treatment "down the road" (for themselves and for their children) or they are saying that they expect others to do what they themselves would not permit to be done to their own body. If everyone felt this way, medicine would simply cease to exist.

Being in a teaching institution has enormous benefits for the patient—benefits of which they are often unaware. In a teaching setting numerous people will be examining, treating and above all discussing the patient's "case." When patients are in a non-teaching hospital, they are taken care of by one physician and errors are far more apt to occur. Someone in a University Hospital must make decisions that in a non-teaching hospital are made in *solo* fashion. That someone making decisions in a non-teaching hospital is generally their regular physician who has treated them before. As good a doctor as he or she may be, they tend, like all humans, to believe their own conclusions. In a teaching setting these conclusions are constantly open to re-examination and discussion and, therefore, errors are more often discovered, discussed and corrected and are, therefore, far less apt to re-occur.

Likewise we have learned an incredible amount from autopsies, and having one's body used in this manner is a vital part of learning. The main purpose of an autopsy is to find out not so much what he/she *died of* but what he/she *died with*. It is here that mistakes may be discovered and, hopefully, corrected for the future. Autopsies are an invaluable part of teaching and learning. But they are a vital part of learning only when good communication between the pathologist and the patient's caretakers exists. The autopsy, when properly done and properly understood by the team of physicians who had taken care of the patient, can be a critically important teaching tool. Teaching physicians to do a better job is, ethically speaking, of primary importance.

The percentage of patients who die in American hospitals and are not autopsied has steadily gone up. Part of this is because the attending physician or resident asks at the most inappropriate time—that is, just after the patient has died. Like advanced directives this question ought to have been discussed long before the occasion. Bringing in the idea that the way we suggest treating the patient is a piece of information gleaned from autopsy data opens the door so that patients feel free to discuss the pros and cons with their family, friends and physician. When the person asking for permission is uncomfortable doing so or if they do not "believe" in autopsies for themselves, body language will soon convey this to the person being asked to sign the permit.

——— CHRONIC ILLNESS ———

One of the triumphs of medicine (and we are not in the least being cynical here) is the creation of chronic disease. In times not so long past, patients with diabetes, hypertension, rheumatoid arthritis, severe congestive heart failure, Alzheimer's disease as well as many other disease states would not long survive, but die either of their primary disease or of some inevitable and then untreatable inter-current illness, usually an infection. Premature children such as those routinely treated today simply did not survive and severely deformed children likewise had a bleak future. When an inter-current illness occurred (generally an infection) patients quickly died. The fact that this does not occur today—or occurs with far lesser frequency—has brought about a large number of quandaries many of which are discussed in other portions of this book.

In this brief section we do not wish to re-discuss the problems posed by not starting, stopping, etc. treatment in inter-current illness in patients whose underlying disease already is only questionably tolerable (to themselves) and who do not have decisional capacity. Rather we want to look briefly at a difference in the physician–patient relationship as it occurs in the course of treating chronic illness over years. It is here that a partnership type of model is the most appropriate. Of course, it is never a true partnership because a true partnership presupposes equal power. Inevitably—and no matter how much the patient reads, informs him or herself on the internet or otherwise comes to know about their illness—they can, in certain ways, never be as informed about the biomedical facts of their disease and what that implies for them as is their physician.

This, of course, is an epistemological question: "what does it mean to know?" Knowing, we suggest, has several components among which "facts" are only one. True knowing has an experiential component as well as a component that we would call "integrative." Persons may be able to describe coronary disease quite accurately but nevertheless be unable to deal with it adequately because they lack experience in doing so and because they lack the knowledge about other organ systems and diseases which must be factoured in when developing a treatment plan. Furthermore, even physician patients who are specialists in their own diseases are ill equipped to treat their own disease: one cannot be objective about an intrinsically subjective matter.

Nevertheless, in chronic disease a well-informed and intelligent patient who is well aware of his own lack of experience and understanding of other organ systems or one who has even more understanding by virtue of herself being in that field can do much to enrich the partnership and deal effectively with that disease. Physicians are well advised to make this partnership model explicitly clear to the patient. Too often today speaking to the patient is much underrated as a therapeutic modality. Despite many of our technical paraphernalia, the relationship between the healer and the patient contributes much to healing and, when "cure" is not possible (as it is not in chronic disease), at least as much to well being. It is, therefore, essential that physician and patient have sufficient time not only to speak with one another about immediate treatment but, likewise, to spend a small amount of time to chat with one another as humans. Our current system of medicine (see sections on managed care as well as Chapter 8) makes this nearly impossible. If physicians are truly interested in practicing ethical medicine—for treating chronic illness ethically requires, above all, time—they are obligated not simply to throw up their hands and blame the system but to unite and do all within the power to change the system within which they perforce must operate.

Patients—especially those with chronic illness—will often be tempted to go to alternative practitioners or try alternative medicine. Insofar as their use of some herbal remedy or incantation is not harmful to them, there is no reason why physicians should not "go along" with this peculiarity as they would with many others. And sometimes things that we do not understand—acupuncture is an example—may, when delivered by persons trained in the procedure, be of benefit. When, however, patients wish either to substitute such medications or procedures for those that are prescribed or to take some medication that is clearly not medically indicated, a physician practicing ethically cannot approve this. When patients demand such treatment the partnership has failed—probably somewhere long before such a demand has even been made. Patients often seek alternative care because the alternative practitioner does one thing that physicians (for whatever reason) often fail to do—talk to the patient as a human being and not treat them merely as the host of a challenging disease.

It is sometimes useful in groups whose culture is totally different to involve the "healer" and to accept him or her as colleague. One case will illustrate this—a child with a Wilm's tumor (whose prognosis while hardly wonderful is far less bleak than a couple of decades ago) should have been operated. The parents refused on cultural and religious grounds. It was suggested that the healer (or shaman) come to a conference where the case was discussed. He received a white coat and was treated like a colleague. X-rays, lab work and the results of other studies were discussed and made clear. At the end of that discussion the Shaman turned to the parents and spontaneously said "they can do better than I; go ahead." The secret here (and we are not implying that it will always work!) is that the Shaman was treated respectfully and as a colleague who, after all, had the same goals as the medical team—the survival of the child. He was not treated condescendingly and the discussion was not allowed to turn into a power play of battling egos.

Managing chronic illness is part and parcel of orchestrating life. It requires more than merely biomedical interventions. People with rheumatoid arthritis, for example, can often be greatly helped by architectural changes (see Chapter 11), diabetic patients may need not only the help of a dietitian sensitive to ethnic habits and preferences but often (especially in their 'teens) the help of psychologists familiar with treating patients of this sort. It must be, in other words and as is so commonly the case in medicine today, a multidisciplinary effort.[8]

Depression—overt or hidden—is a common symptom when dealing with chronic disease. Often it is a perfectly "normal" response to "bad news;" sometimes it is endogenous and can complicate the diagnosis and hinder recovery. Referral to a psychologist or psychiatrist may be of great benefit. Patients—because of an enduring hostility towards and skepticism of psychiatry—may feel abandoned by such a referral. Here, once again, the language is critical. When the physician first speaks of psychiatry, many patients are convinced that their physician believes that "it" is "all in their head." It may be more fruitful to open the conversation by saying "I have a problem fully understanding your case and this referral will help us sort out whether any psychological problems in your response to illness may be adversely affecting therapy."

———— MANAGING CHRONIC PAIN ————

The most frequent reason why patients (other than routine preventive care, eye-glasses and so forth) come to see their physician is "pain." It may be slight; it may be described as burning, as with peptic ulcer disease, but ultimately it comes down to discomfort, dis-ease, etc. And often it is outright pain. Addressing the underlying disease can often and most successfully relieve such pain—the pain of peptic ulcer disease, hernia or bursitis can be well addressed by dealing with the underlying disease in medically obvious ways. While the disease is being dealt with, however, the fact that the patient's complaint was pain must not be forgotten: if relief by taking care of the specific problem can be expected to take more than a short time and if the pain is truly troublesome, then treating the complaint of pain while treating the disease should not be forgotten. That, for medical reasons, may not always be possible; but when it is, even for a short time, it demonstrates to the patient that the physician is at least as interested in the patient who has the disease as he or she is in the disease itself.

The patient may enter the hospital because of intolerable pain from bursitis and during admission a highly interesting (and possible far more important) abdominal mass may be found. Investigating such a mass must be done if physicians are to do their job well. But there is no reason to ignore the patient's chief complaint while investigating the far more interesting and probably far more important nature of the abdominal mass. Indeed, we have seen patients be quite uncooperative (to the point of signing out against advice) because their pain was essentially ignored.

Many of the chronic problems that the physician sees in his or her office are, however, pain from an underlying disease for which no specific curative therapy is possible. Rheumatoid or very severe osteoarthritis, chronic severe back problems not amenable to surgery—there are many such problems. Often physicians can deal with the pain of these conditions with agents that are not narcotic, with physiotherapy, or by helping the patient alter their lifestyle. Simple maneuvers or at times NSAID's may bring the patient sufficient relief and allow the patient's daily activities to continue. But there are a not inconsiderable number of patients in whom this is cannot be done and in whom pain is so severe that function is grossly diminished. Often it is a pain no less severe than that of malignant metastases and yet physicians are often afraid to deal with this type of pain the same as they would with malignant disease—that is, by giving an around the clock dose of narcotic agents adequate to treat the pain but not sufficient to interfere with the patient's daily life. And yet there is ample proof that such treatment improves their efficiency and allows them to enjoy a tolerable quality of life. Adjusting the medication is an art that, in the usual case, is not a difficult one to learn and in the more complicated case sometimes deserves referral to the pain clinic. But a large number of physicians (and this despite the fact that it is well known in the literature) will not prescribe "opiates" which in many cases would be a most helpful thing to do. Physicians' opiophobia can have a number of causes: a fundamental misunderstanding of narcotics, pain and addiction or the fact that government agencies have made the prescription of adequate narcotics for legitimate purposes at times more difficult than obtaining "street drugs."

An experiment done within the last few years ought to be a lesson: A group of persons had their mental capacities and response to threatening situations while driving rigorously tested. They were then divided into four groups: "normal," "nurses coming off an eight hour shift," "patients with well managed pain from malignancies" and "patients whose pain was poorly managed." It is not surprising that those in the "normal" group did best. It is startling that those next best—and rather close— were patients on regular doses of opioids with well-controlled pain and that, those with pain that was only sporadically and poorly managed were considerably worse. It is, however, startling that nurses coming off a long shift were the worst off.[9] If physicians fear that putting patients suffering such pain on adequate round the clock doses of medication will lead to poor functioning they are empirically wrong. If they fear habituation by the patient they may be right in a rigorous sense: yes, as long as the cause of the pain is not alleviated it will interfere with function. Giving them an agent— no matter what called—which alleviates the pain and allows improved function is, in fact, good medical practice.

We are, then, dealing with a multifaceted set of problems—all of them having an ethical component in that inadequate action by a physician or inappropriate laws by a system prevents the proper care of a suffering patient. And it is a problem that, in terms of human suffering as well as a significant decrease in human productivity, is not small.

Pain control, since it is such an important part of medical practice, should not be left to being accidentally learned while treating patients. It should be a disciplined process taught from its basics in medical colleges to its specific application in diseases physicians deal with in their specialty education. When physicians have a basic misunderstanding of narcotic agents, proper use can be learned as can any other medical skill. The fact that effective pain management is simply not taught in a systematic fashion but is left to be "picked up" along the way is indeed an ethical problem and one that medical boards as well as medical educators have neglected. Medical boards, one would think, would consider improper pain management as much a problem of malpractice as they would improper management of pneumonia and would see that remedial educative action (rather than merely, often ineffective punitive action) is taken. Inadequate management of patients' pain should be regarded no differently than is inadequate management of hypertension or pneumonia. At the same time, medical bodies (state societies, specialty societies, etc.) ought, if they are serious about their obligations to the public, bring pressure on the legislature to pass and enforce laws aimed at preventing, as far as that is possible, the illicit use of narcotics while facilitating the proper use of narcotics.

PUBLIC HEALTH

Physicians who choose to enter Public Health may feel themselves in a quandary: on the one side, the benefit or harm done to individual patients, and on the other side the obligation of safeguarding the health of many both locally, nationally and internationally. Part of their obligation is to help make rules or laws that, at times, may conflict with the good of an individual patient. When making such rules one of the essential components is to allow enough "elbow room" to the practitioner within these rules to exercise good medical judgement as well as to provide ready access (for particular reasons in a unique case) to a process that permits setting aside such rules in that instance. Both carry with them the danger of nepotism, which could be partially corrected by mandating an "after the fact" review. While this sounds clumsy, such instances should, in a well-constructed system, be rare.

Public health workers owe their primary obligation to the community at large rather than to the single patient within it. This is akin to the experimenter whose primary obligation is not the individual patient but the experiment and, therefore, the gathering of new "truths." Likewise it resembles the conflict of interest in organ transplantation that has been ameliorated (certainly not "solved") by having two different teams (those who care for the patient and those who will harvest his/her organs) involved.

Ethically speaking the progress made through experimentation is an imperative. We as physicians clearly have the obligation to practice the best medicine possible today and to do all we can to improve it tomorrow. This is the case whether we examine the issue from a deontological, utilitarian, Deweyan or any other perspective. To stop experimentation and innovation is, in a sense, Hubris: it states that our current practice is as perfect as it might be when, in fact, our ability to help the individual patient continues to be limited. Public health workers, furthermore, are confronted with the nature of their constituency.

Obviously local health officials owe their primary obligation to whatever locality they are in and national public health workers (for example the CDC) to the nation. Ultimately, however, their constituency is the population of the world and not merely their locality or their nation. In part this obligation is grounded by the fact of our long exploitation of now developing nations that has left them with deplorable conditions that now threaten all of us—the AIDS epidemic which has devastated parts of Africa cannot be confined to arbitrary borders and other transmittable diseases cannot be stopped by forbidding tourists to bring back food, etc. The fact, for example, that malaria continues to be rampant in some areas of the world because we have failed miserably in draining swamps where we could easily have done so threatens all of us.

—— IMMUNIZATIONS ——

Occasionally parents refuse to have their child immunized. Vigorous immunization laws have essentially wiped out certain diseases. Of late some patients (and in growing numbers) refuse immunizations for their children. This is a classical "free rider" situation. The reason why certain infectious diseases have virtually vanished is due to an immunization program and the fact that the un-immunized today also are unlikely to contract such a disease is due to the responsible behavior of their neighbours. To effectively deal with certain diseases (diphtheria, poliomyelitis, tetanus, etc.) requires a pool of immunized persons, which prevents the disease from establishing a foothold.

The "free rider" phenomenon is ethically problematic for it is predicated on most people "paying the fare." Thus not having your child immunized endangers the child but when the "free rider" phenomenon begins to proliferate it eventually endangers everyone. Autonomy can be carried too far. Certain immunization laws will interfere with the parent's religious or other beliefs. Allowing them to escape a rule that applies to everyone else violates a sense of community. There is, as is so often the case, no truly good answer. Immunizing a child against the parents' will (not allowing him into play-school or school without proof of immunization for example) undoubtedly decreases the parents' "right" to do with their children as they deem fit. Not immunizing because of a parent's idiosyncratic objection, however, sets them apart from the community from which they would (were their child stricken) justifiably expect help.[10,11]

—— "UNNECESSARY" TESTING ——

Many of us were still brought up in an era in which lab tests and X-rays were used indiscriminately and in many instances they still are today. In residency one's patient was essentially "required" to die in electrolyte balance! This fact is an excellent example as to how, before proceeding to rationing, we need to rationalize. When one of these authors trained we were compelled to predict the outcome of tests ordered—to commit ourselves in writing. For if we were wrong this could mean but two things (1) the laboratory had made a mistake (hardly unheard of!) or (2) our evaluation of the case was faulty (also by no means rare!).

In today's arena much of the overuse of technology is due to the time constraints put upon physicians by various forms of managed care. When pressed for time it is easier to quickly order an MRI than it is to take a careful history and do a careful physical examination. The irony is, however, that reimbursing physicians adequately for the time spent with patients ultimately may also be less expensive! Good, ethical medicine above all requires sufficient time to listen, examine and think—not to "plug" the patient into some gadget that we hope will do it for us. If we really have no notion of what we should be asking, we are apt to receive either no answer or one that is misleading. If good ethics starts with good facts—as we insist that it does—then we need to provide physicians with sufficient time to use the most important tools in medicine: listening, examining carefully, reviewing the literature, and thinking.

There is another side to this question. Many advances have been made and much has been learned because curiosity and imagination prompted physicians to order a test or do an examination which standard care does not call for. We do not suggest that elaborate testing should go on without proper indications or where it poses even a slight risk to the patient. But having drawn a tube of blood, getting yet another test not called for by "guide-lines" but prompted by the curiosity of the physician is in the long run beneficial—if only because it keeps curiosity alive!

At this writing and while tests (especially in hospitalized patients) may still be over-utilized, the problem under managed care has been the difficulty that physicians encounter in ordering appropriate tests. Many an examination that should have been done is left undone because "the hassle factor" has caused both patient and physician to forego an examination or not to have a consultation when they ought. This "hassle factor" may well save money in the short term, but has ultimately cost lives.

——— GUIDE LINES VS STRAIGHT JACKETS ———

The use of "guide lines" and of "evidence based medicine" has become very popular in the last decade.[12,13] Unfortunately—if we want to be truthful—both have been largely used as a cost control mechanism rather than as being aimed at benefiting the patient. Guidelines all too quickly turn into straight jackets with the physician apt to be "called on the carpet" if he/she deviates. "Guide-lines" are applicable to a standard patient with a standard disease. A good physician does not treat a standard patient but treats a specific individual with a specific short and long term history and with a particular social or employment background which puts them at greater or lesser risk of particular diseases.

Guidelines can serve well as "check-lists" to make sure that nothing has been forgotten. They tend to be minimalist and to reduce treatment and investigation as much as possible. In being simply a set of standard procedures they are apt to prevent physicians from individualizing care. They are used more and more frequently in investigating a patient's complaints. When a patient comes with a headache or some other complaint he or she is "plugged into a flow chart" or treated as a "case of" X.

The advantage of guidelines is that they may serve as check-sheets to make sure what is minimally necessary has not been forgotten. The danger of guidelines is that (1) they can serve as a substitute for thinking through a problem and arriving at a proper differential diagnosis and a reasonable and well thought out therapeutic approach and (2) they quickly become not the minimal, but the maximal standard of care. "One size fits all" generally works poorly when it comes to clothing, and it fares no better when it comes to taking care of patients.[12]

DRUG COMPANIES AND "FREE" GIVEAWAYS

Physicians as well as other health-care professionals need ongoing education. Much learning takes place informally over coffee-cups. Much takes place in a setting of postgraduate courses. Under our present non-system physicians are supposed to do this unpaid and on their vacation time. This poses a host of ethical problems—not the least of it being that pharmaceutical companies are apt to send physicians to such courses that then turn into 20% learning and 80% leisure activities, etc. Information at such gatherings is not false—but it is, of course, slanted so that a higher sale of whatever product is being sold can be anticipated.

Drug companies do a lot of subtle brainwashing. They also do a bit more than this. The attempt to suppress research findings not favourable to the company having funded such a project is only one of the ways of controlling the ultimate prescription habits of physicians. Physicians may be influenced by being given cheap ballpoint pens or bookmarkers or allowing drug companies to supply lunch at resident meetings. We truly do not perceive this as a major problem—albeit it may be the "camel's nose under the tent" which subtly inclines people towards one rather than another product. As such, one ought to think about allowing or disallowing this practice.

When it comes to funding huge research projects or building research buildings the possibilities of seriously affecting the nature and quality of research and, therefore, ultimately of practice becomes far more serious. And yet, research has to be funded and the funds provided by the government are now minimal in comparison to those drug companies have to offer.

In the last few years drug companies have spent enormous sums of money advertising prescription drugs in lay magazines and suggesting to patients that they bring pressure on their physicians to prescribe such drugs. The monies spent in such advertising are tax write-offs and are, eventually, born by the patient. Physicians, already harassed by time constraints and rules and regulations of their organization, are often inclined to give in rather than to argue with their patients. This, however, is poor medicine. It is again in part the physician's doing but in a larger sense is the doing of a system that restricts the physician's elbowroom of time and possibility beyond reason.

The cost of drugs today—which have escalated beyond reason and which differ drastically from country to country for the same product—are fobbed off by the standard excuse that enormous amounts of money are spent on developing new drugs and on research. Before one can agree or disagree one needs to have a clear definition of what research is and what it is meant to bring about. A slight change in molecular structure of a tranquilizer or loop diuretic so as to get by the patent laws may be research in a limited sense—but it is research conducted mainly to profit the company and not to benefit the patient. Much of what is called "research" has that aim. Drug companies—like all companies—legitimately wish to make a profit and to make as much of a profit as possible. And yet drugs (and medical equipment) are essential to good patient care, which, in turn, is of vital interest to the community and its members. It may well be that the community ought to consider strict communal control of drug companies and regulation of prices.

——— CURIOSITY AND IMAGINATION ———

The beginning of almost all thought is curiosity stimulated by an external or internal stimulus (*e.g.*, a thought, a memory). Scientific as well as other progress is initiated by curiosity and approached through imagination. Here, we think, is the distinctive difference between human and other higher animals. It seems—albeit we can certainly not be sure—that humans wonder not only where a specific light is coming from but, further, also about the nature of light itself. There is no evidence that other higher animals do this—albeit there is no real evidence that they don't. Likewise higher animals other than humans—as far as we know—have no idea of the history of their species nor seem to show any curiosity about this.

All thinking begins with some perception—be it perception of an external event, a memory of a prior occurrence or an imagined event or object. For example, we notice an object and curiosity brings us to look at it more closely and then to use our imagination to present a variety of possibilities of what the object or impression might be. Our ability to think logically and in categories narrows the options until we finally do recognize the perception for what it is. In human thinking perception, curiosity, imagination, reason and emotion interconnect in forming a judgment.

Medicine and health-care ethics would be lost without curiosity and imagination. A differential diagnosis requires curiosity and enlists imagination as it brings what we have learned into sharper focus. In health-care ethics our curiosity (or, if you will, sensitivity) recognizes that a problem exists and our imagination begins to present us with a variety of options as well as the pros and cons of each. It is a sorting and sifting, and in that process (which includes searching the literature and conversing with others about the problem) not only can we handle the problem more readily, but we can also continue to grapple with it to reach a better solution in a similar case the next time around. As we begin to use our reason tentatively to accept or reject what our imagination has offered, our emotions, feelings, past experiences and what we have learned in the past help us formulate a judgment about a particular situation. We may at some point act on such a judgment fully realizing that such acting is just another step in dealing with a problem which has no eternally true answer chiseled in stone.

Our current emphasis on taking care of merely the immediate problem and not "wasting time" by thinking beyond its immediate solution of necessity dulls the physician's mind and is ultimately of great harm to the public at large. Not only does it tend to stifle good patient care for individual patients; it tends to discourage thinking of new ways of dealing with old problems. Physicians ought to be given the time to research the problems they encounter, to look back and see how former civilizations and cultures dealt with such problems and emerge as better physicians, better able to understand—and therefore to cope—with similar problems in the future.

Our current system discourages physicians from allowing their curiosity and imagination to develop. In such a system curiosity and imagination should be tightly reigned in: If the answer to a question does not directly affect dealing with the patient being taken care of at the time, it is not relevant and there is no reason to ask the question. However, if the system (managed care is a perfect example) succeeds in creating the habit of thinking only within a very narrow framework and seeks to keep the physician's nose to the grindstone of immediate practice, it will eventually have done medical progress a great deal of harm. Many—if not most—of the greatest advances in physiology, medicine and biochemistry were made because of a physician's curiosity.[14]

What divides us from other higher animals, we have suggested, is our ability to have a history and to understand it and our ability to wonder not only about a thing, but also about the nature of things. To suppress what is truly human about the profession in order to reduce costs (and increase profits for those outside the patient–physician relationship) seems ethically problematic and in the long run pragmatically counterproductive.

SUMMARY

The ethical problems in every-day practice are rarely discussed in the literature. In part this is because they are considered trivial, in part because they lack the "prurient" appeal of more flamboyant problems and in part because their solutions are at best still unsatisfactory. Many are system related problems in which the fact that physicians should do all they can to change the system is uniquely unhelpful in dealing with the immediate problem. And yet changing the system, keeping involved in bringing about changes, refusing to accept today's shameful lack of care for millions and refusing to allow medicine to be practiced by non-medical people (as insurers do when they "permit" or "disallow" certain tests or treatments for certain patients) is ultimately the only long term and never finished answer for many of these problems. Physicians and other health care workers need to remain involved in fashioning and maintaining a system in which sufficient elbowroom to practice ethical medicine is given. Remaining involved may range from being advisers for the community to acts of civil disobedience when no other course remains open. There is one thing physicians and health care workers cannot, in today's society, countenance: not doing the best possible for a patient in order to increase the profit of managed care organizations.

The most that one can say about any of these is (1) that health-care workers need to think about and then have the courage to apply their priorities to concrete situations, (2) that being the patient's advocate may, at times, entail unpleasantness and (3) that—trite as it may sound—only pushing to create a system which makes proper ethical practice possible can ultimately achieve what all health-care professionals seek: sufficient time and space to deal appropriately with the many daily ethical problems they must face.

REFERENCES

1.   Morreim EH. *Balancing Act: The New Medical Ethics of Medicine's New Economics.* Washington, DC: Georgetown University Press, 1995.
2.   Loewy EH. Justice and health care systems: what would an ideal health care system look like? *Health Care Analysis* 1998;6(3):185–192.
3.   Loewy EH. Physicians, friendship and moral strangers: an examination of a relationship. *Cambridge Q Healthcare Ethics.* 1994;3(1):52–59.
4.   Schafer S, Nowlts T. Personality disorders among difficult patients. *Arch. Fam. Med.* 1998;7:126–129.
5.   Christie RJ, Hoffmeister CB. *Ethical Issues in Family Medicine.* New York, NY: Oxford University Press, 1986.
6.   Loewy EH. Physicians, friendship and moral strangers: an examination of a relationship. *Cambridge Q Healthcare Ethics.* 1994;3(1):52–59.
7.   Loewy EH. Ethical and communal issues in AIDS: An introduction. *Theoretical Medicine.* Sep;11(3): 173–183.
8.   Loewy EH, Loewy RS. *Ethics of Terminal Care: Orchestrating the End of Life* New York, NY: Plenum/Kluwer, 2001.
9.   Strumpf M. et al. Sicherheitsrelevante Aspekte bei Patienten unter chronischer Opioidtherapie. *Anästhesiologie und Intensivmedizin.* 1999; 5: 370.
10.   Froome J, Bedcock K. Should vaccination be compulsory for all preschool children. *Nursing Times* 2003;98(12):26.
11.   Leacock J, Chapman S. The cold hard facts: immunization and vaccination. Australian Press *Soc Science and Medicine* 2002;34(3):445–457.
12.   Loewy EH. Guidelines, managed care, and ethics. *Archives of Internal Medicine.* Oct 14; 156(18) 2038–2040.
13.   Fruti P. Evidence based medicine between explicit rationing, medical deontology and rights of patients. Forum (Genova). 1998;8(4):383–394.
14.   Loewy EH. Curiosity, imagination, compassion, science and ethics: do curiosity and imagination serve a central function? *Health Care Analysis.* Dec;6(4):286–294.

13

Resolving Ethical Problems: *An Introduction to Individual Cases*

— REVIEWING THE FORMAL STRUCTURE — OF THE AUTHORS' APPROACH TO ETHICAL DECISION-MAKING
The Model

In the past we have dealt with a general "travel agent" model that analogized the process of problem solving to the process of asking and answering three questions of seminal importance:

1. "Where have we come from?"
2. "Where are we going?"
3. "How can we get there?"

To reiterate and expand upon our earlier discussions, the first question is concerned with what we cannot change about the case—its history (which also includes the patient's diagnosis). This first question must be answered before moving to the second, which concerns the patient's bio/psycho/social good—part of which (prognosis) often cannot be modified and part of which can be, given a cooperative *milieu* between patient and care-givers. The third question usually has a choice of alternatives—even though sometimes limited or less than desirable—and it can be addressed only when all resources in answering the first two questions have, at least for the present, been exhausted. Because we can't often have the certainty we would like, we will sometimes have precious little to work with—so it is essential we get as much information from the most appropriate and reliable resources as we can.

A Homeostatic Approach,
A Bio/psycho/social Perspective

Though basic, each of these questions helps to identify a very complicated set of concerns having, in turn, at least three distinct parts:

1. The biological aspects
2. The psychological aspects
3. The sociological aspects

Knowing the relevant biological aspects of the case is crucial. Therefore, it is the responsibility of the appropriate caregiver(s) in charge of the patient's case to clarify the patient's biological diagnosis and prognosis. From the standpoint of those doing the ethical analysis, the biological facts of the case are ordinarily treated as 'givens' of the case. For most issues, this is adequate, since the biomedical concern for those focusing on the *ethical* analysis (we refer especially to Bioethicists or to members of ethics committees) is limited to assuring that every usual, reasonable and customary effort has been made to identify them. In other words, every effort is made to assure that the proper specialists have been consulted and that there is enough objective data to warrant consensus among the caregivers about the physical diagnosis and prognosis and the reasonable treatment alternatives.

Knowing the psychological aspects of the case is also essential. The patient's, the patient's significant others' and even, at times, the caregivers' beliefs, motivations, needs, preferences and perceived or factual 'obligations' to others figure prominently and, thus, are critical to understanding the operative dimensions of a case.

Finally, central to understanding the setting within which the problematic case evolves and develops its significance are the sociological aspects of the case. This area includes laws, institutional and cultural mores and social customs of the specific community of which the patient is a part. Any religious affiliation the patient has—and, sometimes, even one a given caregiver may have—can be a significant part of the problem for a patient. Increasingly, as we have tried to emphasize, economic considerations play an important role in the lives of patients and their significant others.

All of these aspects need to be systematically addressed and their ethical relevance assessed, as those first three questions are being asked and answered. The role that diverse values, norms and principles may have played in informing and guiding the unique and complex lives patients have created for themselves (with the assistance—both tacit and explicit—of the democratic community of which they are a part) must never be ignored or denied.

As if this were not difficult enough already, when doing ethical analysis clinicians—just like their bioethical consultants—must always be mindful of the fact that they are _not_ the primary decision-makers. Rather, they are clinical _advisors_ and _consultants_ only. In addition, while the primary role of advisors and consultants in health care ethics is to help patients, and/or their significant others work with the health care team towards a resolution that is the least ethically problematic for all concerned, the patient must never be abandoned. Thus, any ethics consultant's advice should be confined to pointing out alternatives that are not ethically inappropriate, even when some of those alternatives might not agree with the consultant's own *personal* beliefs or choices. This is one of the most difficult intellectual habits for persons involved in the process of ethical consultation and analysis to develop, no matter what their field of expertise.

———— A BRIEF DIFFERENTIATION ———— BETWEEN BIOETHICISTS, ETHICS CONSULTANTS AND MEMBERS OF VARIOUS KINDS OF ETHICS COMMITTEES

———————— Bioethics Committees ————————

In general, ethics committees are constituted to help health care clinicians, patients, significant others and the community to make better-informed decisions about health care options. Although members of such committees need be neither ethicists nor philosophers, they ought to receive sufficient training in the language and methodologies of these disciplines in order to function adequately. As a result, the diverse needs, interests and values of all of those relevantly affected will be better appreciated and more adequately and equitably represented—this includes the recipients of health care, their significant others, the providers of that health care (individual as well as institutional) and, eventually, the community at large.

Generally speaking, an ethics committee's function is threefold: educative, consultative and supportive. Specifically, it addresses these responsibilities in the following ways, and in the following order of importance:

1. In the course of educating itself, it also educates health care staff, patients, their significant others and the community at large about relevant ethical issues and the tools by which such issues might be recognized, critically analysed and resolved

2. It assists with institutional policy review by helping to develop, review and up-date relevant health care policies with an eye to making them more humane and sensitive to the vulnerabilities of those relevantly affected

3. It provides consultation and support for health care staff, patients and their significant others in health care decision-making by helping all parties concerned to develop the sensitivity and insight necessary for a systematic identification and critical analysis of

aspects of individual cases that have become ethically troubling and to construct ethically defensible strategies for the resolution of such cases, viz., strategies that allow the individuals involved to be led by the dynamic of the ethical inquiry into the case rather than simply by their own predetermined expectations and/or interests

We have consistently defended the view that ethics committees, once they are well-functioning and working at peak efficiency, will spend most of their time actively pursuing the first set of responsibilities on the list: their educative responsibilities. This always seems rather surprising—perhaps even a little disappointing—to our audiences! And, most ethics committees that we have observed do, indeed, spend most of their time occupied with the third set of responsibilities. Obviously, when such committees are first constituted, it is the third responsibility—usually in the form of case consultations—that seems to occupy most of the committee's interest, and therefore, attention. Presumably, teaching others about ethical sensitivity, ethical concepts and the philosophical tools available to help them better recognize, discuss, evaluate and resolve ethical issues and cases is not as exciting as rushing in to do an ethics consult—somewhat like superman flying in to the rescue! But if the ethics committee is doing its job properly, the volume of case consultations should actually drop significantly within the first three years of the committee's existence—and this requires sustained and ongoing commitment and hard work: a much more dedicated, effective, though less dramatic kind of heroic rescue.

Institutional Review Boards

Because the goals of therapeusis and the goals of research may differ significantly, institutional review boards (IRBs), unlike ethics committees, focus more narrowly on experimentation and research. Thus, the constitution of the members is generally more homogeneous, viz., researchers of various stripes. To be acceptable, all research must be approved by an IRB. If they are to function properly, IRB's must not only deal with ethical standards or concern themselves about informed consent, but should also ascertain that the experiment is scientifically sound, that it stands a reasonable chance of producing the information sought, and that the experimenters involved are well-qualified. Until such preconditions are met, no experiment can be considered ethically sound.

All institutions conducting human research are expected to have and routinely utilize such IRB's. This is a laudable, but hardly foolproof step forward. IRB's are composed of people, and even at their best, people are not entirely impervious to political pressures or unmindful of the fact that their colleague, whose research they must evaluate today will in all likelihood be the person evaluating their own tomorrow. Furthermore, passage by an IRB does little to solve the quandary in which researchers find themselves when they must look out both for a particular patient's welfare and for the welfare of their experiment. Inevitably there will be times when they are caught between two mutually exclusive—or at least somewhat mutually contrary—goals. Thus, it is unfortunate that, in most health care institutions today, there is virtually no interaction or even an overlap of committee membership between ethics committee and IRB.

—————— Organizational Ethics Committees ——————

On the other hand, yet another kind of ethics committee has more recently evolved: the organizational ethics committee. Organizational ethics is a relatively new phenomenon in health care ethics that questions what it is to be an *ethical* organization. That is, it assumes that there is more to behaving ethically responsibly in the health setting than merely treating patients, their significant others, professional colleagues and other relevantly affected persons with sensitivity and compassion, and it concerns itself with the benefits and burdens that the broader organizational structure itself has on the individuals it is supposed to serve. Hopefully, organizational ethics committees will reward rather than frustrate whole-hearted ethical inquiry and dedicate themselves to promoting a just and equitable ethical environment conducive to good patient-centered care rather than allowing themselves to become merely a handy "hang-out shield" for promoting and protecting the narrower interests of the organization—but that is a chapter that can only be written in the future.

—————— Professional Bioethicists and the —————— Role of Health Care Ethics Consultants

We wish to make only two brief remarks here—first, about the discipline as a profession and second, about the role of the bioethicist as consultant. As to the first: Health care ethics is a discipline that hopefully is on the way towards becoming a profession in the true sense of the word. Professions have their own body of theoretical knowledge and their own literature—which, indeed, the field of bioethics is rapidly developing. But professions also accept certain special social responsibilities (some of which, we have long argued, bioethics has largely evaded), have criteria for and ways of equitable evaluation of who is and who is not considered a *bona fide* member of the profession, have standards of scholarship and practice and have methods of internal control. To date, bioethics lacks at least some of these central features. Furthermore, many whose understanding of the field is meager at best currently claim membership in this "guild." Hence, it is little wonder that many have come to view the "profession" of bioethics with skepticism.

As to the second: Ethicists are not here to "answer questions" in the same way that textbooks or manuals are here to "answer" technical questions of practice. Ethicists help to clarify problematic situations by helping think through questions, analyzing logical process, questioning blindly held assumptions, and asking questions that may help clarify attitudes towards and perspectives about a given issue or problem. As we have reiterated many times before, there are no "right" answers (albeit there certainly can be wrong ones). What ethicists can do is to help work through to a set of not inappropriate courses of action. The final responsibility, of course, remains and must remain with the person who ultimately must act—albeit that all those who advise him or her share in that responsibility.

THE APPLICATION: FLESHING OUT THE FORMAL STRUCTURE

Here we want only to say a few more words about the analysis of the specific cases that follow. How does one go about analyzing an "ethical" problem? Problem solving, in whatever discipline, is, as John Dewey so aptly pointed out, a system of inquiry whose method is not radically different no matter what the problem—albeit it might look quite different, depending on the particular subject matter under discussion.[1] Health professionals are well acquainted with solving problems. In the process of making a differential diagnosis or determining a course of treatment, the same sifting-and-sorting process takes place. It is a dialogical process in which various hypotheses are advanced and tested in relation to the known—and evolving—"facts." As a given hypothesis or "solution" (say, "I think the patient has pneumococcal pneumonia") is advanced, it provokes questions (say, "Was there a shaking chill?" or "Is there rusty sputum?") that the inquirer asks of the database. The answer, then, may confirm or refute the hypothesis: It may show that the "solution" or "answer" is more or less likely to serve under existing circumstances.

Further, the answer may itself stimulate the formulation of additional hypotheses, which must in turn be tested against the database and against each other. The "result" or "decision" is the product of internal and/or external dialogue in which the "conversation" does or may include history and the result of physical examination, reference sources, and the views and opinions of other professionals as well as those of others legitimately concerned with the case. Eventually all of this information (the result of the initial dialogue with others and with oneself) is resolved in an internal dialogue.

Values are not neglected in this process. Clinically, we assign different levels of importance to different hypotheses depending, among other things, on their likelihood and on the necessity for immediate action. (Say, the hypothesis of a dissecting aneurysm, although unlikely, is examined early in the differential diagnosis of severe chest pain, because of the threat that such a diagnosis poses for the patient. Our "value" for preserving life by acting first on what is most threatening has informed our choice of what to ask and what to do.) The questions we ask (be they actual questions or questions we ask by examination or by ordering laboratory tests) and the criteria we choose for judging an answer to be acceptable or not are underwritten by a set of complex and often subconscious values. To the extent possible, the clarification and explication of these subconscious values is one of the necessary functions of problem solving in general and of problem solving in the realm of ethics in particular. Deluding oneself by the claim that "facts" or problems are value-free and not underwritten by values is to misunderstand the whole enterprise of problem solving.

The solution itself is not static; that is, it is not *the* solution for all times or all places. It is one of many or several plausible options that, under existing circumstances, we have chosen to embrace. As such it serves not as the end point for the particular or for similar problems but as a starting point when this problem or other problems like it need to be examined. It is one point in an ever-enlarging chain of inquiry, learning, and growth.

To solve problems, in ethics or in anything else, requires a base of "facts" (knowledge) and a set of criteria in dealing with values. The "travel agent" model we have previously suggested may work well. Deciding, for example, whether a patient should or should not continue to be treated requires a firm basis in facts as well as an understanding of the goals ("ends") such treatment might sub-serve. It requires a lot more than merely that, however. Making ethical choices does not lend itself to a process of decision-making in which the likelihood of various outcomes are simply "cranked" into a formula and whatever outcome is desired by however slim a margin is then chosen. Even here, "the outcome desired" inevitably presupposes values (preferring "cure" to death, which under most circumstances would express a set of values we take for granted).

Solving problems in medical ethics, then, is similar in its methodology to solving problems in other disciplines. Medical ethics is often seen as different because it must deal with extremely complex subject material and options in a setting charged with emotion as well as with a great deal of ambiguity.[2] It rarely offers specific solutions for discrete problems and properly tends to raise questions rather than give clear answers. When asked "What shall we do?" the ethicist properly is prone to first ask a number of questions and eventually suggest a variety of not inappropriate options (none of which are, generally "good") rather than simply saying, "Do this." Furthermore, "solutions" in ethics are at least as prone to error as in more technical endeavors.[3] To ask for certainty is to ask for the impossible in either enterprise. Learning to live with uncertainty and to use error as a prod to further learning and refinement of practice is as necessary in ethics as it is in other fields (see also Chapter 4).

Medical ethics raises problems that cannot be "successfully addressed solely within the confines of philosophy or within the confines of medicine."[4] Finding solutions in medical ethics requires an interplay between an understanding of medical facts and philosophical issues and reasoning. Neither by itself is sufficient to sort out options in difficult real cases.

An understanding of medicine is crucial to the enterprise. And here, we do not mean by "understanding" a necessarily detailed knowledge of the field. The clinical ethicist need not be steeped in the intricacies of differential diagnosis or versed in the interpretation of blood gases; neither is it necessary for the ethicist to be thoroughly conversant with the subtleties of Hegel or Wittgenstein. But clinical ethicists, if they are to function well, must have an appreciation of the technical complexity of both fields. Of sociological and cultural factors, of communication sciences, of law and an understanding and appreciation of what role such disciplines play in decision making.[5] Philosophers who wish to function in this capacity must become thoroughly familiar with the medical process and with the setting in which that process occurs. Medicine is a unique enterprise, and its workplace is a unique setting, one that, like most specific settings (such as tribes or professional groups), has its own, largely unwritten, rules of behavior. Familiarity with medicine, above all, necessitates an understanding of the emotive components of illness: how sick, anxious, and dying persons and their loved ones feel and act as persons who come from a particular culture and who have a particular world-view, as well as what feelings, anxieties, and actions this evokes in health professionals. Physicians who wish to be ethicists need to be familiar with philosophy, with disciplined philosophical reasoning, and with the problems philosophers encounter in their work. The undertaking is a cooperative one: a partnership and a mutual understanding of several enterprises pursuing a common goal.[6]

Moral choices cannot be made without an awareness of the facts of the case. Such facts, first of all, are medical facts. Beyond medical "facts" there are other facts often crucial to dealing with specific problems. These are the "facts" of the patient's personal life, of his or her relationships with others; the "facts" of the patient's prior wishes and worldview; and the contextual "facts" of the hospital setting and its myriad constraints and moral actors, as well as the "facts" of the community and its needs. All of these form the matrix within which the medical facts are embedded. In a very real way these facts are crucial if one is to understand the problem.

Having a sufficient number of facts, then, is the essential condition for making a moral judgment. The facts of the case, furthermore, widen or constrain the possible options. The facts, however, no matter how complete or accurate, cannot solve the problem. Writing or not writing a "do not resuscitate" (DNR) order in a patient who is riddled with metastases depends as much on other considerations as it does on the technical facts; giving or not giving an available organ to one of four possible patients is informed by more than an understanding of HLA loci or even by a knowledge of a given family's circumstances or constellation.

In this process, goals ("ends") must be clearly kept in mind. A DNR order, for example, usually changes the goals of medicine, and acknowledging such a change of goals should come early in the decision-making process. Means and ends are not fixed: What is an end or goal today may become a means to a further end tomorrow. Tentatively establishing the goals at which our actions should aim, and being ready to change the goal in the process of selecting appropriate options or when "facts" change, is an important step in decision making.

Just as in the technical aspects of clinical medicine, all the while facts are gathered, options suggest themselves or disappear. In a true hypothesis-forming way, the options suggest a need for further facts, and further facts change the hypothesis entertained. The options we form in sorting out ethical problems depend critically on a background of ethical principles, beliefs, and presumptions (just as the options we form in making a diagnosis depend on clinical "facts," beliefs, and assumptions). The ethicist (just as the clinician in making a diagnosis) must try to sort these out, make them explicit, and establish hierarchies when principles conflict—while remembering the tentative nature of even the most likely hypotheses we entertain and often are forced to act upon. The ethicist must analyze the problem, making careful conceptual distinctions (distinguishing, for example, between supplying nutrients and fluids to those burdened by the procedure and those not) and showing that, in some cases, cherished distinctions may be problematic or not valid (as, for example, the distinction, under all circumstances, between killing and letting die).

As a background to this activity stand the moral theories that we embrace: The principles, beliefs, and presumptions may differ radically when a utilitarian and a deontologist makes these choices. It is surprising, however, how often the conclusions reached are the same no matter to which ethical theory or deep underlying belief system we subscribe. Persons of good-will often tend to reach the same judgments in concrete problems even when they appear, at first blush, to be informed by widely divergent belief systems. The study of specific cases using principles as guideposts rather than as straitjackets and carefully developing one's ethical judgments from the study of a multitude of cases is essential to clinical ethics.[7,8]

So, in summary, sorting out ethical problems, then, requires at least the following activities:

1. Get the facts (but do not hide the fear of making a decision behind a never-ending quest for more)
   a. Be sure that the "facts" are given and substantiated by those well credentialed in making such judgments
   b. Be certain that the "facts" include the socially relevant facts about the patient
   c. Be sure, ultimately, to examine these "facts" in the context in which they occur
   d. Entertain tentative options in an ongoing manner and let the options guide the search for further (necessary) facts
2. Draw clear distinctions—as distinctions are drawn, other options or the need for further facts may become evident and must be pursued
3. Scrutinize the beliefs and principles motivating the choices
4. Tentatively determine (and do not simply assume!) the goals that are to be pursued
5. When beliefs and principles clash, examine whether they do so in reality or simply because of a misunderstanding
6. Establish hierarchies of principles, duties, and obligations
7. Understand the context in which the problem plays itself out, and take differences in moral views into account—while a narrower analysis may be useful, removing problems completely from the context is artificial and may distort them in peculiar and unexpected ways (what *is* an analysis completely bereft of context anyway?)
8. Be aware that any "goal," "solution," "answer," or series of "options" is purely tentative and merely a step in further decision making

CASES FOR ANALYSIS AND DISCUSSION

These cases are for analysis. The comments following each case are meant to serve only as a skeleton for further discussion, not as a solution to the problem presented. Once again, it must be remembered that there is not any single answer that is "appropriate" but rather that there are several answers that may, depending on how the vagaries of the case are clarified and the values subtended, be not inappropriate. This approach is not meant to suggest that "anything goes." Just as there are many answers that are "not inappropriate," there clearly are also answers that are not appropriate whatever the circumstances may be. The reader will notice that as we proceed through the cases, we become less discursive, positing instead the sorts of questions that the reader should begin to be able to raise and explore.

CASE 1

You are a third-year medical student newly arrived on the ward. Your resident informs you that all members of her team (students and otherwise) will be introduced as "doctor" to the patient. After rounds Mr. Jones, who is one of your patients, confides to you that he is happy not to have students to contend with and says that he would not ever permit a student to draw his blood or to do other procedures. At this point your resident asks that you do a necessary test on this patient. You explain your reluctance to her, but she insists that you perform the procedure.

This is a tough one. Medical students are vulnerable for several reasons: They are the lowest on the totem pole; they are the least knowledgeable both about technical facts and medical protocol; and they are afraid that anything that displeases their attendings or residents may eventuate in a bad grade, which they can ill afford. The pressure to comply is therefore great. But medical students, in part by virtue of their vulnerability, also have several advantages: They have the possibility of appealing to a higher authority and to have their appeal heard, and, because of their inexperience, they can ask questions that might otherwise seem unallowable. Those are slender advantages, but they are nevertheless very real.

First of all, there is the question of being introduced as something one is not but is in the process of becoming: a doctor. In today's hospital being introduced as something one is not is acknowledged not to be permissible. (In former days, this was different. Students were introduced routinely as "doctor," and the patients knew, understood, and sometimes joked about it. That certainly did not make it "right." But since everyone, or virtually everyone, was aware of the practice and understood its implications, it was a far different matter. It is an example of cultural change and with it of acceptable ethical practice). It is not permissible today because, whatever else may be said about it, it is a form of deception and therefore an assault on another's dignity. So this matter should have been nipped in the bud at the very beginning when the resident stated that this was the way in which she would introduce her students. Talking to the resident (either the student by himself or herself or, preferably, a group of students acting together) might have taken care of the matter. When all other appeals fail, the fact that the guidelines issued by the American Association of Medical Colleges is emphatic on this point may be persuasive.

Secondly, after it was regrettably but understandably not nipped in the bud (who wants to get off on the wrong foot with one's resident?!), it should have been addressed when Mr. Jones made his very pointed comment about students. A simple remark such as, "Well, Mr. Jones, all students on this team are called `doctor,' and I am a student" followed by an explanation of the student's role (and supervision) might have done the trick.

Thirdly, the moment of truth comes when you are told to do a procedure on someone whom you know to have been deceived as to your actual role—and, furthermore, that you were a part of that deception. At this point, you have little choice: Either continue the deception or refuse to play along. The initial deception was of the more tacit kind: You permitted another to deceive, and you failed to respond when the patient made his statement. But now deception is active: Knowing how the patient feels about students, and knowing that he has been deceived, you, nevertheless, actively participate.

Deceiving persons is considered to be a form of lying, of not giving the truth to someone entitled to the truth and who asks for it. If one is to respect others, lying or deceiving persons is at least a *prima facie* wrong: *prima facie* because some would argue that lying or deception under some very exceptional circumstances is either permissible or at least unavoidable. (Of course, Kant would hold that it is never, and under no circumstances, permissible.) Deceiving patients under these conditions is especially unacceptable: The professional–patient relationship (and students by virtue of being students in the field partake in this relationship) as ordinarily perceived is based on trust. Furthermore, patients, by virtue of their being patients, are in a situation of reduced power *vis-à-vis* the medical team and are exceptionally vulnerable. That, perhaps, does not give them any special rights compared to one less vulnerable; but it does give a special obligation to the one holding power by virtue of such vulnerability: Taking candy from a baby may, strictly speaking, not be any more wrong than taking it from a trained sumo wrestler, but it is considerably more reprehensible.

Students do have at least a quasi-physician–patient relationship with their assigned patients. Learning to accept and to feel comfortable dealing with the physician–patient relationship as well as relationships with other healthcare professionals is one of the important (and often forgotten) aspects of the clinical training years. It is part of the necessary socializing process accompanying training in any field, and it is very much what "becoming a doctor" is all about. The student–patient relationship is, however, somewhat more complex. First of all, it involves on the "physician side" others who have considerable power over the student's fate. Secondly, it is an assigned role and one not chosen by either of the participants. Thirdly, it is a transitional one: transitional because the student is neither quite a lay nor quite a professional person. The transitional role, incidentally, often gives the student a unique opportunity. Not rarely patients, because they too perceive the student as standing between doctor and layman, will turn to the student with questions and requests they would hesitate to broach to others on the health care team. The student becomes *their* student, one who, perhaps, can better understand their fears and hopes than can (or do) the other members of the medical establishment. The student, as it were, becomes the critical missing link. Ultimately, the student confronted with such a problem will have little choice but to solve it either by changing the resident's mind, appealing to higher authority, or confronting the patient with the truth. Perpetuating deception, the only other alternative is, under almost all circumstances unacceptable.

There is one other critical ethical problem, one we often tend to neglect: Those of us engaged in teaching students in any field and at whatever level bear an ethical responsibility to students. This responsibility and the whole nature of academic ethics is one to which all too little attention has been paid and about which all too little has been written. Teachers have a professional relationship with their students that bears similarity to the relationship healthcare professionals have with their patients. In this relationship, students are more vulnerable than their teachers and are therefore deserving of a great deal of respect, consideration, and effort: Belittling students, short-changing them when it comes to the time students need for help (be it with their thesis or with patient problems), coercing students, or exploiting and manipulating them to serve one's own ends denies the respect due others and due especially others who are dependent and vulnerable. Beyond this, teachers are (for good or for bad) role models whose behavior their students are apt to emulate or, at least, to consider as acceptable. The way teachers deal with their problems and with their colleagues as well as the habits they exhibit in daily life (rigor or sloppiness, courtesy or rudeness, honesty or deception) inevitably influence not only the student but also, in the aggregate (since students are tomorrow's professionals) condition the future itself.

CASE 2

You are a first-year resident on the surgical service. Your patient is advised by your attending that he urgently needs an operation. In fact, you yourself feel that the procedure is not as absolutely necessary (and that a less radical one might serve equally well) as your attending thinks.

Consider two different scenarios:

1. You discuss the matter with your attending, who disagrees with you, maintains that the operation is absolutely needed, and explains his reasoning to you. You are not convinced. On the afternoon before surgery, the patient asks your opinion.

2. You discuss the matter with your attending. He agrees with you that the surgery is not as necessary as he has said but says that he put the matter in the strongest terms because he knows that the patient would otherwise refuse this type of surgical intervention. The surgery, while not absolutely necessary to save the patient's life, is still probably the most advisable in the long run. Your attending asks that, for the ultimate good of the patient, you support him in his stance.

In part, this case again is one that involves deception. Physicians often feel that deception is permissible in "furthering the patient's good." One cannot truly state that this is "always" and under all conceivable circumstances wrong: Scenarios in which deception may be arguable can certainly be constructed. But deception is, at the very least, a *prima facie* wrong; that is, it is wrong unless powerful reasons why it may not be wrong under a given circumstance can be marshaled. And in medical practice instances in which such arguments can be successfully upheld are rare. Deception "for the patient's good" is, even under the best of circumstances, an act of crass paternalism, and acts of paternalism must be justified. In stating that we are doing a given thing "for a patient's good" when the patient himself or herself could consent but has not consented is to say that we either know how the patient would define his or her good (which ordinarily is impossible) or that we think that the patient's definition of his or her own good is plainly "wrong." In the competent patient, this violates all notions of respect.

There are really two problems here. The first, and perhaps more important one, is the problem of deception. The other is the problem of interdisciplinarity: that of the proper integrative functioning of a ward team and the mutual discipline necessary to its proper function. A team, by necessity, must have someone in charge who is, in the final analysis, both responsible and accountable for what happens. Such a person, however, is not above reproach and, thus, his or her unethical or technically incompetent behavior cannot be tolerated. When the head of a team makes a technical decision (say when he or she decides whether to operate or to pursue a given course of therapy), a decision that by reasons of training and experience he or she is most capable of making is made. In making this decision, the head of the team may be wrong, but he or she is less likely to be wrong than are those with less training and less experience.

When, however, a moral matter is at stake, things are not quite that simple. The head of the team may have had the most experience with the technical aspects of the problem (he or she may be most aware of the likely outcome of doing or not doing something), but the head of the team, like every other member of the team, is a moral agent who must live according to his or her lights. One may claim that the head of the team knows the patient best and therefore knows what the patient "would want," but that is a claim shown to be wrong much of the time. Although one may expect that, in technical matters, the team will ultimately comply with the decision of its head, even here, one would hope that when time is available discussion and learning will take place. When it comes to matters having a heavy ethical content, discussion becomes all the more important. Members of the team who meekly comply with what they consider to be ethically inappropriate demands of a "sovereign" member are not living up to the demands of their moral agency.

True emergencies or even extremely urgent situations are a somewhat different matter. In an emergency or extremely urgent situation there is little time to reflect and discuss. Hesitation may, at such a point, do irretrievable damage. Except under the most crass circumstances, one must assume that the person in charge is handling things properly. This "benefit of the doubt" is based on the immediate threat as well as the probability that the best trained and most experienced (and, therefore, hopefully the one in charge) ultimately will "know best." But this benefit of the doubt does not imply that, after the fact, doubts and questions do not need to be aired, discussed, and addressed. Not airing and discussing one's doubt would itself be morally problematic since it would make the person in doubt a party to possible future wrongs.

Most situations like the situation in the case under discussion are neither emergencies nor so urgent that discussion and dialogue cannot proceed. Whatever else is done, legitimate doubt necessitates a legitimate and courteous question and deserves a well-thought-out and courteous answer. All parties, furthermore, should be ready and willing to engage in such a dialogue and must, in the face of sufficient evidence, be willing to alter their opinion or to change their mind. Furthermore, it is equally important to involve others who are concerned with that particular patient's care. One of the advantages of working as a team is just that: Members, provided they are willing to do so and have not stubbornly invested their ego in their own particular point of view, can learn and profit from each other's experiences and alternative points of view.

In the first scenario, the resident would be well advised to insist on discussing the matter with the attending. If unconvinced by the attending's arguments, the resident would have to weigh his or her own ability to make a technical judgment against the technical ability and experience of the attending. When asked by the patient, the resident might well choose to say that while he or she was personally not convinced, the greater experience of the attending inclined him or her to support the decision. The matter here is a predominantly technical one and, therefore, one in which the chance of the attending's being correct is significantly greater than is that of the inexperienced resident. Supporting the attending seems the reasonable thing to do.

In the second scenario, there most probably is no technical dispute. The attending acknowledges that he or she is deceiving the patient (even if that deception is believed by both the resident and the attending to be for the patient's "good"), and the resident has the choice of either going along with the deception and, therefore, becoming a party to it, or refusing to do so. Prior to such a decision, a dialogue with the attending in which both come to state their points of view and search for shared values is essential. Hopefully one may convince the other: The attending may convince the resident either that the patient is not competent to make a meaningful choice (in which case courses of action other than deception are possible) or that the technical facts of the case are other than the resident understood them to be and that surgery is, indeed, urgently needed; the resident may convince the attending to do other than deceive the patient. Whenever possible such a dialogue should also involve the other healthcare professionals caring for the patient—especially if they are expected to play a role in the patient's ongoing care. If, however, the persons involved cannot convince each other, the resident is left with the original quandary: Participate in the deception, leave the case once adequate patient care has been assured, or tell the patient the truth.

There is one other ethical problem that bears brief mention: The smooth functioning of a team—since in today's world it is critical to an optimal outcome—is something that, in and of itself and beyond the particular relationships of the particular persons involved, has ethical standing. In a sense it constitutes the necessary condition for all else that can or will be done for the patient. All members, within the limits of reason and ethical probity, are ethically compelled to do as much as they can to promote and are certainly advised to refrain from disrupting its smooth function.

CASE 3

You are a nurse working in a large hospital. The patient, whom you have come to know fairly well over the last few days, has just been diagnosed with cancer of the lung with metastases to the liver and brain, and it is a cancer that is unlikely to respond to any form of therapy. The patient has repeatedly asked you for the diagnosis and you have repeatedly suggested that he take up this question with his attending physician. When the patient continues to inquire, you ask the resident what you should do. The resident explains that the attending physician does not wish to "give the bad news to the patient" and that he or she has simply told the patient that he had a "small growth" and that he would be treated. Confronting the attending, you are told that he or she "knows the family" and that it is their wish that the patient not be told.

Again, the problem here is one of truth telling, deception, and trust. Since the patient is obviously continuing to ask and is asking you, the nurse, instead of turning to the attending or to his family, it seems very likely that the patient realizes that he or she is not being told the truth. But even if this were not the case (even if the patient really did believe the physician but simply wanted to talk or to be "doubly sure"), would that alter the basic ethical question?

Many factors are involved. The patient's treatment is obviously not only in the hands of a single professional (as he or she might conceivably be in the privacy of the physician's office—and even in the office there are almost always others involved) but is in the hands of a team headed by that particular physician (see Case 2). Such decisions, therefore, ethically should involve all members of the team.

You as the nurse have taken the first appropriate step: You have been sensitive enough to recognize the problem and to informally discuss it with the resident on the floor. That this has led nowhere does not speak against such an initial first step: You could easily have learned things that you did not know before and that might very well put the matter in a different light. Your next step, it would seem, would be to confront the physician and to discuss his or her point of view. Again this is likely to lead nowhere, but, even if only in fairness, it ought to be done. Beyond this you might wish to talk to the family and attempt to persuade them that their fears and therefore their actions were not well founded: You could, for example, suggest that they read Kübler-Ross, consult with their clergyperson, seek help from an ethics consultant, or take the case to an ethics committee. Depending on hospital structure, the next step you could (and should) take is to discuss the matter with your head nurse, with your supervisor, and ultimately, if need be, with the ethics committee. There are, as in most ethical problems, several courses of action open to you as well as several possible courses of action you could not ethically defend. Lying to or deliberately misleading the patient is one of the courses of action that, under most circumstances, one would find hard to defend. On the other hand, going directly against the attending's wishes is apt to shatter the team, in itself a problematic step to take. In the particular case and if all other courses of action are exhausted, you could ask to be removed from the case.

But beyond the case, it seems obvious that there is an institutional problem of truth telling. Such a case—as a generic and, perhaps not as a specific, case and also not necessarily to point fingers of blame—might very fruitfully be taken to the ethics committee for deliberation and finally for establishing a policy or, at least, for rendering an opinion on such a course of action. An ethics committee might, for example, decide that deceiving patients under most such circumstances was considered to be improper behavior and go on to suggest that when deviations from this policy appeared to be in the patient's true best interest, consultation with the committee or with a consultant would be advised.

One of the problems health professionals other than physicians (and more and more physicians also) confront today is that they are employees whose livelihood is dependent on their "on-the-job" function as judged by their superiors. Such a position, inevitable though it is, lends itself to power plays and coercion. As with all else, persons must seek their own "bottom line:" the point beyond which they simply cannot go, and still maintain their personal integrity. All too often these "bottom lines" are obfuscated and the actions one takes are rationalized instead of being rational. Such a course of action only adds hypocrisy to the morally already tainted act and all too easily leads to the kind of self-deception that makes repetition of the same and similar courses of action easier.

When nurses complain of being "put in the middle," it is often a quite justified charge. On the other hand, it is "a middle" from which they could in many circumstances avoid or, by forthright action, remove themselves. Not every dispute should be taken elsewhere or to a higher authority: Most, in fact, should be resolvable and resolved within the structure of the team itself. When they cannot be resolved, however, forthright action (instead of complaining, while complying) may help resolve the problem and incidentally, by creating mutual respect, head off such problems in the future.

CASE 4

You are a practicing specialist who is asked to see a patient in consultation. The physician asking you to see the patient is one who refers a large number of cases to you. The patient is seriously ill. After a careful evaluation of the case, you conclude that the treatment given by his physician is less than optimal and that another approach would be more advisable. You write an explicit consultation note to this effect. When you see the patient the next day, none of your advice has been followed, and the patient is worse. The family sees you in the hall and asks about their father and husband. You speak to the physician, but he or she states that he or she disagrees with you and is not ready to use your approach.

This case highlights a situation that, while not frequent, is by no means rare. When consultants see patients, a new professional–patient relationship (that between consultant and patient) is established. The ethics of referral, however, are not that simple. The relationship is not purely between consultant and patient: The referring physician as well as the other members of the healthcare team must also be considered.

To make things more complicated, the consultant depends on consultations for his or her living, and the referring physician's continuing reliance on him or her is economically either essential or, at the very least, highly desirable. In our fee-for-service entrepreneurial economy, the decision to have serious disagreement with persons who refer a large number of cases may have dire consequences. Not only is it likely that referrals from that particular source may dry up; it is also not entirely unlikely that others who refer patients may turn elsewhere. Serious economic consequences may attend the consultant who quarrels with referring physicians. One may say that such considerations are irrelevant to making a moral choice. But choices are composite, and ethical considerations are only a part of the choice eventually made. It is unrealistic to claim that anyone can be oblivious to his or her own interests. Loyalties, therefore, tend to be complex.

As usually understood, medical etiquette discourages the washing of the profession's dirty linen in public. Criticizing a colleague is "not playing the game." Medical etiquette ought, perhaps, not be a consideration when it comes to making decisions critically affecting a patient's life, but medical etiquette is, nevertheless, a powerful tacit as well as explicit force. It cannot be entirely ignored. As in a Venn diagram, medical ethics, medical etiquette, economics, personal feelings, and many other considerations often overlap.

Often, when physicians differ, there is no absolute or even relative standard of truth to which one can appeal. Saying that the consultant has, by virtue of his or her being a consultant and a specialist, more intimate knowledge of that which he or she is being consulted about than does the referring physician is usually true. But that statement fails to take into account the fact that more than just expertise in a particular field is involved. The primary care physician who asked for the consultation is, unless he or she withdraws or is dismissed from the case, the physician of record and therefore in charge. Furthermore, he or she may know something about the totality of the case that the consultant may not and may be more acquainted with the patient's values and wishes. More than just the organ system in which the consultant is expert may be at stake. The referring physician may, for various reasons, disagree with the proposed course of action.

In this case, adequate communication is the primary requirement and may by itself resolve the problem. One or the other of the physicians may come to understand the problem from an aspect not seen before, and disagreement may therefore disappear. When the problem remains unresolved, the consulting physician is left with a variety of choices: In essence, he or she can continue to try to persuade the attending physician while publicly supporting him/her and showing a "united front," or he or she can inform the patient or the family (if the patient is unable to "think straight") of the disagreement. If time allows (which is by no means always the case), the consultant would be well advised to communicate with the referring physician, and suggest that an additional opinion be obtained. In addition and when all else fails, such a dispute can and should be taken to the respective chiefs of service. Prior to confronting patient or family and essentially asking them to make a choice and a judgment for which they are ill prepared, every reasonable avenue must be exhausted.

Coming to terms with this problem requires an examination of a hierarchy of values in which loyalty to one's colleagues, self-interest, medical etiquette, and, ultimately, the obligation to a critically ill patient vie with each other and need to be resolved. Consultants who materially disagree with the referring physician's course of action but who nevertheless "go along" against what they perceive to be the patient's best interest participate in a form of deception. Consultants who "blow the whistle" in a sense violate etiquette and may do themselves personal harm. Both the consultant and the referring physician are enmeshed in a patient–physician relationship, which under ordinary circumstances takes precedence over all else.

Such disputes, furthermore, do not take place in a vacuum. Other members of the team inevitably become involved and almost as inevitably begin to take sides (and may even, regrettably, come to enjoy the "show"). Such disputes, therefore, are highly injurious to the smooth function of the healthcare team. Since the smooth function of the healthcare team is essential for proper patient care, the integrity of the team is therefore a proper moral concern.

CASE 5

Ms. Jones is a 26-year-old unmarried lady who is now about 8 weeks pregnant. She went to see her physician with vague abdominal complaints 5 weeks before. Not knowing that she was pregnant, her physician ordered an abdominal CT scan, which was performed and was negative. The patient is now in the hospital with a urinary tract infection from which she is recovering well. She is being taken care of by two physicians: an internist, who is handling her infection and who was her prior physician, and an obstetrician, who is overseeing her prenatal care. You are an ethicist who is called in to see the patient because the two physicians disagree: One feels that this patient should be aborted because of the exposure to radiation, and the other feels that this should not be done. How do you go about "sorting out" this problem?

This case is one of those in which a technical problem masquerades as an ethical one. There simply are not enough available facts to make a proper decision. How much radiation did the uterus receive, and what is the effect of this amount of radiation on a 3-week fetus? Is the risk of serious fetal malformation 2% or 90%? Answers to these questions are available but may require some research. No rational answer can even be attempted before the facts are established. (If they cannot be, then the absence of factual information has to be dealt with.)

At any rate, abortion is a decision for the patient and not one for the doctor to make. Physicians can give their patients facts, can tell them what they do and do not know, and can counsel them about risks and benefits, but ultimately it is the (rational and competent) patient who must decide. How the physician personally feels about abortion is not the issue here. Physicians who are morally opposed to abortion cannot, of course, be expected to participate, but by the same token, those same physicians cannot hold other moral agents hostage to their own belief system. Failing to supply patients with full information (for example, hiding from them the fact that abortion is an option even if an option repugnant the physician or misleading them by pretending to knowledge physicians do not have) is deception and makes the patient a hostage of the physician's idiosyncratic belief system or to his overly developed ego.

If the facts are not known and are not obtainable (say, if no information as to the percentage of risk to the fetus from such exposure exists), the patient is entitled to be told about this lack of information. Physicians must learn to share their ignorance as well as their knowledge with patients. Failing to do so is, likewise, withholding pertinent information: information the patient needs to assess risks and benefits in his or her particular case and circumstances.

CASE 6

Mr. Smith and his wife have been your patients for some time and you know them well. The marriage has not been a good one. About 7 months ago you found Mr. Smith to have cancer of the colon. The lesion was resected, but metastases to the liver were discovered at surgery. The patient refused further anticancer treatment. In many talks with the patient and his wife, you all agreed that treatment should be limited to promoting the patient's comfort. Specifically, the patient asked that intercurrent illnesses should be treated vigorously only if such treatment would promote his comfort and not merely prolong his life. Some of these conversations were held in front of the patient's wife, and you have noted the substance of this decision in your records.

You are called to see Mr. Smith in the emergency room where he has been brought after sustaining an inferior wall myocardial infarction. At the time of examination there is complete heart block, the heart rate is 25, and he is unconscious. You explain the circumstances to the wife and state that you will admit the patient to the regular hospital floor, keep him comfortable, and allow him to die. She asks what you would do if he did not have metastatic cancer. You explain that ordinarily one would insert a pacemaker and admit the patient to the coronary care unit and that there would then be a moderately good chance of survival. She insists that you do this. You remind her of the agreement between you and the patient and remind her that she was privy to the agreement. She agrees to the facts as stated by you but says that she has changed her mind and wants "everything done." Furthermore, she threatens to sue you if you do not do "everything possible to save my husband's life."

This is a most disagreeable case. Many factors enter into the decision: factors such as the personal relationship of the actors prior to this event, the fact that the physician inevitably feels empathy and loyalty not only toward the dying patient but also toward the wife, the fear of being sued, and the knowledge that, when all is said and done, the easiest course would be to do as the wife wants. The physician knows quite well that the chance of this patient's living, even with a pacemaker and even in the best of hands, is poor. Most likely if the doctor follows the easier course, the outcome will be the same. The patient will die, and no one will be the wiser.

On the other hand, there are the troubling promise and the implicit contract. Even if not formally signed by the patient, the decision is recorded in the chart and not even disputed by the wife. This is not the time to consult a lawyer, and, besides, the lawyer here can only advise as to the legal problem and not as to the ethical question. What if the patient were to live and then challenge the doctor for breach of promise?

The conflict here is not so much one of loyalties (the loyalty to the wife, while existent, is not of the same order as it would be were she herself the patient). Without a doubt the physician's first obligation is to the patient: This can be justified by an appeal to the obligations inherent in the physician–patient relationship or by appealing to fidelity and promise keeping. A utilitarian argument that failure to live up to the obligations of the professional relationships and failure to keep promises eventually dilutes trust in medicine and brings more harm than good can likewise be made.

But that does not settle the practical issue completely. Physicians do not practice in a vacuum. Institutional and other social pressures may be brought to bear and, while, puristically speaking, they are not relevant, they do often play a crucial role. In doing what they perceive to be "right," physicians must be fully aware of these pressures and must, when possible, learn to cope with them.

In this case, the physician has the option of doing as the wife wishes and treat the patient in a vigorous manner (thereby breaking a promise made to the patient), or the physician may choose to honor the promise and allow the patient to die in comfort (thereby risking the wife's anger and possible suit). The option to pretend to treat but to do so only half-heartedly and ineffectively or to lie to the wife and deny that effective treatment exists, is, of course, a practical option: practical in the sense of being achievable. It is, however, an option that adds deception to promise breaking and therefore one that would be even worse than either of the others.

If at all possible, the physician should attempt to persuade the wife to go along with the patient's wishes and with the promise given to him. This can be done by searching for and appealing to a set of shared values: such values as trying to alleviate suffering, not prolonging dying, and keeping trust with the patient's wishes come to mind. Other shared values, values that may be peculiar to the case in question, must likewise be sought. When time and opportunity allow, enlisting others who were perhaps close to the patient and his wife should be sought. Here, in persons who are religious, the clergy can be of great help. Ultimately and finally, the promise given to the patient will have to be honored even when doing so may be a risky thing to do.

[*Addendum*: This is an actual case well known to the author [EHL]. The patient's wife, after all efforts at persuasion failed, threatened to sue and walked out. With considerable misgivings, the physician did as he had said he would: The patient was admitted to a ward room, kept comfortable, and allowed to die. Mrs. Smith remained the attending's patient and never again mentioned what had occurred. In thinking about the whole affair, one wonders if the wife's action was not an elaborate and undoubtedly unconscious way of alleviating her own sense of guilt. The marriage had been poor, and she may have feared censure (by her own conscience as well as by his relatives) if she acquiesced to what she knew would be her husband's death.]

CASE 7

The parents of 6-year-old Jimmy bring him to you for care. You have taken care of the child for about a year and have noted that the parents are very devoted to their child. The family members are staunch Jehovah's Witnesses of long standing and have informed you of their refusal to take blood or blood products.

Consider these different scenarios:

1. Jimmy, who has been entirely healthy, is involved in an accident and is brought to the emergency room. He is bleeding profusely and needs blood. His parents refuse.

2. You find that Jimmy has acute lymphatic leukemia and is severely anemic. The child is in urgent need of blood transfusions and chemotherapy. There is a good chance that with proper treatment this child's life can be prolonged for some years and a fair chance that a long-term remission or cure can be achieved. You know that, should the child be transfused, the parents' attitude toward the child may change dramatically.

3. Jimmy, who is now 15, is involved in a scenario similar to either scenario 1 or 2 (above). You have talked to the boy on many occasions and know either that he shares or that he does not share his parents' beliefs when it comes to accepting blood. How would you handle this problem in any of these events?

4. Jimmy has grown up and you continue to take care of him and now his family. He is 30 years old and in need of surgery. Jimmy is eager to "get it over with" and asks you to schedule him as soon as possible. The surgery is one in which the need for blood could easily come up, and Jimmy reiterates that he would not accept transfusion even if that refusal should cost him his life.

Let us first consider scenarios 1 through 3. There are a number of interesting problems here. The first is the problem of what has and what has not been said by you and by the parents over the year that you have taken care of Jimmy. Did you or did you not know the parents' feelings about blood transfusion? If you did not know this, and if this fact was "sprung" on you at the time of the emergency, it is quite different than if you had known of their refusal right along and had tacitly gone along with it.

Let's assume that you didn't know. In the first instance, (1) you are suddenly confronted with a dying child who almost certainly could be saved. Your action or inaction will depend on how you feel toward the "rights" of parents to their child. If you feel that these "rights" are almost absolute (if, in other words, you feel that parents are more like owners than stewards of their child), your actions will be far different than if, on the other hand, you feel that children are vulnerable members of the community that are given to the parents in stewardship. If you transfuse this child first and then get a court order, the court will undoubtedly back you up (provided, of course, that reasonable grounds for assuming that the child would die without transfusion were indeed present). An overwhelming number of ethicists feel that in this eventuality the good of the child overrides the parents' wishes and that transfusion is (almost) ethically mandatory. Whether you could be sued if you failed to transfuse is a moot point and one largely irrelevant to the ethical judgment made. (Or is it?)

Let's, however, assume that you did know. In that case (2) you have known about these parents' feelings right along, have perhaps tried to persuade them (and failed), but have, nevertheless, continued to treat the child. There are two possibilities: (a) It is possible (but unlikely) that you told the parents that in the unlikely event that the child urgently required blood in an emergency you would transfuse and appeal for a court order. This seems an unlikely scenario because parents who are Jehovah's Witnesses would have been most likely to object and change physicians under such circumstances—though, for interesting and complex psychological reasons, in some cases they do not. And they did not. Here you have dealt with them honestly and your course of action is clear (and is, incidentally, a course of action the parents in a sense colluded in). On the other hand, (b) it is likely that you never informed the parents and gambled that the problem would never come up. And you have lost. Now you are in a fix: By not informing the parents of how you would act in a straightforward way, you have inevitably conveyed the impression that you would acquiesce with their wishes. In this sense you have colluded with the parents. Now transfusing and turning to the court is far more ethically problematic and far more of a breach of faith than it was in the previous scenario.

In (a) there is little doubt of the lesser problematic course of action: A critically ill person who lacks decisional capacity and who has not made clear provisions to the contrary (which a child cannot do) and who without treatment would be sure to die must (under ordinary circumstances) be treated. Physicians who fail to treat under such circumstances ethically become a party to the patient's death and may, although there has been no case like it ever brought, possibly be legally liable. In (b) arguing on the basis of promise keeping, it seems clear that the physician is in a situation to a large part of his or her own making. Conversations with patients and their relatives above all have to be honest: The parties are entitled to know where each one stands and what they can expect. By your silence acquiescence was, at the very least, suggested.

One could argue on the basis of ethical consistency that since this is the case you are bound to respect the parents' wishes. The only argument to the contrary would have to be that in a hierarchy of values, treating the patient (and perhaps, but with no very certain chance for success, saving his life) has primacy over the tacit promise given. In other words, the good brought about by breaking the promise would be greater than the good brought about by keeping it. This utilitarian argument, however, certainly is arguable: It hinges on what is seen, under these circumstances, as the "good" of the patient. Rule utilitarians, furthermore, could still invoke the rule against promise breaking and have us much in the same fix.

In scenario 2 above, the resolution is far from clear. The child with proper treatment may live, but proper treatment is long-term and may require several transfusions distributed over a long stretch of time. Further, the chances of this child's long-term survival are, at best, far from splendid. In many circumstances transfusing the child will distort the relationship between parent and child. (That is true in the first example also; but in that example the child would be healthy after transfusion and not in need of prolonged care and support during a critical illness.) Is it in the best interest of the child to allow him to live out his time surrounded by the warmth of a loving home, or is it in his best interest not to have the chance of long-term survival denied? (Again assume, as is probably the case, that you can get a court order to proceed.) In such a situation physicians would be well advised to enlist the help of social services, psychologists, and other members of the healthcare team in investigating the actual situation and arriving at a reasonable conclusion.

In the third scenario, the matter is even more problematic. Jimmy is still legally a child, but 15-year-olds are not quite the same as 5-year-olds. If the child, when privately spoken to, insists that he shares his parents' belief, then overriding his belief comes dangerously close to the paternalism that would override the belief of an adult. Not all 15-year-olds are the same: Some are more, some are less, mature. Further, the social and parental pressures may be such as to render the decision only questionably autonomous: How much coercion (internal and external) is at play here? After all, Jimmy's worldview has been strongly molded by home and church. It is therefore really *his*? (And could this not, in the final analysis, be said of all of us?) If the child does not share the parent's belief, one can breathe a sigh of relief and transfuse; but one must be aware that coercion can be a multifaceted thing. How coercive is the hospital or you? Inevitably physicians (like all persons) will feel more sympathetic toward those sharing their own system of values (and more sympathetic toward decisions emerging from the same system) than they will toward those with whom they materially differ. Even if physicians cannot help feeling this way, they must be aware of their feelings and take them into account.

If discussions have taken place and promises have tacitly or explicitly been made between you and the parents, scenario 3 still presents a severe problem. If Jimmy continues to share his parents' belief (and if he truly shares their belief and is not simply forced to agree in their presence), he essentially agrees to and becomes a party to the promise. If, however, he does not, then it is difficult to argue that a promise made by one party to another as critically concerns a third is binding. (Promising my wife without my consent that you will not treat me when I become critically ill does not appear to be a binding promise.)

Jimmy grown up in scenario 4 is quite a different problem. By now you ought to know that Jimmy and his parents are Jehovah's Witnesses (if you don't, it doesn't say much for the kind of relationship that you have had over the years!). In past conversations you may have agreed to take care of all eventualities and to refrain or not to refrain from blood transfusions. If so (and you have probably agreed to abide by Jimmy's wishes since patients who are Jehovah's Witnesses in general will, if they have a choice, not stay with physicians who do not give this promise), the current situation has merely forced Jimmy to cash in on your promise. Physicians confronted with patients who demand a promise that they, the physicians, are unwilling to give (say, not to give blood should the need arise), either give it against their will (and then are ethically "stuck") or see to it that the patient receives other competent care. Physicians who are "the only doctor" in a reasonable area may then face a difficult choice: They can either make their refusal known early on in the relationship and thus put the ball in the patient's court, or comply with the patient's wish. Hoping that the situation will never come up is whistling in the dark: Inevitably the situation does come up.

Jehovah's Witnesses, since their beliefs are so firmly held and yet seem so "unreasonable" to most healthcare professionals, are a source of great frustration to healthcare providers. When dealing with such persons, firmness together with compassionate rationality is the order of the day. The parents of such children, after all, do not have it easy: Watching one's child die is a most terrible thing, and watching it die when it need not die is especially difficult. The parents are ground up between deeply held convictions and deeply felt love and affection. Healthcare professionals must be aware of this and, even while they are doing what they need to do to save the child's life, must do all in their power to act with compassion and understanding toward the parents.

CASE 8

The patient is a 78-year-old single lady whose only known relative is an 80-year-old sister who, while spry and bright, has had no contact with the patient for over 40 years. The patient has a history of chronic depression and has been in a nursing home for some years. There are no other relatives and no living friends. Two months ago the patient refused any further investigation of a probably cancerous breast mass. At the time, her physician judged her to be competent enough to refuse such investigation and treatment. Following a heart attack, the patient is in an intensive care unit. Attempts to treat her severe heart failure have been unsuccessful. You are called because the question of using more aggressive and invasive means (an intra-aortic balloon pump or angiography and possible angioplasty and/or surgery) has come up. The patient is stuporous and incapable of making any decision. Her sister states that she does not

know the patient well enough to participate actively and leaves the deci-
sion up to the medical team whom you are called to advise.

This is a difficult case because the physician at first thinks that there is no way of knowing the patient's wishes. If this patient either had not been previously held to be competent to make an important decision or if she had never had to make such a de-cision, the matter would be much more difficult.

When it came to her breast mass, the patient, despite her psychiatric history, was allowed to refuse further intervention. It seems clear that the patient did not want any invasive or extensive intervention then, and it seems not unreasonable to believe that she would feel the same under existing circumstances. It can, of course, be argued that a breast mass, which may or may not be cancer and which may or may not require surgery, is quite a different thing from an imminently life-threaten-ing occurrence. Perhaps the patient would feel differently about the temporary use of an intra-aortic balloon pump or about angioplasty than she would about having a breast removed. (After all, a mammogram would either mean surgery, which the patient might not want, or that nothing need be done, in which case, and in retro-spect, why get it?) Can we really know how the patient would feel about aggressive temporary measures of this sort? And if we can't, on what basis do we choose ei-ther to act or not to act?

The question of the patient's competency (or, more properly, decisional capac-ity) in the first place is an equally interesting question. Her physician either concluded that she was competent to make the decision (or at least not incompetent enough that her choice could be easily overridden) or, perhaps, the physician's decision was based on a composite judgment: On the one hand, the physician may have felt that surgical intervention would, for whatever reason, be undesirable and, on the other, that she was competent enough to refrain from challenging her (especially since her refusal coincided with the physician's judgment of the case). Now, however, the physician may feel differently because without intervention, the patient will shortly die. The emotive, even if not necessarily the ethical circumstances are different. Should this make a difference?

Was the sister right in refusing to make the choice? What if she had known her sister well but still refused or made the choice that the decision was up to the physi-cian? If the patient was indeed judged to have decisional capacity, should her refusal to allow further investigation of her breast lesion have given rise to a frank discussion as to her wishes in other eventualities? Is appealing for public guardianship here ei-ther required or ethically advisable? Remember, once public guardianship has been granted, limiting therapy is often extremely difficult if not almost impossible so that, in a sense, appealing for guardianship creates a self-fulfilling prophecy.

CASE 9

The patient is an 86-year-old man, recovering in the hospital after an acute infection. He has been confined to a nursing home for some years, is badly disabled by arthritis so that he can hardly get out of bed, and was taken to the hospital when unconscious. Although his infection clears with treatment, he refuses to eat or to take adequate fluids. Unless tied down, he attempts to remove the nasogastric tube. A psychiatrist has seen the patient and has found him to be competent. The patient's physician wishes to insert a gastrostomy tube for feeding, and, when the patient refuses, you are called in to advise the physician. In talking to the patient, you find him to be entirely clear. He states that he has no relatives, is unable to pursue any of his interests, and is miserable. He says that he has lived long enough, has had a good life, and now does not wish to continue to live.

There is no doubt that unless the patient changes his mind, he will die. Under ordinary circumstances patients who refuse reasonable medical treatment in the hospital can be asked to discharge themselves against medical advice. In a patient completely confined to bed, this is not an option, and nursing homes will not ordinarily accept patients under such circumstances.

The question here is one of decisional capacity: The patient appears to know the options, the outcome of the option he has chosen as well as of alternative options, and he is able to articulate rational reasons for his choice. He is not uninformed, has deliberated, does not appear to be under coercion, and, as far as one can tell, his choice is one consistent with his worldview. If this patient did not know what the actual consequences of his choice would be (if, for example, he claimed that despite not eating he would not starve to death) or could not articulate a persuasive reason for his choice, matters would be quite different, and, while feeling uneasy about doing so, we might have a sounder argument for interfering.

The fact that this patient has been judged to have decisional capacity and the fact that, deep down, we sympathize with his decision are important. Physicians who feel compelled to support life under all circumstances (and there are not many of those today) and who feel that keeping alive is a "moral duty" will, of course, be inclined to doubt the patient's competence. This is because reasoning in this fashion is part of their particular definition of incompetence. In the present case, however, competence has been established and is therefore one of the "givens" of the case. Physicians who hold suicide to be immoral will likewise have difficulties with this case and may act quite differently from those more willing to accept suicide or, at least, to look at the matter on a case-by-case basis.

Competent patients, who have reasonable decisional capacity, are held to be ethically (and legally) free to accept or reject medical intervention. Forcing such patients to undergo a procedure they refuse is coercive and, ultimately, constitutes legal assault and battery. When it comes to fluid and nutrition, strangely enough, we tend to adhere to other standards. Why is this the case, and is it justifiable? Is there a distinction between forcing fluid and nutrition (say, by a nasogastric tube) and doing a gastrostomy for the same purpose? Allowing patients to starve themselves to death in an institution, moreover, may affront both institutional sensibility and the moral views of the health professionals involved.

A resolution of such a case, like a resolution of most cases, lies in an attempt to discover an underpinning of shared values between those who would and those who would not condone such a practice. What might some of these shared values be? If a decision to allow the patient to have his way is made, does that end the obligations of the healthcare team? If not, what are these obligations, and why? What if those who must care for the patient (physicians, nurses, and ultimately the institution) are ethically unable to allow a patient to starve before their eyes even when doing so accords with the patient's wishes? If you must advise the healthcare team, what principles would you invoke?

There is another problem: Many nursing homes (as has been said) will not accept patients under such conditions, and hospitalization in an acute care hospital likewise will not be covered in such a case: There is, after all, no need for "acute care." This is, like so many other problems in healthcare ethics, an institutional and ultimately societal one. It is not one that a single or a team of healthcare professionals or an administrator can solve. Such a state of events once again underlines the fact that healthcare professionals need to be active in bringing such problems to the attention of the community and its elected representatives, who alone can deal with the basic nature of this generic problem.

CASE 10

The patient is a 5-month-old girl. She has one healthy older sister. The parents live in a basement apartment and have a marginal income: The father is disabled after a head injury, and the mother has been caring for the family. The child has been definitely diagnosed with Werdnig–Hoffman disease, a condition in which there is relentless deterioration of motor function so that, ultimately, breathing becomes impossible. Death usually occurs somewhere about 1 year of age. Intellectual and sensory functions are preserved; that is, such patients can feel and think in the same way that healthy infants would. The child is comfortable, but her ability to breathe is deteriorating. You are called to advise as to "how far one should go."

In dealing with patients like this, physicians bear a relatively lighter burden than do the other healthcare professionals, who must spend longer hours with the patient. It is here once again that a well-functioning team that talks over their problems among themselves can prove itself of great value not only to the patient but also to everyone within the team.

What makes this case so very agonizing is that intellectual function and sensation persist intact while muscle power inevitably falters, leaving the patient unable to cough or, eventually, breathe. It is this feature that makes the case critically different from, say, Tay–Sachs or other conditions in which both sensory and intellectual function regress. The options are to treat all intercurrent conditions vigorously and to support the infant with parenteral nutrition and the use of a respirator until death, inevitably, ensues or to abandon some or all of these efforts. Do we have all the data we need to have to get on with trying to reach a plan of action?

If a decision to treat all intercurrent disease fully and to ventilate and artificially feed as required to sustain life is made, what could it be based on? If, on the other hand, it is decided to forego some or all of these measures, what justification could be invoked? Is there a difference in means or in goals (ends) between these two courses? If a decision to abandon the treatment of intercurrent infections, for example, is made, is there any reason why ventilation might still be appropriate (remember that the patient is able to feel and think and that air hunger is awful)? What about a feeding tube?

What if the parents disagree (with you or between themselves) with your proposed course of action? Are there several acceptable options or only one? If several, how and why would you rank them, and in this particular patient, what would motivate you to choose one over the other? If one were to treat one but not another intercurrent illness (or decide to use one but not another therapeutic modality), on what basis would such choices be made? What shared values could be appealed to? What role do the individual members of the healthcare team play in the decisions made about the patient? What might be the role of an ethics committee or consultant?

Do the economic and social circumstances of the parents properly play a role in the decision? If, as is probably the case, the treatment of this case would use up resources so that they were unavailable for other uses (say, continue an immunization program or expand the hospital's outreach to the community), would that have any effect on what might be suggested? Are such considerations ethically permissible ones?

CASE 11

Your hospital asks you to help them formulate a policy for allocating funds. Specifically, the hospital is located in an area in which many automobile accidents occur, and, therefore, a fair number of organs become available for transplantation. At the same time, the hospital adjoins a poor area where many are without proper medical care. Limitations in funds have made it necessary to choose between (1) establishing an active transplant program (which would enhance the hospital's prestige) and (2) ex-

panding a completely inadequate outreach program to provide medical care for those without it. The hospital administrator points out that the hospital cannot do both. Further, he points out that, once established, albeit at great cost, a transplant program would begin to "pay for itself," while the outreach program would remain a constant drain.

The "you" who are asked to help formulate this plan is an important consideration. Ideally, the "you" does not have an axe to grind or is one on whom the hospital or its administrator can exert undue pressure. Assume that it is the case that you are a relatively free agent and that you have at least as open a mind about such things as anyone.

What are the considerations here? The economic reality seems to be that starting up a transplant program holds out a very real hope of self-sufficiency in a short time. The transplant program would not be an eternal well into which money would disappear. In addition, the prestige (and, therefore, referrals and eventually money) that would accrue to the hospital might well make this a very real economic asset: one that could even conceivably fund the outreach program!

On the other hand, transplantation is a rather sophisticated need *vis-à-vis* basic healthcare. It directly profits only a very few patients as compared to the many that an expanded outreach program would benefit. Does the fact that the outreach program already exists, whereas the transplant program has to be newly established, give any grounds for preferring one to the other on a moral basis? Does the fact that patients can go elsewhere (even if with some inconvenience) for transplantation, whereas the poor living in the adjacent houses do not, as a practical matter, have this opportunity, matter? Does the fact that the hospital is near a highway and, therefore, has many potential donors make a difference?

On what basis can this matter be argued? Are there shared values that underwrite the decision to establish either of these programs? Are there morally acceptable ways that might make both transplantation and outreach possible: say, agreeing formally ahead of time that a given percentage of the profit from transplantation would be used for the outreach program? Should one count in the indirect profits? In that such a plan of action would force the transplantation group to give up moneys it earned for an enterprise in which it might have little interest, would it be morally defensible? Why? Who should ultimately decide which course of action to follow? Who should be responsible, and why?

CASE 12

The patient is a 56-year-old married lady with long-standing chronic pulmonary disease. She agreed only reluctantly to be placed on a venti-

lator 6 months ago and then only because she thought that she had a fair chance of "coming off it." Home care was never very successful and was interrupted recently by an intercurrent infection. Since in the hospital (and now with her infection cured), she has persistently asked that the ventilator be discontinued and that she be allowed to die. She has been seen by a psychiatrist and judged to be competent. Were the ventilator to be removed, the patient would inevitably die in the course of 2 or 3 hours. The family members, who are very close to the patient, support her in her decision.

The difficulty here, as contrasted to Case 9, for example, is that this patient's death, once the ventilator is discontinued, will be agonizing. It is generally agreed that slowly strangling to death is one of the very worst tortures to befall sentient entities. Our 86-year-old gentleman in the example above could at any time decide to start eating; this lady would not have such an option.

What are we, in fact, saying when we refuse to disconnect her ventilator and override her wishes? Are we inappropriately or appropriately exercising our greater power? On the other hand, if we disconnect the ventilator and allow her to strangle, are we demonstrating callousness? Can one moral agent force another (consent or not) to torture him or her? Is the family truly aware of the consequences of either course of action when they "support her decision"?

Is there a resolution to be found in seeking for shared values? If so, what might these be? If a decision to continue ventilating the patient against her wishes is made, how can this be argued for or implemented? If, on the other hand, the decision to discontinue is made, how can this be done? Is the agreement (or, at least, the absence of strong disagreement) on the part of the rest of the treating team crucial here?

Is a middle road (say, sedating the patient to obtund her) an acceptable option? And if not, why not? If so, how can such a course of action be defended? To what extent would such a course depend on agreement of the entire treating team? of the patient? of the family? What if the physician feels strongly that he or she cannot be a party to discontinuing ventilator support? What if one of the nurses assigned to this patient felt strongly opposed to such a course of action?

CASE 13

Mrs. Jones, who is 18 weeks pregnant, has been in the ICU for 3 days. She was taken to the emergency room after a head-on collision, and, over the last few days, brain death has been established. There did not appear to be any abdominal injury, and the pregnancy does not appear to have

been disturbed. Mr. Jones is a laborer, and there are two other children.
You are called to advise about continuing to support Mrs. Jones so that
the pregnancy can continue.

First of all, as in Case 4, we need more technical information to arrive at a rational decision. The current state of the art is that at 18 weeks a fetus is far from being viable. Many more weeks would have to elapse before viability could even be reasonably possible. Supporting a brain-dead person for that length of time, while possible, is far from easy: It requires the use of sophisticated apparatus in an intensive care setting and is enormously expensive. Furthermore, while one or two cases here and there have used a brain-dead mother as an incubator for her developing fetus, there is not even a moderately reliable track record. Under these circumstances, each such venture is a trip into the unknown.

What are the physician's obligations under these circumstances? If the husband should wish to continue his wife's support under these circumstances, is it proper to continue to use up the resources of a busy ICU (built to take care of critically ill patients for their own good) for such a purpose? On the other hand, if the husband does not wish to continue his wife's support under these circumstances, is it proper to abandon all chances for the fetus' survival and to go along with his wishes?

Would the patient's lacking or possessing insurance coverage be a legitimate consideration? If the patient were maintained, a delivery finally effected, and the infant were severely damaged and required lifelong support, what would be the ethically acceptable course: making the father responsible? making the community responsible?

What if the father had not wanted continued support but continued support had been maintained and a damaged infant had resulted? Who should be responsible for the costs and, ultimately, for the infant's ongoing care? If not the father, how does this differ from a "Baby Doe" case in which treatment is given against the parents' will, and the parents are then held responsible for the further care of the child they had not wanted? Does stopping support here constitute abortion (the deliberate killing of a fetus)? If not, how does it differ? If yes, why? If the mother starts to abort spontaneously while the issue of brain death is being settled, is there an obligation to try and halt the abortion?

CASE 14

Your patient is a 92-year-old gentleman whom you first saw in the ER
when he was unconscious with a heart rate of 24. After a pacemaker is
inserted, the patient recovers fully, and you find him to be a very pleasant,

fully alert man who readily consents to having a permanent pacemaker placed. The pacemaker is placed. All goes well until the second day, when the leads become dislodged and the patient's heart rate drops to 32. He is stuporous. You are about to reposition the leads when his 62-year-old daughter, who is in the hospital when this occurs, intervenes and states that she does not want "anything more" done. She threatens to sue you should you proceed despite her refusal.

It might be well to review and think about your responses to Case 5, in which the issue hinged about the right of competent patients to make their own decisions and to have their wishes carried out even when they are not conscious. In Case 5 a patient no longer wished to have measures to prolong life taken, and his wife, once he was unconscious, demurred; here a patient presumably wishes to live, and, after he is unconscious, his daughter objects to further treatment.

The case would have been slightly—but not severely—stickier if the daughter had objected when the patient was first brought into the emergency room. Even here, barring evidence of the patient's stated desire to the contrary, treatment despite the relative's objection would have been given. It would have been given because the patient's relationship to his daughter was unknown and her having the patient's best interest at heart could not be assumed in a situation in which the patient's symptoms could very likely be reversed. In that eventuality, the situation could have been reassessed.

Communication is, once again, essential. Prior conversation with the daughter might or might not have forestalled the situation but, at any rate, should have been attempted. In accepting or not accepting a relative's decision about a patient's care, a decision as to the relative's motives must be made. Even when such motives are almost never entirely clear and are undoubtedly not entirely known to the person himself or herself, a reasonable assessment should be attempted. Situations in which serious doubts remain should be resolved in favor of acting to sustain, rather than in acting not to sustain, life. Is this presumption in favor of life a reasonable one, and if so, why?

(*Addendum*: This is an actual case known to the author [EHL]. Every attempt to convince the daughter to change her mind was made, but she remained steadfast in her opposition and continued to threaten suit. When all else failed, she was asked to leave the hospital or to stop causing a commotion in ICU. The pacer was uneventfully repositioned, and the patient left the hospital a few days later. Conversation between him and the physician revealed that the patient and his only daughter had not been the best of friends, that he was rather well-to-do, and that, as the only living relative, she, at least until this time, stood to inherit his entire fortune.)

CASE 15

Mr. and Mrs. Prenkovich have had a good marriage. Mr. Prenkovich was recently found to have an inoperable carcinoma of the pancreas. After biopsy has established the diagnosis, Mrs. Prenkovich informs you that she would "not allow you to tell Mr. Prenkovich." She feels that he could not "take this" and that it would shatter all hope and spoil his last few weeks or months. Mr. Prenkovich has always been the sort of patient who asks little, waits for you to volunteer information, but then seems grateful to you for having done so.

This is a very frequent problem and one that rests on a frequent misunderstanding of a critically ill patient's needs. Mrs. Prenkovich may think that her husband's last weeks or months would be hopelessly darkened if he knew his prognosis, and she wishes to spare the man she loves that pain. In addition, Mrs. Prenkovich may feel that she herself could not cope with her husband once he knew, and she may, often without knowing it, want to spare herself this pain. The motives are complex and generally rest in a misconception. Kübler-Ross, in her book *On Death and Dying*, has done all of us a great favor by exploding this myth: Not only are patients with a fatal disease quite capable of coping with such knowledge, they generally are quite aware of their prognosis (or, what may be worse, terribly afraid of what they surmise but do not know) anyhow. Not telling patients about their true prognosis, furthermore, cuts off communication at the very time that communication is essential: communication between doctor and patient, staff and patient, family and patient, and, above all here, between Mrs. Prenkovich and her husband. In his last days on Earth, Mr. Prenkovich is likely to be forced to "play games" instead of communicating and receiving what solace he can.

Physicians confronted with this problem often acquiesce to the relative's desire and lie to the patient. Often this is because physicians are most reluctant to discuss the poor prognosis of death with their patients and are only too glad to have an excuse for not doing so. Occasionally, it may also be due to the fact that physicians are loath to alienate those significant others who survive the patient. The problem, once again, is one of truth telling. The physician, first of all, is obligated to reason gently with Mrs. Prenkovich and try to convince her that lying and subterfuge are not in her husband's true best interests. A promise to stand by, to help, and not to abandon not only therapy but also more human support is essential. This sort of dialogue takes time: time to hold and time for what has been said to be absorbed and digested. There is no rush: The decision does not have to be made today.

If Mrs. Prenkovich (as she well may) persists in her attitude, what is the physician's obligation to his or her patient? If the decision to go along with Mrs. Prenkovich is made, how does the physician deal with the patient? How does he or she handle the unasked question? How the direct one? If the decision to tell the patient the truth is made, how does the physician go about telling the truth, and how can be or she involve Mrs. Prenkovich? Is there a ground of shared values between the physician and Mrs. Prenkovich that might help in the resolution of the problem? Are there others who might help? Are there ever situations in which a Mrs. Prenkovich might be right?

So far we have spoken mainly about the physician's role. This has been one of the difficulties in arriving at a solution. The entire healthcare team should, as it properly should in most of these decisions, be involved in the dialogue from the very beginning. It is quite likely that they know the patient and his wife from quite a different angle than does the physician. If no clergyperson has been involved, it might be well to involve one—especially one with clinical pastoral training who had expertise and understanding for dealing with such problems, provided patient and wife agree to this. Compassionate rationality (compassion for the wife's grief, the patient's fears, and the needs of both, and reason to sort out the various aspects and dispassionately present them to all involved) can go far in helping the team discuss and ultimately come to meaningful grips with the problem.

CASE 16

You are taking care of a professional pilot who was brought into the hospital after having sustained a seizure. Workup has revealed no specific cause, so a diagnosis of epilepsy is made. Your patient beseeches you not inform his company because this would result in the loss of his job.

This problem has many guises: the husband who refuses to tell his wife about a communicable venereal disease, the psychiatric patient who tells the psychiatrist that he or she plans to harm someone, the patient with any disease that may threaten others but that he or she wishes to hide. The physician is obligated not only to his or her patient but also to the community. Neither of these obligations, however, is absolute. Or are they? Is the obligation of confidentiality an absolute or a *prima facie* obligation (one that "on the face of it" is binding but that may, under special circumstances, have to be ignored)? What are the circumstances that would permit a physician to breach confidence? Why?

In this case, once again, communication is essential. If at all possible, the pilot must be induced to cooperate, and the physician must undertake to help the pilot reach an equitable agreement with his company. If the pilot is obdurate, what is the physician's obligation? May the physician coerce the pilot and thus obtain a grudgingly given consent? (For example, could the physician inform the patient that come what may, he or she will report the condition, but offer, if the patient finally consents, to try to mediate so that reassignment to non-flying duties rather than losing a job would result?) Should the physician evade the problem by relinquishing the care of the patient to another physician? And if not, why not? What are the limits of communal responsibility today?

If, say, a new law were to require physicians to report the names of all HIV-positive persons so that they could be published in the local paper, should physicians cooperate? If so, what are the grounds on which such a belief could be sustained? If not, what are the grounds for non-cooperation? What if drug addicts had to be reported to the authorities so that they could then be jailed? How do you come to terms with such conflicts?

CASE 17

PBC is an eventually fatal disease. Current studies show that treating this disease either with steroids or with another immunosuppressive agent prolongs life (but does not cure the disease). A new agent may also be efficacious but has not been tested. It is suggested that a double-blind study comparing this agent to no treatment be set up.

1. You are on an institutional review board (a board that must approve the protocol of all studies in an institution). Do you favor this study as proposed?

2. You are a physician whose patient has PBC. You have the option either of enrolling the patient or of not enrolling this patient in such a study (assuming that it has been approved by an IRB). What are your options and obligations?

Testing new drugs is something that needs to be done. When it is done, it must be done with as fully informed consent as possible. The very fact that the venture is experimental means that we are far from certain about the outcome or the attendant risks. The patient, then, essentially consents to take a trip into the unknown with us in which he or she runs all the risks while we watch. While we are far from certain about the outcome, and while we are unaware of all the risks, we do know some things: We know, for example, what animal or computer models have shown, what experience has been had with similar substances, and what treatment offers or does not offer in the same disease today. Is it acceptable to compare a treatment with a non-treatment group in a disease in which treatment is known to benefit the patient? Is there an alternative? Are there ways other than such experiments to find out how patients who receive no treatment will do?

If you have a patient who might be enrolled in such a program and you have no qualms about it, you do not have much of a problem in suggesting enrollment to the patient. If, like we, however, you have serious reservations about a protocol that fails to treat a treatable disease in order to establish a "no-treatment" control (and think that a protocol in which one treatment is compared with another would do just as well: after all, finding out how the new stuff works relative to the old stuff is far more useful than finding out how it works as compared to no treatment at all), you have a problem. The new treatment may be far superior, and each patient has a 50% chance of getting it! On the other hand, the patient also may receive no treatment at all when some treatment has been shown to be beneficial. Under these circumstances, should you simply not mention the protocol and treat the patient in the established way? Should you mention it to the patient, not express your opinion so as not to influence the patient, and let the patient decide, or should you mention it to the patient, make your recommendations, and then abide by the patient's decision? What arguments might support any of these courses of action?

CASE 18

Mr. B. developed myocarditis during her pregnancy. Although she eventually delivered a premature, otherwise healthy and viable infant, her cardiac function deteriorated until a transplant was the only way of saving her life. Eventually a properly matched donor heart became available, and she underwent an uncomplicated transplantation. Within hours thereafter and while still in the recovery room, cardiac function deteriorated and it was obvious that she had undergone a "hyper-acute rejection" (a process whose etiology is poorly understood and which, in about 95% of cases, leads to repeat rejection should another transplant be attempted). Ms. B. was placed on temporary pump support. She was awake, able to speak, and rational. There were two other potential recipients awaiting transplantation when a heart that would be a good match for Ms. B. as well as for the other patients became available. You are asked to advice about attempting another transplantation for Ms. B. or giving the heart to one of the other two waiting candidates.

The questions here point up the difficulty physicians find themselves in when making allocation decisions involving specific patients. Who most "deserves" this heart? What are the criteria on which such judgment can be made? Under these circumstances, does the surgeon who operated on Ms. B. have a special obligation to her that he or she does not have toward these other waiting patients? Would such an obligation be changed by the fact that the surgeon has also already come to know the others? Does the fact that the chance for a successful transplant in the other waiting patients is very much higher than it is for Ms. B. militate against giving the available organ to Ms. B.? If so, why? And if not, why not? Should Ms. B. have any say-so in the matter?

One way of conceptualizing such a problem is to consider what would happen if the option of re-transplantation were not to be made available after an event of this sort had occurred. Physicians or surgeons feel (understandably and arguably rightly) that they have a special obligation to a patient once treatment by them has been started. This would be an obligation to do all that is needed to save that particular patient's life regardless of (virtually) all other considerations. The one is an identified life and an identified life in which one has become deeply involved; the others are either not identified or have not had the degree of surgeon involvement as has Ms. B. And yet these two other lives are just as distinctly individual lives (and are just as "identified," even if by someone else). One way out of this impasse would be to make a communal judgment (perhaps by the professional community with the help of ethics committees) that patients who had sustained hyper-acute rejection would be barred from receiving another transplant. Such an event would then become simply one more of the risks a patient knowingly undertook in undergoing transplantation.

Can one justify such a way of proceeding? If not, what other ways of dealing with such a situation should be made? Should this point of view be enlarged so that those with little chance to survive would be barred from intensive care units, especially when there was a shortage of beds? Or should a strict first-come, first-served policy be more acceptable?

CASE 19

A 66-year-old diabetic woman was admitted to our hospital for a non-Q Wave myocardial infarction. Soon thereafter she had combined coronary bypass and mitral valve surgery, tragically complicated by sternal wound infection, renal shutdown, and multiple cardiac and respiratory arrests which left her neurologically profoundly impaired and has made her dependent on renal hemo-dialysis, tracheostomy and ventilator support, tube feedings, and total (24 hour a day) nursing care. In the course of her subsequent fifteen months hospitalization on the telemetry unit, she has suffered multiple respiratory arrests, episodic cardiac arrhythmias (including ventricular tachycardia) for which a pacemaker has been placed, multiple episodes of aspiration pneumonia, urinary tract infection, systemic septic shock, erosive stage IV sacral decubiti, malnutrition, anemia, problems with control of her diabetes, hypotension, congestive heart failure, shunt and permacatheter clotting. Over the past several weeks, progressive ascending wet gangrene of the right lower extremity following unsuccessful antibiotic therapy of an osteomyelitis of her right foot has developed. *Throughout this long ordeal, she has remained non-verbal, though she responds to the staff with facial grimaces and can withdraw her right leg, evidently in pain, when dressings are changed or the leg is moved. The leg itself is malodorous, filling the room air with the odor of rotting flesh in spite of the liberal use of various de-odorizers.*

Her husband of over 50 years is a faithful attendant at her bedside and her room is freshly and repeatedly filled with cards and flowers from her family and friends. He was reluctant to consider a No Code order for his wife, whom he deeply loves, continued to hope that she might be weaned off the ventilator and be taken home. He finally and only very reluctantly consented to a no-chest compression order (all other resuscitation measures to be used) because he recognized that the extirpation of her sternum (following a wound infection) made compression useless and possibly harmful. He is constantly talking to her, singing to her, and interpreting " what she wants" to the staff. He tells the doctors, nurses and other support personnel about her beauty when they met, her sweetness and generosity to others, her (and his) deeply shared Christian beliefs, and their long connection to one another. He also appears to be communicating effectively with her, and she is clearly different in her responses to him, turning her eyes toward him and smiling when he speaks to her as she does to no other speaker. He still expresses the hope that she will eventually be weaned from the ventilator so that he can "take her home and care for her there."

The husband is very concerned about the ascending gangrene on her leg, her pain as well as the odor of death in her room. He had been told that amputation of the leg could be done, though the surgery might kill his wife, but that it might also relieve her of pain and would certainly ameliorate the dreadful smell. He would like the surgeons to proceed with the amputation, and assures the staff that this is also the wish of his wife.

The vascular surgeon consulted to perform the amputation refuses to do the procedure because he is "morally opposed" to what he believe is futile surgery in this uncommunicative, catastrophically injured and clearly dying woman. He thinks her husband wrong to prolong her suffering in the forlorn hope of saving her life. The surgeon does not wish to operate because he feels that it would be morally wrong to do so because "it would be futile."

This case which was suggested to us by our colleague, Dr. Faith Fitzgerald, has many facets. In the absence of an enduring power of attorney for health-care matters is the husband the proper surrogate? If so or if not so, why? The husband is supposed to decide on the basis of the "patient's best interests," setting his own interests aside. Is that proper? Is it possible? When the surgeon declares that he is "morally opposed" to amputating this leg, why does he invoke futility? What does futility denote here?

━━━━━━━━ ◆ REFERENCES ◆ ━━━━━━━━

1. Dewey J. Logic, *The Theory of Inquiry*. New York, NY: Henry Holt, 1938.
2. Clouser KD. What is medical ethics? *Ann Intern Med.* 1974;80:657–660.
3. Loewy EH. The uncertainty of certainty in clinical ethics. *J Med Humanities Bioethics.* 1987;8(1):26–33.
4. Gorovitz S. Moral conflict and moral choice. In: Gorovitz S, ed. *Doctors' Dilemmas: Moral Conflict and Medical Care.* New York, NY: Oxford University Press, 1982.
5. Loewy EH. Teaching medical ethics to medical students. *J Med Educ.* 1986;61:661–665.
6. Thomasma DC. Medical ethics training: a clinical partnership. *J Med Educ.* 1979;54:897–899.
7. Toulmin S. The tyranny of principles. *Hastings Center Report.* 1981;11(4):35–39.
8. Jonsen AR, Toulmin S. *The Abuse of Casuistry: A History of Moral Reasoning.* Berkeley: University of California Press, 1988.

Index